TO BE OR NOT TO BE PHILOSOPHICAL

A Tiptree Inspector Decides

Noel Boulting

Introduced by Carl Hausman

UPFRONT PUBLISHING
LEICESTERSHIRE

TO BE OR NOT TO BE PHILOSOPHICAL: A Tiptree Inspector Decides

ISBN 1 84426 198 0

First Published 2001 by
MINERVA PRESS

Second Edition 2003 by
UPFRONT PUBLISHING
Leicestershire

TO BE OR NOT TO BE PHILOSOPHICAL

A Tiptree Inspector Decides

With his permission, this book is dedicated to the philosophy and life of Charles Hartshorne (1897–2000) who not only became the first philosopher to live in three centuries, but who also published his last book, The Zero Fallacy and Other Essays in Neoclassical Philosophy *in his hundreth year.*

ACKNOWLEDGEMENTS

In the *Nicomachean Ethics*, Aristotle remarks: 'the philosopher, even when by himself, can contemplate truth, and the better the wiser he is; he can perhaps do so better if he has fellow-workers, but still he is the most self-sufficient' (1177 a 33–5).[1] Not satisfying that last condition, even if still striving to meet the former, I have been most fortunate in having many fellow-workers who have made the writing of this book possible. I am most deeply indebted indeed to Maggie Lycett, Richard Farrow and Pauline Stevens – whose book on the history of Lower Halstow was recently published – for giving me a tough time in making sense of the first eight chapters. I am even more indebted to my wife Pat Hall and to Jennifer Smith for making suggestions in the writing of all thirteen chapters.

Two people suggested writing a book on Peirce's philosophy: Carl Hausman of Penn State University at a NOBOSS meeting and Michael Krausz, of Bryn Mawr College, at a Conference of the Society for *Philosophy in the Contemporary World*, though neither must be held responsible for the way this book came to be structured. NOBOSS is a philosophical society founded by Cyril Hodgkinson and myself over twenty-five years ago and conversations I have had with its members, no doubt, have played their part in the way arguments are forwarded in some of the book's chapters. I am also indebted to suggestions from George Goodwin with respect to the book's Preface; Valerie De Furrentes and the late Dorothy Gibson in relation to chapter I; Dr Martin Bertman – editor of *Hobbes Studies* – in the case of chapter IV; Jon Taylor of Great Falls University, Ann Everett, Doina Cornell, Sue Tappin, Tom Mayhew, Maureen Leather and Pat Turner with respect to chapter VII and Andrew Smith, Stuart Lewin and Dr Julius Tomin for their remarks on chapters VIII and XIII. Numerous conversations were enjoyed during the intervals between sessions at the Tiptree Inquiry. Memorable discussions, which, no doubt, influenced what appears in the book, were enjoyed with the late Bill Davis, Nicholas Nardecchia, Dick Harman, Jeannie Taylor, Laina West and Mary Tappenden; the last person making considerable efforts to ensure I was properly informed as to the case for the appellant. None of the persons mentioned in these acknowledgements is to be credited with any factual or interpretative errors which may have found their way into the text.

I am most grateful to John L Craig of the School of Environmental and Marine Sciences, Auckland University for sending me copies of his articles on ‘Conservation’ and on ‘Managing Bird Populations’. I appreciate, too, permission granted by J F Perry (editor) allowing me to refer to my articles published in the *Interdisciplinary Studies in the Philosophy of Understanding Journal: Ultimate Reality and Meaning*, as well as Joe Frank Jones III (editor) enabling me to refer to my publications in *Philosophy in The Contemporary World: An International Journal Sponsored by the Society for Philosophy in the Contemporary World* especially my ‘Aesthetics of Nature’ paper. These publications are listed in Appendix III.

I would also like to thank the RSPB for granting me permssion to quote from, and refer to, material published by this charitable organisation.

NOTES

[1]Trans by W D Ross.

FOREWORD

I would like to share two things with you. One is a practical problem; the other is more theoretical. The practical problem is what are we to do, what are we to think, what should we feel when we learn that there is a proposal to alter the landscape, a landscape which provides a setting for our ordinary lives? The theoretical issue resides in the question 'Can any kind of rationale be justified for such a change?' And if such a rationale can be offered, what rationales might be considered for opposing it. In this book an attempt is made to relate such a practical problem to this theoretical issue.

The practical problem arose when I discovered in 1995 that a range of hills behind a nearby village – Lower Halstow – was the site proposed for generating a waste tip. The most obvious response could have been NIMBY: not in my back yard! But a waste tip may be necessary somewhere. Waste from London has to be accommodated in the Home Counties and waste is exported out of Kent. People living in highly concentrated areas won't tolerate poisonous gases descending upon them from incinerators. People's rubbish bins have to be emptied each week and proper arrangements for waste separation for possible recycling have not been organised nationally, as they have been in Germany for example.[1]

The theoretical issue arises in the way I interpret the philosophy of Charles Sanders Peirce.[2] It does so because his philosophy does not focus, as the works of so many philosophers do, only upon what we might think in relation to such a practical problem. It also throws light on the way we may feel and act in relation to what we can experience. Most important of all, of course, is how feelings, thoughts and reactions can be related to each other. Philosophical inquiry, in a Peircian spirit, not only shows how they can be so related but also suggests how different perspectives on theoretical concerns – those from *ecological ethics, environmental aesthetics and so on – can be brought together. However, because we are dealing with human responses, must a human being's interest in the environment just be human-centred, rather than related to the environment as a whole? Much of the writing of this book is haunted by this very dilemma.

The reader is introduced to the book's main issue in chapter I: the case for making yet another waste tip in a Kent landscape. Arguments forwarded against such a landscape change are considered. They can be grouped in three classes: the *aesthetic, the *instrumental and the *scientific. These are then connected to Peirce's three categories which relate to how experiencers' feelings (*Firstness), actions (*Secondness) and thoughts (*Thirdness) can refer to something outside the self. A fourth category – the unconcerned – is then delineated. Given the appeal against Kent County Council's rejection of allowing the construction of such a tip, forwarded by its proposers, and given these four categorial standpoints which can be adopted towards such a proposal, must the standpoints of the protesters not be *anthropocentric in character, tied only to a human concern for the environment? The argument that the scientific attitude, which provides knowledge of the natural world, must lead to a *non-anthropocentric perspective is considered in chapter II. Its connection to a moral point of view is also examined. The *cognitive pitch of the book is raised in this chapter. Even if a reader does not follow all the arguments within it, its conclusions are important since these are carried in the rest of the book.

Chapter III narrates the opening of the inquiry and develops themes from the previous chapter. Attention is given to the different ways in which scientific evidence is treated by the participants in the inquiry: in one case dispassionately and, in the other, as a way of forwarding an instrumental concern for economic gain. The question of morality once more reappears. To deal with that issue, the way this evidence is handled is related back to the basic attitudes already indicated. Finally, other possible attitudes towards nature, still anthropocentric in character, are delineated.

The philosopher Charles Sanders Peirce was always suspicious of mixing theoretical inquiry with practical applications. The result can be that the latter renders the former farcical in generating make-believe *reasoning – *rationalisations. That possibility was engendered at this inquiry by one of the participants seeking to put what was called 'a semblance of rationality' on their case. In chapter IV, by using Peirce's categories again, *sham reasoning is distinguished from rationalisations and these two set apart from the activity of giving reasons as well as from the idea of being *rational. The thickets of Hobbes's philosophy are entered in dealing with such questions as 'Aren't all attempts to be rational, rationalisations?', 'How can we be rational?' and 'What grounds the activity of being rational itself? This latter question is important in deciding how far any argument, in relation to what is to be done to the environment, has or has not to be anthropocentric in character.

This concern about the relative importance of the anthropocentric point of view is advanced in chapter V when it is shown why the appeal had to be abandoned temporarily. Different types of reasoning are considered in the light of Peirce's three categories: Firstness, Secondness and Thirdness. After relating these to the significance of his *phaneroscopy or *phenomenology – the study of anything which may appear to *consciousness – the issue as to why we should reason at all – an issue raised in the previous chapter – is re-explored. Now a scientific way of knowing is contrasted with *instinctual or *common sense forms of awareness. But either approach still leaves Peirce's phenomenology anthropocentric in character; doesn't it?

This latter question is taken up in chapter VI in considering the case for the importance of the site with respect to habitats for birds. Now Peirce's three kinds of reasoning are related to each other through his logic of scientific investigation, illustrated by the way an observer seeks to identify a bird. But in relation to the issue of altering the environment, why should birds be considered and why should we think it necessary to study them? These questions raise the issue of the importance of bird songs both to humans and to birds. Use of *analogical reasoning suggests that birds might enjoy each other's songs aesthetically rather than just *functionally. If so, and by employing once more Peirce's logic of scientific investigation, it is shown how a strict anthropocentrism might be undermined.

The former chapter concentrated on Peirce's category Firstness, in indicating the significance of something non-anthropocentric in character. After examining the documents used by KCC's landscape witness, a concern for what exists – Secondness – now becomes the focus, issuing in an examination of the landscape issue. By showing how Peirce's categories relate to his theory of signs in chapter VII, landscape is defined as 'a human perspective from some place upon a physical environment capable, in being appreciated panoramically, of yielding an *iconic experience'. Strictly anthropocentric approaches to the question of defining landscape – whether in terms of Cosgrove's iconic or Hirsch's cultural approach – can then be undermined. Peirce's categories are developed helping us to appreciate how the environment can be altered whether *unintentionally, *causally or *intentionally by man's activities. This trichotomy of distinctions is then used to answer the question 'What gives landscape its value?' Again the values, which emerge, transcend the kind of anthropocentrism dominating earlier chapters. In the next chapter a fictional reply to what has been claimed is forwarded. It sets out to provide an anthropocentric rationale for taking a *'specular view' of the landscape. That view was adopted by the appellant's landscape witness at a further inquiry, called to decide whether or not Tiptree Hill was worthy of a Local Landscape Area (LLA) designation. Thereby it can be seen as a defence of Cosgrove's position.

The opposition between the anthropocentric and the *ecocentric standpoints is reinforced in chapter IX by examining the issues as to what makes a landscape significant – scenic quality – since this issue proved important when the inquiry resumed. Because the debate about this criterion raises philosophical issues, it runs into reflections on the narrative in this part of the work. The rationale for the appellant's landscape witness – set out in chapter VIII – is shown to make sense of the Countryside Commission's documents, used by him at the inquiry with respect to seeing a stretch of land either aesthetically, instrumentally or scientifically. This examination leads to the formation of three philosophic questions which drive the rest of the book: i) must the discourse of the different parties at the inquiry remain *incommensurable? ii) how can the *Fact/Value dichotomy in making landscape judgments be overcome? iii) how can the six criteria used in landscape evaluation be justified?

The contents of chapter X show how utilitarian arguments dominated the final speeches at the inquiry's last day. Heffernan's principle is introduced to deal with the clash between the utilitarian or anthropocentric case and the *ecological or ecocentric argument, though once more the issue of defining a *'community' is raised. More consideration might be given to this principle if the hill was awarded an LLA designation, for, in that case, the landscape and environmental features would be recognised as significant, thereby ensuring the future of the biotic community inhabiting the area. But the possibility of Tiptree Hill acquiring such a designation leads to re-examining the scenic quality criterion, and this examination is unpacked by employing Appleton's *prospect–*refuge theory. But that move might imply the importance of living as an 'insider' within what is evaluated. A defence is thereby mounted from this perspective over the idea of taking a 'bird's eye view' as an 'outsider' towards this landscape, which is just what forwarding the 'specular view' entails. But are these two discourses – that of the 'insider' and 'outsider' – incommensurable with each other? The answer to that question is explored through a return to Peirce's basic attitudes towards nature now shown to make sense of the work of Porteous, an environmental aesthetician. The possibility of this incommensurability problem being overcome raises the spectre of Hobbes's philosophy once more, as that was explored in chapter IV.

The idea of resolving a conflict through the mediation of an impartial authority is re-examined in chapter XI, but even if its decision is fair *procedurally, would it prove sound *substantively? And if so sound, would that result throw any light on solving the Fact/Value dichotomy? After all, it was indicated in earlier chapters that once one is committed, as a matter of fact, to inquiry then that commitment of itself ushered in certain value imperatives. Criticisms of this *logistical way of grounding values are considered. Another route for solving the problem is pursued: that value judgments as to what is good or bad can be grounded upon the aesthetic dimension. In addition an appeal is also made to *casuistry to overcome the Fact/Value dichotomy. This latter approach would make sense of Peirce's *critical commonsensism as well as his conservatism in such matters, but the many differences in the various attitudes held by the participants, mapped out in chapter X, foreclose such a move. This chapter ends by suggesting a tentative solution to be explored next.

In chapter XII, certain legal anomalies emerging in this inquiry are drawn out which can induce immoral consequences. But once more the question, 'What makes a decision moral?' is raised. A model, based on the suggestion at the end of chapter XI, is presented showing how moral judgments can be grounded. Its roots lie in the idea of something being valued for its own sake. What does that mean and how can that claim be justified?

In the book's conclusion, a fictional argument occurs between key figures at the Tiptree Inquiry who debate the inspector's decision. Different ways in which arguments about the manner in which the visual impact and other landscape considerations can be viewed are raised, but this time from an aesthetic standpoint. In order to do this, the different ways aesthetic viewing can be grounded are delineated: through *specularism, by means of scientific objectivity or through lived experience. It is suggested that it is only the latter which can give authority to what exists, its due. It thereby provides a way in which anthropocentrism can be overcome, to allow for a more ecocentric point of view to emerge.

Earlier I spoke of interpreting Peirce's philosophy in a certain way. The way in which I interpret his philosophy is not entirely in accord with that of Carl Hausman, whose Introduction to this book I deeply appreciate. We have been arguing about how Peirce's philosophy is to be interpreted over the last ten years or so. His alternative approach, which he sets out in the last two paragraphs of this book's Introduction, is best appreciated after the book has been read and attention given to Appendix III: A Tribute to Fathers? In the book's Bibliography, my own publications –

articles – are listed and they are also indicated in the endnotes of each chapter. In addition, in Appendix I, there is a glossary of philosophical terms to give the reader a sense of how they are used in some philosophical circles and how they may have been angled in the interests of making issues clearer in this book. These terms are preceded by an asterisk within the text just as they have been presented in this very Foreword. The point of the second glossary – Appendix II: An Index of Thinkers – is to list the names of thinkers so as to indicate something as to their ideas, particularly in relation to their relevance to questions raised within the text of each chapter.

This book is dedicated to Charles Hartshorne, who was one hundred years old on 5 June 1997, becoming the first philosopher to live in three centuries.[3] He died on 9 October 2000. His work has not been regarded very highly on this side of the Atlantic probably because of his advocacy of the writings of Charles Peirce and the English philosopher Alfred North Whitehead, who was a teacher of Bertrand Russell just as Peirce was a teacher of John Dewey. In addition, Hartshorne is thought of not only as a philosopher of religion but also as the world's first chief 'ornithologist' to explore the nature of bird songs. His famous *Born To Sing* (1973) has been republished recently. His ideas, particularly about aesthetics, haunt the writings of my present book.

NOTES

[1]Emma Tucker, 'Where Rubbish is a Way of Life', *Financial Times*.

[2]N E Boulting (1994), 'Peirce's Idea of Ultimate Reality and Meaning Related to Humanity's Future as Seen through Scientific Inquiry' in *American Philosophers' Ideas of URAM*.

[3]His last book, *The Zero Fallacy and Other Essays in Neoclassical Philosophy* (ed. with an Introduction by Mohammad Valady, Chicago, Open Court) was published in 1997.

ABOUT THE AUTHOR

Noel Boulting studied under Richard S Peters at the London Institute of Education to obtain his academic diploma in the Philosophy of Education; under David Hamlyn and Stuart Brown at Birkbeck College, London, to obtain his first degree in philosophy; and under Imre Lakatos and John Watkins at the London School of Economics to obtain his mastership in the Philosophy of Science. He has taught philosophy for the Extra-Mural Department, University of London; Philosophy of Education at Trent Polytechnic and Educational Studies at Mid-Kent College of Higher and Further Education. His philosophy club, NOBOSS, was formed in 1977 on the basis initially of forwarding an interest in the philosophies of A N Whitehead and C S Peirce since such interests could not be pursued in English universities outside theology departments at that time. NOBOSS meets at least twice a year, and professors of philosophy from America and Germany have attended its sessions. His publications include articles on C S Peirce, Edward Bullough, Thomas Hobbes, Aldo Leopold, Jean Paul Sartre, Simone Weil and the Aesthetics of Nature besides those on nationalism, ecological responsibility and an examination of the concept of action and the justification of value judgements.

PREFACE

> Philosophic thought has to start from some limited section of our experience – from epistemology, or from natural science, or from theology, or from mathematics. My own belief is that at present the most fruitful, because the most neglected, starting point is that section of value-theory which we term aesthetics.[1]

In a Preface, rather than in a book's Foreword or its Introduction, an author gets the chance to say what he thinks is important about his composition. This book is driven by one central concern: to show the place of intellectual activity in everyday life. Isn't expressing an idea, forming a critique of some prevalent thought, reflecting on one's experience or trying to make sense of one's life in our present culture as or more important than earning a wage, shopping or watching some favourite TV show? But that latter question upon examination, yields four subsequent ones. How are intellectual activities to be pursued? How can they be characterised? How do they relate to what might appear not to make use of them? Why are such activities important?

Within the society in which we presently exist – whether in the UK or in the USA, countries in which the writer has taught, lectured and lived – the first question has a clear answer. Institutions have evolved in which intellectual activities are pursued. A number of unfortunate results follow their development. Those who don't enter such institutions may be deprived of knowing how significant such activities are: those who enter can come to think that such activities have nothing at all to do with the business of everyday living or, are concerned only with the business of earning a living. But this tale of woe does not end here since it has consequences for our second question: how are intellectual activities to be characterised? Intellectual practices become divided into different forms, each sustained by an old-fashioned, unionised mentality which regards one kind of intellectual activity as best forwarded in a particular department. So the examination of some *cognitive interest, as that which emerges say in 'the fortunes of some institution or person',[2] belongs to the history department; engagement in empirical research into features of the physical world, to a science department; the study of the behaviour of different kinds of social groups in response to various types of social processes, to the sociology department. But when it comes to the geography department things can become more problematical: should the geographer be interested in what it is that gives rise to certain kinds of vegetation and forms of agriculture some given land mass makes possible – in view of its geology and the kind of rainfall it enjoys? Or should s/he be interested in the significance such a land mass can have for its inhabitants in terms of its footpaths, fences, homestead provisions and so on?[3] And disputes as to whether or not such disciplines have been rightly characterised in the way they have been presently articulated to you – the reader – belong, it might be thought, to the *philosophy department. We can now see why philosophy could be regarded as a *second-order activity – a *meta-narrative[4] to employ a term in current usage. It can, for example, focus upon the logic and upon lines of *demarcation that separate one kind of intellectual activity from that of another.

All this may strike a possible reader as needlessly theoretical, so let's consider a practical example. A layperson is interested in the subject 'landscape'. S/he goes into a library or scans the internet after typing in that term on his or her computer. S/he may do this for all manner of reasons. Where did the term arise? What is it? Why is it important? How does the term enter ordinary ways of speaking about living in some environment? On what grounds are some landscapes thought to be more significant than others? Where might an inquirer find something about the kinds of landscape which seem similar to the one s/he faces every day? And so on. In answer to such questions, s/he would soon find that they are touched upon in resources found in the geographical, anthropological, historical or psychological sections, to name but a few of the possibly relevant disciplines. But no one source attempts to bring these kinds of inquiry into a single domain; similarly for the study of conservation management. Each discipline singly does and must have much to contribute to our imaginary layperson's questions about either the topics 'landscape' or 'conservation management', but no one source brings all the relevant *cognitive contributions together in either case.[5] With respect to the issue of landscape, this book attempts to do so through a gradual introduction of the reader to Charles Peirce's philosophy.

For the purist – those persons whose life centres on one kind of intellectual activity – such an outcome hardly appears promising since it can always be said that his or her cognitive interest can't be handled at least satisfactorily or, more seriously, rigorously by such an attempt. But this book is not for the purist only. It is directed towards a lay person with little in the way of knowledge about any of those named specialisms.

The third question, concerning the way intellectual activities are to be characterised in this book, goes to the heart of the matter. Through a case study which seeks to be factual in character, I try to show that the really important question about a decision – in this case what should be done in relation to a specific part of one's environment – raises issues which may be either overlooked or ignored in common sense rationales for doing something. So the central issue involves the narration of a decision process: whether or not to construct a waste tip in the vicinity of the writer's home. There is then a narrative sustained throughout the following chapters examining how a particular decision was reached. A reader might be interested simply in how that decision emerged and regard the exploration of the rationales, thought to be pertinent to it, as of only passing interest. On the other hand, an interested purist might be engaged more by the logic behind such rationales and less in the actual nitty gritty of the specific details involved in the making of this decision, in this case at a public inquiry. But both these ways of viewing the book do it a disservice, since its aim is to reject that very split, this *bifurcation, between a narrow concern for actual practicalities on the one hand and/or theoretical matters on the other. Indeed a mere concern for the actual facts underpinning the practicalities without any concern for the theoretical issues involved is blind: an obsession merely for an exploration of the theoretical issues devoid of their practical application can be simply empty.

Our fourth question concerns the importance to be attached to intellectual activities. Through these activities we can come to gain knowledge about the natural world, other people and ourselves. Earlier, it was pointed out that discourses dealing with these areas have become divided into different disciplines and that the demarcation lines between them, their respective logics and the way they can be integrated with each other are matters involving a second-order or metadiscourse called philosophy. But increasingly during the twentieth century these issues have been drawn back into these very disciplines so that activities sustained under the name of the philosophy of science, the philosophy of the social sciences or the philosophy of art become regarded as particular aspects of the main specialist discourse itself, whether in a science department; a sociology or psychology department; an arts and creative design department. What role then remains for philosophy?

Traditionally, the pursuit of philosophy was seen as leading to wisdom through investigating man's nature and his relationship to the world. Indeed, the ancient Greek philosopher Socrates might have been happy with the idea that philosophy concerns inquiry so as to ascertain what the things we think, in themselves and by themselves, can really be. Within this Preface it has been claimed that there is no single discipline any longer, which deals solely with such an enterprise, but, rather, that different disciplines have evolved and continue to do so in the development of human inquiry. But besides being a possible cognizer of the world, as the human subject engages in inquiry, within one or more of these disciplines, that cognizer remains a citizen. To be a citizen is to employ a language which does not articulate the basic concepts of one discipline alone. Rather language has evolved in response to ordinary experiences of human beings down the years. In addition, in the contemporary world, there are 'experiences' – meaning screen mediated experiences – of contemporary human beings and the taken-for-granted ideas that are thought to have derived from the special disciplines already indicated. I say 'thought to have derived' because it is a matter of dispute whether the ideas, taken to have arisen from inside a specific discipline, are represented accurately within a citizen's culture. So, at a particular level, each of us is aware of how some new idea is 'hyped' and its discoverers then have to explain how their discovery has not been articulated accurately by its reporting.

At a more general level, there are disputes arising from the question as to how far a discipline taken as a whole, is represented in a culture. Is it really a distorted version of its true nature, according to its practitioners? Much debate, sparked by the writings of the philosopher Karl Popper for example, has focused upon whether science is an activity which can establish once and for all what is true or false, which is how it is regarded outside scientific circles. Or, to put the matter crudely in the eyes of a purist, is science to be understood as constituted by bundles of theories? None of these theories can be thought of as true. Rather each can be regarded as acceptable in view of present-day evidence but which may fall in the light of further inquiry. The issue of what makes a question a scientific one is a problem which arises within this text. It does so because it is often claimed that to be rational is to adopt the scientific turn of mind.

Having established the existence and importance of what can be called ordinary language and drawn your attention to Popper's idea of science, why not adopt a view which might not have displeased him: 'Philosophy deals with problems the solution of which enlightens common sense'.[6] Philosophy would then have a proper role. But philosophers are the kind of people who are never really satisfied; they always seem to be nit-picking! So it will come as no surprise to you to realise that this writer has reservations about such a view. For those who are practically or scientifically minded, life indeed may be constituted by problems. But those of a more contemplative disposition may be fascinated by the question 'What do these things – whatever they may be; human expressions, existing things or someone's remarks about something – what do they mean?' A second problem with Popper's approach is similar to a difficulty one could have with a definition which might not displease Oxford philosopher Geoffrey Warnock: 'Philosophy is the acquisition of systematic conceptual knowledge through the investigation of ordinary language'.[7] The investigation either of problems or the nature of different forms of conceptual knowledge can be sustained from within the different disciplines in which such problems or different forms of knowledge arise, a possibility already articulated.

So instead, then, of speaking about specific problems or different kinds of knowledge, let's remain with Popper's and to a degree Warnock's insight regarding the examination of what might be taken to be common sense as expressed in ordinary discourse. We can do this by turning to Whitehead's contribution in all this: '...the very purpose of philosophy is to delve below the apparent clarity of common speech.[8] That is exactly what *To Be Or Not To Be Philosophical: A Tiptree Inspector Decides* delineates. But that kind of 'delving' has to be characterised by a way of handling what is delved into, a kind of *methodology if you like; a certain kind of spirit. Here Charles Peirce puts the matter exactly: 'Philosophy is the activity of discovering what really is true; but it limits itself to so much of truth as can be inferred from common experience' (Peirce: vol. 1, para. 184).[9]

Besides delving for the truth in a spirit concerned for what is true, there are two other features in this case study approach which constitute this author's manner of engaging in inquiry. The first is indicated by Hegel's remark that: 'Philosophy moves essentially in the element of universality which includes within itself the particular'.[10] Without endorsing his way of doing philosophy, this approach indicates the way in which philosophy can point to general truths arising from reflections upon particularities. Once enunciated these general truths can be employed again in a further specific or particular case in the future. Reading this book, for example, at a practical level, a reader can learn something as to how one can go about using arguments in the defence of one's own piece of landscape which may be under the threat of developers. Indeed I, myself, have re-read sections of this book more thoroughly as a result of being made aware of another proposed landscape development closer to home! Anyone reading this book at a theoretical level may come to appreciate the way such arguments can be justified or be liable to criticism.

The second feature of the philosophising that takes place in the development of this case study approach is indicated by Wittgenstein's remark about philosophy: 'Philosophy is a battle against the bewitchment of our intelligence by means of language'.[11] His aphorisms are, of course, subject to much scholarly debate as to their meanings but for this book's purposes, it is assumed Wittgenstein *might have meant* that ordinary language is something that has to be offset, put out of gear, regarded from a critical stance. In this way we can prevent ourselves taking ordinary, everyday assumptions for granted. One way of doing this is to focus upon particular moments of experience where we are at a loss to know what to say. Such moments or 'drops of experience', to quote William James, may be moments of intensity in the contemplation of nature, some artefact which we come across quite suddenly by accident – the Vietnam War Memorial in San Antonio, Texas, not mentioned by the bus tour guide, Deborah Butterfield's scrap metal construction of a horse, *Palma*[12] or Sherlock Holmes's statue, standing outside London's Baker Street station – or a turn of phrase someone employs, which might be regarded as peculiar, quite shocking or very endearing.

With the exception of a dream sequence at the start of chapter I, each chapter begins with a piece I have written in the light of conditions as described. Many of them were composed in acts of contemplation of the natural world and could be described as attempts to reverse what Whitehead called 'the slow descent of accepted thought towards the inactive commonplace'.[13] Because of their negative character and because ordinary common sense beliefs are negated or suspended, such pieces, which tend to set the tone for what follows in the rest of the chapter, could, following Adorno, be described as 'evanesces'.[14] That possibility is rejected because his reflective pieces are not tied particularly to specific places where they were composed. Such compositions could be called *reveries but that term recalls Rousseau's *Reveries of The Solitary Walker* and the study of his reveries shows that he used the term as Peirce did later, to indicate ruminations or daydreams which were triggered by the sight of something, *not* in relation to something's particular existence or its qualitative features. Perhaps such pieces are best called *aestivals, a term 'ugly enough to be safe from kidnappers' (Peirce: vol. 5, para. 414), as Peirce once wrote of his term *'pragmaticism'. After all the term 'aestivation' in zoology refers to the idea of being opposed to hibernation and, as a noun, is used in botany to refer to the form in which petals fold over one another in the flower bud. Of course, the term 'aestival' is not a noun, but an adjective meaning 'pertaining to the summer' and most human beings still seem to take delight in a warm summer's day. The 'noun' aestival then will be used to refer to being fully awake in the act of forming some expression whose meaning can unfold under examination, or lead to further elaboration. Such an aestival can bring delight, at least to its composer, since it characterises exactly what was experienced in a particular place at a certain time of day.

To speak in this manner is to refer to the idea of something being valued for its own sake, something's *intrinsic value. We are, then, referring to aesthetic experience for it is aesthetics, as Santayana once put it, which is 'concerned with the perception of values'.[15] Recall that the very question we are now considering, our fourth, was 'Why are intellectual activities important?', what gives them their value? Given that to engage in them one must know how to inquire, how to think – the subject matter of logic – two different kinds of inquiry are essential in philosophy. They have and always will be central in its pursuit: ethics – the discourse explicating the grounds we have for acting morally[16] – and aesthetics, the discourse that concerns the examination of experiencing something for its own sake.

An adult student, a car maintenance teacher, once said to me he found philosophy difficult because 'you are always wanting to tell me a story about something and the point of a story is that it isn't true!' He had, then, a particular conception of what the truth was. He was adopting a philosophical position: what was true was what corresponded with the brute facts of the matter where they can be clearly established. His observation incites two remarks to close this Preface. The first is that I doubt that he would be very satisfied with it since he, as a possible reader, has been doing philosophy with the writer. That is to say, he has been presented with a narrative as to what constitutes the nature of philosophy;

he has been offered a story as to what this writer thinks is the point of philosophical activity. Yet he might be pleased with what he finds in this book because each chapter has plenty of brute facts within it, which could cause him to reflect. It is hoped that his reflections upon them may be guided by what he finds in the third part of each chapter. It is also to be hoped that he finds they are written in accordance with the spirit of a philosophy defined in the following terms: philosophy is the critical interpretation of experience so that our normal or more conventional ways of rendering it meaning can be offset to reveal experience's redemptive, felt, yet fleeting character.

NOTES

[1]Alfred N Whitehead, 'Remarks' (1936) in *The Interpretation of Science: Selected Essays*, p.211.

[2]George Santayana, 'The Sense of Beauty: Being the outlines of Aesthetic Theory' (1896), in *The Works of George Santayana,* vol. II, p.90.

[3]Jay Appleton, *The Experience of Landscape*, p.8.

[4]Jean-François Lyotard, 'Science has always been in conflict with narratives. Judged by the yardstick of science, the majority of them prove to be fables. But to the extent that science does not restrict itself to stating useful regularities and seeks the truth, it is obliged to legitimate the rules of its own game. It then produces a discourse of legitimation with respect to its own status, a discourse called philosophy. I will use the term *modern* to designate any science that legitimates itself with respect to a metadiscourse of this kind making an explicit appeal to some grand narrative, such as the dialectics of Spirit, the hermeneutics of meaning, the emancipation of the rational or working subject, or the creation of wealth.' 'The Postmodern Condition' in *The Postmodern History Reader*, sect. 2. The term metanarrative refers to this 'metadiscourse', this 'grand narrative'.

[5]Consider conservation management. 'The current structure of education divides knowledge into separate disciplines and this makes the development of integrated management difficult. For example, management practice and economics are taught by business schools, policy formulation is rarely taught explicitly, ecosystem structure and function is the domain of science and conservation biology a newly emerging discipline. Unfortunately the separation of science into the classical disciplines of botany, zoology, chemistry, physics, etc., means that the physical, chemical and biological components of ecosystems are rarely linked. The consequence of the power divisions in academia means that the subject of conservation management, which involves integrating knowledge from all these areas, is rarely taught and where it is, few government agencies allocate sufficient funds to train their staff in the necessary skills for their jobs.' John L Craig,

'Managing Bird Populations For Whom And At What Cost?', *Pacific Conservation Biology*, vol. 3, pp.172–182, **p.176**.

[6]Karl R Popper, *Objective Knowledge*, chapt. 2 sect. 2.

[7]Geoffrey J Warnock, *English Philosophy Since 1900*, pp.158–159.

[8]A N Whitehead, *Adventures of Ideas*, p.285

[9]C S Peirce, *Collected Papers of Charles Sanders Peirce*, vol. 1, para. 184.

[10]G W F Hegel, 'Preface: the Scientific Cognition' in *Phenomenology of* Spirit, p.1.

[11]Ludwig Wittgenstein, *Philosophical Investigations*, part 1, para. 109.

[12]Deborah Butterfield's construction Palm – a named reference to Picasso's daughter Palma – was displayed at a Paris Gibson exhibition in Great Falls, Montana during the summer of 1996. It gives the impression of 'a horse – perceived to be leaning over some non-existent field gate – whose physical presence can be felt in a human shoulder whilst failing to make contact with it, within a spatial proximity normally reserved for she or he who would take such intimacy for granted.' (From my *The Art of Montanian Communication*, 30 June 1996).

[13]A N Whitehead, *Modes of Thought*, p.237.

[14]Theodor Adorno, *Minima Moralia*, p.16.

[15]George Santayana, op. cit., p.14.

[16]For a striking analysis showing how philosophical activity is tied to values, particularly moral values, see Julius Tomin's, 'Pursuit of Philosophy' *Hist. of Pol. Thought*, vol. V, no.3, Winter 1984, pp.527–542.

CONTENTS

INTRODUCTION

Noel E Boulting's introduction to philosophical thinking has a distinctive vitality arising from but not limited to his own personal concerns about an issue literally 'close to home'. He presents a story about a practical and protracted controversy that occurred in the geographical area where he lives. The controversy is over whether environmental-*ecological-interests have priority over economic social ones. This is the inspiration for far ranging philosophical reflection. The account of the proceedings is complex and filled with many details about the natural setting of the land subject to a commercial proposal. Boulting brings the controversy to life through the details of the pros and cons offered by both sides. He also introduces a dialogue with the reader once he turns to the application of the story to its philosophical assumptions, which are viewed in the light of the thought of the American philosopher, Charles S Peirce.

The account of the positions taken for and against the proposals are also embedded in some fascinating literary descriptions that serve an aesthetic interest. The aesthetic quality Boulting lends to his story seems to me not only to enhance his account, but also to link the story with one of the more fundamental purposes of his book, in addition to offering a case study as to how the decision whether or not to pursue the commercial interest emerged, which is to provide an introduction to philosophy through the framework of the American philosopher, Charles S Peirce. At the same time, it is the *aesthetic interest that plays the role of the final goal which Boulting finds could underlie the opposition to the commercial proposal. The literary quality of many of Boulting's descriptions is not surprising to me, because I know him to be a sensitive and enthusiastic admirer of nature as well as a person with an extraordinary commitment to the value of both historical and philosophical approaches to vital issues – to the 'importance of vital topics', if I may here introduce one of Peirce's own ways of referring to one of his writings. It also seems to me that there is a message coming from a broader, although Peircian, perspective. This is the promotion of our practising *critical common sense when addressing vital issues. The book is a call for critical reflection with common sense passion.

The Peircian basis adopted by Boulting is not intended as an exposition of Peirce's philosophy as a whole. Instead, it focuses on Peirce's three categories, which he developed in terms both of a conception of logical relations and a *phenomenological description of the most pervasive features of experience. Boulting attempts to relate these two in his 'A Tribute to Fathers?', his Epilogue to the book in Appendix IV. Boulting begins with and sustains a phenomenological understanding of the three categories in his application of them to the arguments offered by proponents of the two sides to the controversy. However, Boulting's book does not simply assume his interpretation of Peirce's thought without a contribution of his own. Boulting introduces a fourth category, which he adds to Peirce's short list of categories. I mention the short list because philosophers are prone to identify categories that are limited to specific aspects of what we experience, such as the category of things that are related by cause and effect or the category of essential nature and accidents or unnecessary qualities that may qualify a thing's essential nature. Here I would qualify Boulting's use of Peirce's categories because it seems to me that the original three can still accommodate what Boulting identifies as a fourth category. However, this is not the place to pursue this point. The main thrust of Boulting's argument stands regardless of whether one accepts the proposal of a fourth category. Indeed, the priority of the aesthetic quality intrinsic to ecological processes is Peirce's category of *Firstness, which Boulting discusses at length and places as crucial to his support of ecological interests, so important within present public debate.

What is of most interest to me is a fundamental philosophical issue raised by Boulting and the way Peirce's philosophy contributes to one way of resolving it. The issue is whether there is an objective grounding that may counter the relativism and *subjectivism seemingly rampant in many contemporary views of knowledge. This is to look beyond the *anthropocentric limits of human perspectives on the world. Much of my writing about Peirce has centred on this issue.[1] I do not claim that Boulting agrees with my way of moving beyond subjectivism, since he seems to regard Peirce's 'immediacy of feeling' to be identified with the phenomenological experience of human beings in relation to an object (see last chapter). I too believe that Peirce's ideas about aesthetic intrinsic value, 'the Admirable per-se', do not locate beauty or immediacy within mere subjectivity. Immediacy of feeling, as one way of identifying the category of Firstness, is neutral with respect to its *existential location. What can provide a move beyond subjectivity is a difficult issue, however. I would like to offer a suggestion of my own.

Peirce clearly recognised the diversity of cultural contexts and semiotic systems distinguishable even within the same culture. He insisted on the *fallibility of his own hypotheses and the radical conceptual revolutions that contribute and have contributed to the growth of scientific theorising. He also proposed his *metaphysical views tentatively in the sense that they were necessarily hypothetical proposals rather than demonstrative theories. However, Peirce also provided an account of how semiotic systems are subject to constraints not completely dependent on finite linguistic or conceptual frameworks. There is resistance that does not reduce to the linguistic and conceptual pressures of any historical or cultural period. These restraints are experienced under the category *Secondness to which Boulting calls attention and explains, especially in his discussion of the idea of landscape in chapter VIII. Instances of experienced restraints, of seconds, are, in semiotic terms, effected by virtue of the interactions of interpretations with what Peirce referred to as the Dynamical Object, or the object that functions in experience so that interpretation is invoked. The Dynamical Object is the object of interpretation. The object interpreted is the Immediate Object. There is always a residue, something not perfectly subjected to the interpreting process, and this residue lies in the extra-subjective Dynamical Object. There is inevitably more to be subjected to interpretative intelligence, more to a sign than its symbolic and qualitative aspects – to its *Thirdness and its Firstness. I believe that attention to the resistance of the Dynamical Object could contribute to Boulting's arguments in the final chapters of his book where the point is made that our responses to nature are not wholly subjective.

It seems to me, then, that Boulting's conferring of priority on aesthetic quality does not in itself provide the *objectivity* of nature that he insists on in response to the finality of the perspectival relativity commonly attributed to aesthetic value. He does argue that the priorities of nature are in some respects resistant to the feelings and attitudes of human perspectives. However, I would like to see more attention given to the condition of the resistances in observing and appreciating nature, which would imply giving more attention to Peirce's interesting, although difficult, view that there is always a Dynamical Object underlying the results of interpretation. This resistance is experience of constraints that pressure theory and aesthetic responses, pressing science through surprising observations and insights that sometimes occur when there are innovative, creative acts involved. In science, the condition of these reactions to the world yield new theories and result in evolving laws of nature. Similarly, there are the constraints of the Dynamical Object on interpretations and appreciative acts of artworks and of beauty in nature. The constraints of course can be expected to vary according to historical contexts and cultures, but those constraints, on the view I see suggested by Peirce, are also dependent on something external, some condition that is wholly independent either of the whims of subjective responses or the contingencies of cultural likes and dislikes. This suggestion needs much more discussion, which I have offered in the works mentioned in my previous footnote. But whether one claims such a suggestion seems to raise a mystery rather than a resolution, that suggestion is still well worth considering.

Carl R Hausman

NOTES

[1]Charles S Peirce, *Evolutionary Philosophy*, Cambridge University Press, 1993; 'Metaphorical Reference and Peirce's Dynamical Object', the Appendix to *Metaphor and Art*, pp.209–231; 'Charles Peirce's Evolutionary Realism' in *Classical American Philosophy*, and a manuscript to be published, *Charles Peirce's Process Realism*.

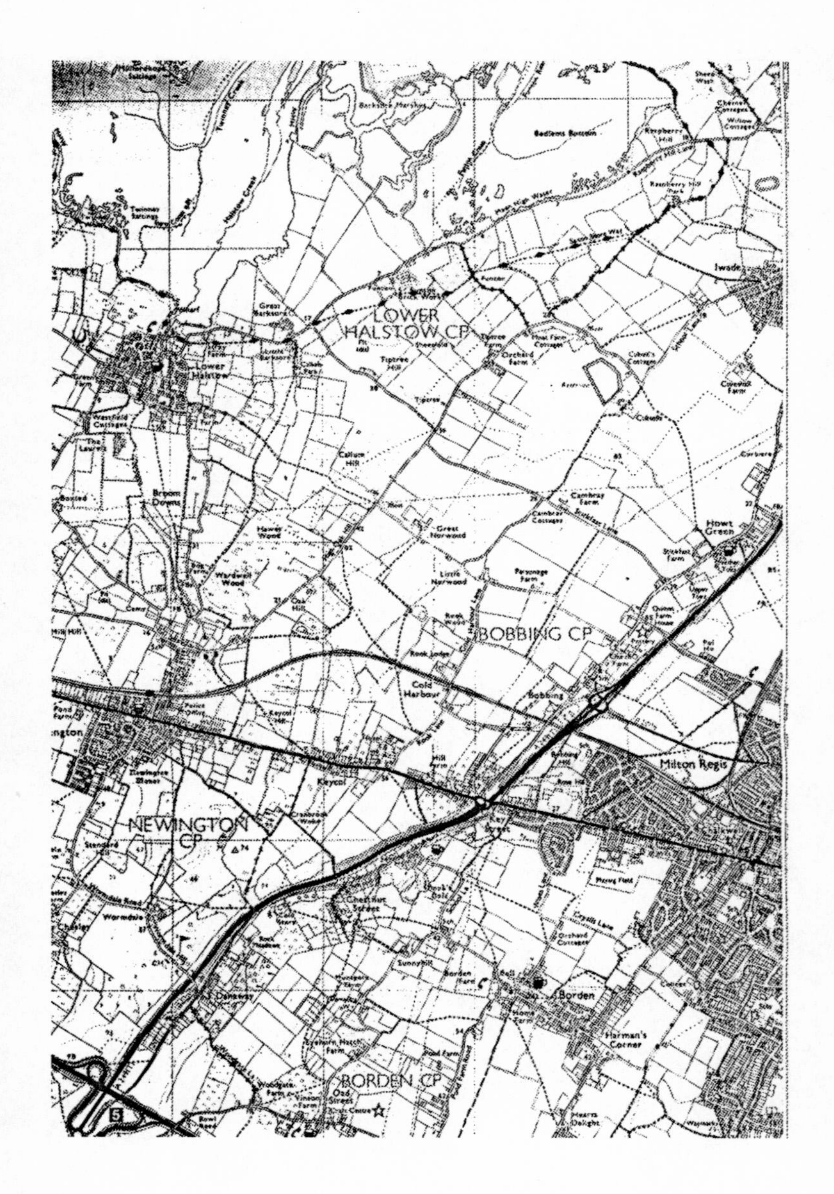
LOWER HALSTOW CP
Lower Halstow
BOBBING CP
Bobbing
NEWINGTON CP
BORDEN CP
Borden
Milton Regis
Iwade
Howt Green
Cold Harbour
Keycol
Harman's Corner
Broom Downs
Great Norwood
Little Norwood
Badlesmere Bottom
Raspberry Hill
Sunnyhill

PROLOGUE

Imagine travelling from London to Canterbury as if the M2 had never been constructed. The A2 main road can be used, the old Roman Road – Watling Street. The Medway Towns will be traversed before passing through Gillingham and Rainham. Beyond Rainham a garage and a Little Chef Inn will appear on your right. A left-hand turn leads into Oak Lane where one passes under a railway bridge rebuilt after having been struck by a V1 rocket during the Second World War. Just before the golf club, a left-hand turning into Canterbury Lane will be passed, where a two-seater Bleriot monoplane with a tabby cat and two Frenchmen aboard had to land on 12 August 1910 because the aeroplane had run out of fuel. Having reached a high viewing point, the road drops suddenly into the village of Upchurch. Here a church appears where Sir Francis Drake's father was a parson. The road continues towards Lower Halstow. Parking your car in the village will enable you to walk north along the wharf side opposite the Norman church. Tiptree Hill stretches north-eastwards behind this building and can be seen as you walk along the wharf. As this is done you will be traversing what is called the Saxon Shore Way.

The proposed site at Tiptree Hill, the focus of the present book's inquiries, lies within a mile, east from Lower Halstow as you pass through the village to go to the Isle of Sheppey. On the Ordnance Survey Landranger 178 – The Thames Estuary – Map, its grid reference is 874674. It is positioned in a prominent part of the highest section of land in a flat landscape west of Iwade village, recently granted a motorway bypass, and north of Newington, where pilgrims used to stay on their way to Canterbury. It overlooks Minster and the Isle of Sheppey to the east, the Medway Estuary to the North, and Gillingham and beyond to include the Medway towns to the west. Three miles to the South East is the town of Sittingbourne.

Continuing from Lower Halstow, you can turn right after Barksore Hill, following the sign to Sittingbourne to climb Tiptree Hill[1] rather than continue to the Isle of Sheppey. The site is now on your left-hand side and Callum Hill on your right towards the south. Once on the top it does not seem unreasonable to say that you will see how this line of hills provides views as much enjoyed by locals as visitors. Not only can the river with its low lying marshland and adjoining creeks be seen but also a port and industrialisation at Thamesport across the Medway on the Isle of Grain to the north-west. The agricultural land, dotted with farm buildings, is designated Grade 3 standard. It supports spring blossoming fruit trees, beef cattle, sheep and horses, the latter used for recreational purposes. On the site's northerly and westerly sides there are a considerable number of very old oak trees. A small mixed woodland stands on the western boundary of this site. Two small ponds are situated on the site's southern boundary.

Having enjoyed the magnificent views of the Thames and Medway estuaries to the west of this site, you will see, as you proceed east, beyond the steep north facing slope, that the land falls gently in a southerly direction. Suddenly Rushenden Marshes on the Isle of Sheppey will appear as well as Iwade. Eventually Ridham docks on the Swale to the north-east of the Isle of Sheppey can be seen.

The hill rises up to 58m AOD (above ordnance datum) since it is a ridgeline regarded as a London Clay outcrop. A bridleway passes from the east of Funton Brickworks at the north corner of the proposed site south to School Lane which runs from Newington village to Iwade. The main path, a section of Kent coast's long distance footpath – part of the Saxon Shore Way[2] – crosses the proposed site from east to west.

So much for geographical matters; you are now equipped to begin understanding some of the issues in the events about to be narrated to you underpinning this – a philosophical – inquiry. Of course, the reader must be on his or her guard about the possibility of the *fallibility* of the descriptions of these events, firstly, because they are written to give a sense of their immediacy and, secondly, because they are written from a layperson's standpoint!

NOTES

[1]To local people Tiptree Hill is referred to as Basser Hill, the term Basser being a corruption of Barksore; the hill I referred to as Barksore Hill is in fact named Little Basser!
[2] The Saxon Shore Way is a footpath 140 miles long which runs alongside the River Medway. It begins in Gravesend and takes in Kent's scenic coastline (on its way) to Rye in Sussex.

Chapter I

HUMAN INVOLVEMENT

At the side of the road by which I walked, a large pool of water came into view. It emptied into a stream further along the road ahead of me. This pool was very inviting; clear light shone right through it. Since I became aware of just how warm and comforting the water was, did I bathe there quite nakedly? I felt I rolled and frolicked in it. It was so very beautiful, for the light was in and around me wherever I turned. Or was it that I was filled with a desire to lie there as I found myself standing once more on the side of the road looking into it? It was indeed a magical pool.

A small, smiling, black-haired girl may have been playing in the water too, but I only became aware of her presence on turning round to see her standing at the top of the stream which ran into this pool. The upper stream fell like miniature falls gently into this shiny-surfaced water. The vision of her inclined me to share this experience with others, hoping that they might like to play in the pool too. I went to the village to find them.

When I returned we were all disappointed. I visited the farmer nearby, passing through his home to reach him at the back of his house. A red-haired woman was driving a tractor. She said the water had been diverted onto the farmland, as it was needed to irrigate the ground. The soil was very parched. However, she said that the water's flow would be redirected to its normal course once they had finished with it. We would have to wait. In the meantime she displayed its magical powers by showing me what happened as it was splashed onto the dry soil. Immediately potatoes grew and she was able to pull the tractor over the plants so that a great yield was produced. I returned through the farmer's house, but paused in his drawing room, a beautiful room surrounded on all sides by French windows enabling me to see Kent's rolling countryside. As I watched the tractor I wondered whether the pool would be restored or not.[1]

Large black letters appeared on a bright, golden-yellow poster. It read, 'Say No to Tiptree Tip'. It hung in more than one window in Upchurch. Seeing that poster was my first introduction to the idea that there was a place called Tiptree. Later, in our local magazine, details were given of a meeting in Sittingbourne, at the Wyvern Hall on Thursday, 27 July 1995. Having cycled to the town I entered what is called the Swallows Leisure Centre to examine, outside the entrance to the hall, large maps and photographs explaining the nature of the proposed Tiptree Landfill development.

The meeting had been called to consider an application by a company – South Eastern Minerals Ltd – to extend what had been a borrow pit, a clay extraction area, on Tiptree Hill adjacent to the road up Barksore Hill. The site would be extended from this pit north-eastward to include the whole area between Barksore Hill Road, the Sheppey Road and School Lane creating a zone the size of the financial city of London. The site would then be dug to remove London Clay to a such a depth that even a building as tall as Big Ben could disappear within it without trace! The space created would then be infilled with rubbish. The meeting had been called, as its chairman explained, to enable Kent County Council's planning sub-committee officers to hear the views of some five hundred local residents, who filled the hall and other interested parties in relation to this application.

The applicant's representative spoke of the feedback the applicant – South Eastern Minerals Ltd – had received following a consultation exercise involving three exhibitions. He explained that three million cubic metres of clay would be excavated over ten years. A further three years would be necessary to remove an extra one million cubic metres. He tried to reassure his audience by saying that before any clay was removed sixty thousand trees would be planted and bunding (a bund I take to be an embankment) five metres high would be erected around the perimeter of the site to screen it. This screen would provide protection for walkers along the path – part of the Saxon Shore Way on the north-west side of the site, parallel to the Sheppey Road – as well as local residents living along School Lane and Barksore Hill.

To the accompaniment of some noise in the hall, he claimed that, on the basis of the information received from the County

Planning Authority, the National Rivers Authority, the Medway Ports Authority and various contractors, he, on behalf of the applicant, believed that over the next five years there would be a shortfall of clay of between 1.5 and 2.4 million cubic metres. Covering that shortfall, besides providing clay reserves over the long-term, was the point of the present application. In addition, London Clay provided a good base for rubbish infilling. This site, he argued, was well placed to satisfy the need for clay in the north of Kent and adjoining areas for sea and river defences, construction, land reclamation and so on. In addition, from an ecological, agricultural or landscape perspective, the site was not subject to any kind of significant environmental or planning constraint. Clearly the local roads would have to be upgraded to achieve a suitable standard, especially Stickfast Lane and its Sheppey Way junction, to accommodate the movement of forty vehicles per hour. As far as the new housing development for Iwade was concerned, that was 1.8 kilometres of open countryside away from the site, and the building of such houses was to start as late as 2001, perhaps even 2006, some six to twelve years after the quarry had been excavated.

A local county councillor pointed out that this countryside was rich in bird life (the Habitat argument) and added that there was no need to extract the clay (the Clay Need argument) The Road Danger argument was also put: how could cyclists and pedestrians use these lanes any more if so much traffic traversed them. To widen them would affect Grade one agricultural land (the Agricultural Land argument) A neighbouring Liberal county councillor reinforced these arguments and then added another: the development and associated traffic would cause great damage to the locality and its environment.

The majority leader of the Swale Borough Council – another Liberal – stated that the proposal was at odds with the ongoing developmental plans forming the new Borough Strategy. Although the area was covered by the Thames Gateway Planning Framework, initiated by the Conservative Government at Westminster this involved no requirement for clay. Indeed clay was not a good material for land reclamation and it would be a criminal act to destroy the Tiptree Farm site since it commanded fine views in all directions over the estuary and beyond (the Landscape argument).

Besides challenging the claim for the need for clay, the borough's Chief Planning Officer listed the various plans this application infringed. A clear distinction, he said, had to be drawn between a supposed demand and an actual need for clay. He also explained how it offended against the Draft Waste Local Plan (the Waste Disposal argument) since the infill method was no longer regarded as the best way of disposing of waste. In addition, he underpinned the Landscape argument by saying that under the developing Swale Local Plan E68 it was proposed to make the site part of a Local Landscape Area, thereby excluding it from such developments as the applicant desired, since it served as an important backdrop for so many locations in this area. He raised, too, the issues of the new housing development planned for Iwade, excessive traffic movements and the significance of the Saxon Shore Way. In addition, he continued, although the conservation group English Nature had withdrawn its objection to the proposal, the Kent Trust for Nature Conservation had an objection and the Royal Society for the Protection of Birds had yet to make its position clear.

The chairman of the Lower Halstow Parish Council gave the most entertaining speech. Once he had thanked the KCC Planning Sub-Committee for arranging the meeting he focused on the central issue as he put it: the Need for Clay argument. If there was no need for clay, then there would be no hole and thereby no waste disposal site. Before he set out that argument, however, he did reinforce the Landscape argument in pointing out that the only objection to Swale Borough Council's draft local plan for making the site part of a Local Landscape Area arose from those who were the representatives acting for the applicant.

He launched into the Clay Need argument by pointing out that the Kent Minerals Local Plan regarded permitted sources of clay sufficient to meet present-day requirements. To start with, a local site, supposed to need clay, was already receiving dredgings and materials from London's Jubilee tube line extensions and this demand would end in December 1995. The Sheerness Dockyard – a listed area – hardly needed London clay since that shrinks and cracks as it dries out, making it of no use, and in addition London Clay was hardly suitable for capping other waste sites.

The Wyvern Hall echoed with laughter when he pointed out that the applicant had stated it would need 500,000 cubic metres of clay to infill the Sheerness dry dock. Since the dock measured 90 metres by 30, it would have to be over 600 feet deep to take the clay. Clearly the actual figure had to be 50,000 cubic metres at most. Other inert materials taken from elsewhere would serve the purpose. Obviously then this application had to be refused.

The chairman of Iwade Parish Council argued that Iwade residents would have to endure noise, dust and smell for many years to come, besides suffering from the fact that lorries would clearly use the quickest rather than the prescribed routes. Hence, children would be at risk from road accidents and risked illnesses from air pollution. Again, how would local residents be compensated for the loss of value in their homes? Chairmen from Upchurch, Bobbing and Newington parish councils supported these views and underlined other objections expressed at the meeting, particularly the consequences of heavy traffic use on other adjoining roads.

The chairman of the Tiptree Action Group (TAG) explained how the group she represented had formed to fight this application and keep members informed as to its 'progress'. She claimed that the application was being forwarded at the wrong time, at the wrong site, using the wrong access point and was being foisted on the wrong community. It was common knowledge that the Secretary of State, John Gummer, was opposed to landfill as a form of waste disposal. What wasn't good enough for him shouldn't be good enough for KCC either! Methane gas oozing from the site on its restoration in years to come would be a pollution issue (the Environmental Quality argument). She then raised the specific issue of road accidents and their implications for residents.

Another TAG member, after pointing out he had won a KCC award for promoting Environmentally Sensitive Tourism, drew attention to the rich potentials for overseas visitors in relation to wildlife, stressing, in reiterating the Wild Habitat argument, that at this site four of the five British species of owl nested near by. A healthy range of predatory birds indicated there was abundant wildlife. Although he did not specifically draw attention to the Scientific argument – that the proposed site is adjacent to a Special Protection Area as a Site of Special Scientific Interest (an SSSI site) – he argued that a more thorough ecological survey was needed than had been provided by the applicant so far. If the present site was indeed developed, it would take at least one hundred years for it to return to its present ecological condition.

In her summary of TAG's case, its chairman asked how, if the applicant company had only been in existence since November 1994, could any track record be assessed and proof of any kind of effectiveness in relation to its long-term management be tested?

A report was presented by the British Riding Schools' Association's representative on behalf of the Padbrook Meadows Livery and Riding Centre at Little Callum Farm located at the bottom and on the other side of Tiptree Hill Lane. It was not only the danger from accidents which was worrying about this development but sudden noises which could cause horses to throw their riders, besides the sudden appearance of large vehicles (the Countryside Protection argument). Vermin from the tip could well arrive at the stables. The school's work, funded through the Department for Trade and Industry (DTI), supported by the borough council and its Economic Development Unit, would be undone by this application.

The land manager of Millread Homes then underlined remarks made by previous speakers about the prospect of 'a buying blight' for the homes that were to be built in Iwade. He too doubted the need for clay and pointed out that the only kind of experience the applicant seemed to have had was in quarrying ragstone. The representative for Ward Homes supported this view, drawing attention to 'creative invention' in relation to the applicant's claims about a clay need.

After pointing out that infilling as a means of waste disposal was being replaced by the use of constructed waste transfer stations, the representative for the council for the Protection of Rural England also rejected the applicant's use of the Clay Need argument. Moreover, he underlined the Landscape argument by pointing out that this site also acted as a 'buffer zone' between different areas of development. Then a spokesperson for the Ramblers' Association, though stressing the affects of the application on the Saxon Shore Way, developed a point made earlier by another TAG member, 'What type of waste would be used for infilling at the site?' She was opposed to the application since she had noticed that some lay-bys had suddenly appeared in Stickfast Lane and that the volume of Heavy Goods Vehicles (HGVs) had increased.

The borough councillor for Hartlip and Upchurch stressed once more the traffic implication, particularly in relation to an already overcrowded A2 (Watling Street). A local resident advocated the idea that the site be turned into a countryside park so that everyone could enjoy local views, and a neighbouring county councillor said, 'Everyone should be committed to opposing the "desecration of the North Kent countryside".'

The meeting's chairman thanked those who had attended and remarked that, despite the fact that feelings were running high in regard to this application, he thought that the meeting had been a constructive one. The next step would be the county planning officer's professional recommendation in relation to this application. In the light of this meeting the KCC Planning Sub-Committee would then consider this application at a meeting where anyone, besides the interested parties, was welcome to county hall to listen to the debate. As I cycled home that evening it seemed to me cut and dried: there was no chance that permission would be granted to extend the borrow pit at Tiptree for clay extraction and waste infilling purposes! This pit, from which clay is sometimes 'borrowed' – as Scottish folklore supposes that March borrows its last three days from April[2] – 'borrowed' to be used for sea defence purposes, would remain as it now was.

In the dream – 'Dreamt-Pool' – narrated at the beginning, there seems to be a concern in relation to two distinct kinds of attitude. One attitude implies that the pool be seen as inciting pleasure, *aesthetic[3] pleasure or, alternatively, its waters could be seen as *instrumental, say, in the growing of crops. For the moment no specialised use of language is intended here; checking a dictionary will confirm that the term 'aesthetic' is connected to the idea of appreciating something, usually as beautiful, and seems related somehow to principles of art creation. Perhaps the novelist Charles Dickens was referring to this completely non-instrumental perspective when he remarked of the River Medway: 'Running water is favourable to daydreams and a strong tidal river is the best running water for mine'.[4] On the other hand most dictionaries tie the term 'instrumental' not to the activity of *contemplation – appreciating something for its own sake – but rather to the activity of doing something in which some object is used as a means, not as an end in itself, for producing something else. Does this distinction between the aesthetic and the instrumental throw some light on the proceedings of the Wyvern Hall meeting?

You might wish to object to this suggestion. Clearly the Landscape argument, and, perhaps, with some strain the Wild Habitats together with the Countryside Protection arguments might veer towards the aesthetic. And perhaps, too, the Need for Clay argument and the case for Waste Disposal can be covered by saying that both provide instrumental reasons for opening up a quarry at the Tiptree Site. But can the Agricultural Land argument, and – as part of the Environmental argument – the increased Traffic and Pollution arguments, as three arguments, be brought together by claiming that all three are simply bad consequences of the instrumental use of one part of the land. The first of the three – the Agricultural argument – represents the undoing of one instrumental use by another. The second – the Traffic argument – represents a direct attack on the quality of life to be enjoyed or suffered in the area. The third – the Pollution argument (of air or water) – could have wider implications. And what about the *scientific argument, which hardly seems to fit any of these categories?

In order to resolve these difficulties the writer suggests that reference be made to the greatest American philosopher – perhaps greatest world philosopher – in the nineteenth century. His name is Charles Sanders Peirce, who was born in 1839 and died in 1914. In 1896 he wrote the following about the scientific attitude:

> If we endeavour to form our conceptions upon history and life, we remark three classes of men. The first consists of those for whom the chief thing is the qualities of feelings. These men create art. The second consists of the practical men, who carry on the business of the world. They respect nothing but power, and respect power only so far as it (is) exercised. The third class consists of men to whom nothing seems great but reason. If force interests them, it is not in its exertion, but in that it has a reason and a law. For men of the first class, nature is a picture; for men of the second class, it is an opportunity; for men of the third class, it is a cosmos, so admirable, that to penetrate to its ways seems to them the only thing that makes life worth living.[5] (Peirce: vol. 1, para. 43)

You could argue against what is said in this quotation right away: after all, what Peirce says or at least implies about there being only three classes of men, never mind three basic attitudes, seems at least somewhat preposterous! It has to be admitted that there are some difficulties in understanding this passage, even if it does delineate three ways of regarding nature – aesthetically, instrumentally or scientifically. To distinguish only three such attitudes does seem to be a bit odd; this writer will suggest at least four! But before this passage is dismissed, three points in defence of it must be borne in mind. Any writer spending long periods of time on his work without direct contact with other writers[6] might enjoy his moments of self-amusement. Indeed, elsewhere Peirce does speak of inciting a sense of 'buffoonery' on the grounds, as he said in 1903, 'that a bit of fun helps thought' (Peirce: vol. 5, para. 7). Because it seems so absurd, the reader is not likely to forget his claim about 'three classes of men'. Secondly, this division into three categories is very important for Peirce's own ⋆methodology – how he goes about doing philosophy – the significance of which we will need to explore subsequently. Thirdly, this quotation clearly implies something about what it is to be a human being: that human beings are feeling creatures capable of action who can think or reflect about the two other dimensions, or at a subsequent moment about thinking more generally understood! Peirce refers the term ⋆Firstness to ⋆feeling, ⋆Secondness to action or what can be brought into *existence through action of some kind, broadly interpreted, and *Thirdness to *thought.

What has just been written reflects a basic presumption in this writer's approach to *philosophy: because all three features are central in the experience of human beings, all three realms – those of feeling, action and thought – have to be addressed in any kind of philosophic discourse, even if much contemporary philosophy is prepossessed with what can be called thoughts about thoughts. And any education for any kind of learner, if it is to do justice to these three capacities, would have to be measured in terms of its effectiveness in relation to these three dimensions: feelings, doings and thinkings. It was John Dewey, Peirce's student, who, in his life as an educator, demonstrated the importance of these ideas, even if his educational philosophy is not so fashionable today.

It is quite clear that in relation to these three attitudes some kind of awareness, even *cognizing, knowing or understanding – but not of the same kind – is involved. Elsewhere, Peirce, in 1903, pays tribute to the artist as being capable of 'seeing what stares one in the face, just as it presents itself'. The ordinary man at the sight of snow, sees 'white, pure white, whiter in the sunlight, a little greyish in the shadow', whereas the artist 'will tell him that the shadows are not grey but a dull blue and that the snow in the sunshine is of a rich yellow' (Peirce: vol. 5, para. 42). And this tribute to the artist in representing nature accurately in his activity is quite distinct from the artist's use of his or her imagination in creating a work of art which enables the spectator to appreciate something different about nature altogether! Of course, Peirce does praise imagination, but he is more interested in scientific imagination rather then 'mere artistic imagination' or 'mere dreaming of opportunities for gain. The scientific imagination dreams of explanation and laws' (Peirce: vol. 1, para. 48).

As has already been implied, the phrase 'opportunities for gain' refers to the *practical and knowledge here has an instrumental value: it can be applied in a means-end sense by someone 'for some ulterior purpose such as to make money, or to amend his life, or to benefit his fellows' (Peirce: vol. 1, para. 45). If South Eastern Mineral Ltd had forwarded some benefit to the Lower

Halstow people by offering, for example, to provide a well-furnished playground for children, resources to improve learning conditions at the school or a forum to decide what might be done with a grant to improve the village's quality of life, such a gesture might have undone opposition to the Tiptree development. Such a desire to benefit others in an instrumental sense, however, was never demonstrated! Finally, there is the kind of knowledge which most interests Peirce – the *scientific. Pierce defines this in terms of inquiry 'into truth for truth's sake', when someone's activity is controlled 'by passion to learn' arising from 'an impulse to penetrate into the reason of things' (Peirce: vol. 1, paras. 43. 44).

By now you may have anticipated a fourth possibility. Peirce clearly does not consider the idea that in a given human being's life there might be no 'chief thing', no activity which 'seems great' in itself which has any kind of *being-aware content in it at all. In other words, there could be a class of attitudes which do not register an awareness in themselves, attitudes which negate or are entirely opposed to the three delineated so far.[7] So, for example, someone can be lying in the sun on a beach just for the sake of 'getting a tan'. Whilst lying on the beach there might well be no concern for the experience of any object for its own sake, thereby ruling out any kind of aesthetic dimension. Could this activity be subsumed under the practical, because s/he may think that the sun's rays are doing him or her some good? If questioned – surely s/he might think this activity a bit odd, especially if the interviewee didn't know there were such people who asked strange questions such as philosophers are thought to do – answers might be readily given: 'I find it enjoyable', 'Look what other people are doing' or 'My mates at work will ask "Where did you get that tan?".'

Now you may have detected a chink in this argument. What could be more purposeful than lying in the sun to get a tan to impress your mates? It would not count as working or labouring but it would be doing something purposeful wouldn't it? Two arguments can be used to strike against this objection. Firstly, all instrumental activities are tied to something which exists and such objects sustain salient features of that thing, allowing for its purposeful use in our active concerns. Lying in the sun is merely a passive state. Of course, the sunbather's mind might be active: the sunbather could be thinking about what to do with the rest of the day, reflecting on some cognitive problem or, indeed, recalling the memory of someone who once enjoyed the sunbather's company. None of these possibilities, however, define or are necessarily connected to the state of simply lying in the sun.

A second argument develops this one: you don't need, and indeed it might be safer not to use, the sun for sunbathing purposes. A sunlamp could be substituted for the same effect and may well not diminish your enjoyment in getting the tan. Speaking more broadly, the case for pleasure, for self-satisfaction alone can take us away from the *intrinsic properties of things in the world we use to trigger them. Consider, for example, the significance of virtual reality experiences, which lie in their capacity to trigger pleasurable sensations, substituted for objects in the world normally associated with such satisfactions. Consider how a child, for example, in using a slot machine, can simulate racing a car as opposed to really driving one!

None the less it does need stressing that the search for pleasure is usually associated with an attempt to escape from work. Consider the life of those who are in employment: many travel to work peering through a car screen, are occupied at work by a computer screen and then entertain themselves at home by turning on a TV screen. The latter activity, to judge by most of the programmes available, are designed to provide an escape or a distraction from any other awareness which may have arisen at work. Indeed, that last sentence used a term which could serve to name the kind of attitude Peirce does not consider – the *distracted.

Before you reject what has just been suggested, let's consider why this idea of the distracted might be useful in relation to this meeting's proceedings. Clearly not every resident in every village attended. Was this because it was not at a convenient time? Not every window in the area displayed a bright, golden-yellow poster either. Surely it cannot be assumed that thereby support for the Tiptree development is being indicated by their absence? After all, some people might object to having any kind of poster in their window and might even regard seeing them displayed elsewhere as an aesthetic obscenity! But no posters have been displayed indicating support for the proposal. Is it really reasonable to assume, then, that all the non-displayers are too frightened or not sufficiently interested to indicate some kind of support as opposed to a rejection of this proposal? Surely not.

You may well have been angered by what you have just read in the name of an activity which purports to be called philosophy. Firstly, are not value or *normative judgements being passed on people's lives whose views are not taken to fall under the three categories Peirce delineated. Secondly, it is an *empirical question as to what attitudes people actually have, which means that the writer would have to do some studies or try to gather evidence in order to establish his claims as to how people regarded the proposed development. Thirdly, much of what is assumed here does rely on the idea that the focus of people's lives is provided by what they do for a living. Is that true? Finally, surely it's bad enough trying to understand the use philosophers make of such terms as 'aesthetic', 'instrumental', 'scientific' and so on – without, in addition, trying to understand the particular jargon Peirce employs: Firstness, Secondness and Thirdness for example.

Your first objection can be met by agreeing there is always a danger that in *describing something – saying what it is, giving an account of something – it is easy to be *prescribing something, saying what should be so or commending it. Clearly the meeting's proceedings were described, whilst in saying that a particular speech was 'entertaining', that speech was prescribed or recommended to you. If one does admire Socrates – judged by many to be the first philosopher, since he asked questions in an ongoing discursive activity – and prescribes his philosophy, then one may well adopt the view that the unexamined life is not worth living! To the obvious retort 'Who says so?' it can be replied that we are not seeking an *authority to settle this question for us, but rather what the *argument is for showing that the unexamined life is more worthwhile to live than the examined one, an argument Peters, for example, explored.[8]

In the light of this admission, you may seek to press the writer on the grounds for his claim that earlier he was doing a descriptive job in relation to certain people's attitudes in our culture. In reply, just as reference was made to the quality of television programmes, for which 'ratings' are available in terms of their popularity, so it can be pointed out that the millions who read the tabloid the *Sun* every day, or for whom TV commercials are regularly made, such persons are indicated a form of life which involves thoughtless consumption, whether that be in terms of food, drink or having a good time on holiday. Indeed, such commercials can be 'normative' themselves: one seen at a cinema recently narrates very graphically the danger to natural life itself which Western man's present existence threatens. Suddenly it finishes with the remark, 'Forget it and come down to us, your local travel agent and book your holiday now', to the accompaniment of appropriate images!

You might reply that this move, in assuming what people understand of cultural phenomena is in danger of landing its writer in a *fallacy, an error in thinking. You might wish to claim that the writer has assumed that the way cultural messages are produced and transmitted determines the way they are in fact received by their consumers.[9] Many people might regard the *Sun* as something to be read as a comic! Such claims then raise complicated questions as to the relation between what people say and what they really believe! It might be better to rely more on direct experience of what people actually say in relation to something specific, for example on the nuclear weapons issue, membership of Greenpeace or the acceptance of a Tiptree poster! 'Not just now, I'm busy'; 'It has nothing to do with me'; 'There's nothing I can do about it' are not isolated comments. They suggest that people, for one reason or another, might not be able consciously to address an issue; they are distracted by other problems. Difficulties with children, earning enough to live or concern with their own personal problems spring to mind. They are in a state of distraction.

Until now we have distinguished only one noncognitive attitude, namely the distracted which, although appearing analogous to the aesthetic in being passive, is opposed to it since it is not concerned with contemplating something specific. The term 'escapist' does not seem to be an unfair way of characterising this attitude. Taking drugs, consuming alcohol, spending hours watching TV uncritically are activities which may be chosen to escape from some specific state of affair. But there are, of course, activities which can engage us to take our minds entirely away from the broader cognitive issues which are the focus of this chapter. Indeed, in talking about the human response to technological advance, it has become fashionable to distinguish 'back viewing' – passively watching TV for example – from 'forward viewing', interacting with what appears on the screen. If the term distracted relates to the former, the term the *distractive can be used to characterise the latter. Take, for example, that remark uttered when a resident opened the door and in response to the idea of taking a Tiptree poster to hang in a window replied 'Not just now, I'm busy'. So many jobs are waiting to be done, especially at the weekend: cleaning the house, cutting the lawn, taking the children out and so on. Unlike the case of being distracted, there are activities which are not passive: they can generate a general sense of busyness and they involve work, unpaid or shadow work if you prefer to put it in that way. And to make that point is to say something about the third objection, since busyness at home can mirror a sense of what is done at the place of employment!

To this third objection, then, against the claim that the focus of most people's lives is centred on the need to work – as though we now live to earn rather than earn to live – this writer is saying that such a view is accepted as *axiomatic in our culture. The focus in education – as full-time work opportunities shrink by the day – is upon vocational training, so that schools can be regarded as providing young people with requisite skills helping them to do jobs to enable the country to be competitive in the modern world. Go to any social gathering and people will ask you what you do in terms of earning a living. In addition, those out of work seem to feel the need to explain why they are no longer working, unless they are of an age where one can be proud to say one has retired, in which case they can now be busy around the house and the garden!

So far we have delineated only two noncognitive attitudes in relation to the concerns which were the subject of this public meeting held to discuss the Tiptree development. The first was the distracted, held by those seeking to escape the worries of their world. The second was the distractive, which characterised the attitude of those too busy in relation to the work demands of their job or home to be able to focus on local environmental developments. A third can also be distinguished: the *enured.

The word inured means accustomed or habituated, perhaps hardened. For better or worse, we may have become too used to hearing about earthquakes in different parts of the world, killings in some emerging nation state or ecological disasters in an environmentally sensitive area of our globe. Earlier I cited the example of the cinema commercial. Most of us do not explore the cognitive awareness which may explain how these events arise, what they actually involve nor their long-term consequences. They are simply held in consciousness whilst reported. We may send a donation to help ameliorate those consequences, speak to someone else about the horrors involved or fret about them in the middle of the night. But by day our cognitions are elsewhere engaged.

I will use the term 'enured' to refer to a noncognitive attitude in relation to cognizing. This attitude is used or practised habitually in relation to such reporting, conveyed by what we read, see on TV or hear on the news. Normally we absorb arguments with respect to the kinds of things already indicated. We make up our minds or form some cognitive decision about them simply on the basis of 'the spin', the presentation or the 'packaging' of the points of view which are set before us. We hold no specific cognitive attitude ourselves because we have not done the necessary inquiry into such matters in order to be able to take a secure cognitive position in regard to them. This attitude is given more consideration in chapter X, but for the moment we need to bring the ruminations of this chapter together.

To make sense of the proceedings of that July night meeting in 1995, it was claimed initially that one of three attitudes could be taken towards the developments proposed at the Tiptree Site. Concern for its affects on the perceivable features of the landscape for their own sake alone, would reflect an interest tied to what has been called an aesthetic attitude; seeing the site as a valuable resource for the production of clay and its use for infilling represents an instrumental attitude; a third would be one manifested in a concern as to what will happen to undermine the ecology of adjacent sites of special scientific interest. Finally, a fourth class was delineated. It was made up of three further attitudes and a person sustaining one of these three might be quite uninterested in any of the issues raised by the contents of this chapter. The distracted was a term for a passive attitude where a person might simply want to collapse, perhaps out of sheer exhaustion, in the face of all the things they see as needing to be done. The distractive was a word used to characterise a state of being busy at home or work possibly to the point of being under stress in relation to the tasks one is concerned to complete. The enured referred to a noncognitive attitude with respect to something cognitional. Consider a character in a rather low-spirited play by Alan Bennett – *Getting On* – who remarks that her passion for Coleridge and his poetry was successfully undone by her university education so that she would be immunised against such intellectual enthusiasms for the rest of her life![10]

What has not been drawn out sufficiently so far are the implications of the way arguments were used at the coming inquiry. If claims in relation to the noncognitive attitudes are suspended for the moment, it is clear that instrumental arguments really dominated the debate at the public meeting. These were supported to a degree by the aesthetic whilst the scientific hardly emerged at all. Is this point unconnected to the idea that we live in a work-orientated culture? That issue will require further exploration.

Another issue, not really examined so far, has been the claim that there can only be these kinds of attitudes in relation to things external to ourselves. If there are more, they will have to be combinations of those already delineated: consider the way scientific theories are used for practical purposes or the affect of an aesthetic concern upon an instrumental attitude. If there are further attitudes to be delineated, what these are, will require further examination.

Finally, it could be argued that all of these attitudes are held by human beings and concern their interests alone. That being so, are we locked into a kind of prison by our use of language to such a degree that it is impossible to view the issues raised by this chapter from a different point of view? Consider one of Ludwig Wittgenstein's aphorisms: 'A *picture* held us captive. And we could not get outside it, for it lay in our language and language seemed to repeat it to us inexorably'.[11] How far are we imprisoned in 'a picture', which relates to us only the concerns of human beings through our language use? This is a subject to be taken up next.

Behind all these issues is a more central one, that is the status of philosophy itself. If philosophy is a term used to describe an activity whereby the self attempts to establish a relationship between itself and the world surrounding it, or, indeed, between oneself capable of acting and the self who reflects,[12] then that activity must concern values. Such values are embodied in the attitudes we bear towards the world, and on the evidence of this chapter it would appear that the more usual attitude, sustained by human beings, is the instrumental. Otherwise the particular attitude we bear, whether we are aware of it or not, is constructed from one or more of the basic kinds established in this chapter. Indeed, you as the reader, may have felt challenged to consider whether there are further ones and if there are, whether or not they can in fact be reduced to or be encapsulated by the concerns of the attitudes so far delineated. This thesis – that in a basic sense, there are only the three fundamental ones along with their negations – will of course be tested in subsequent chapters. If it holds, then we would be provided with one reason at least to take Peirce seriously when he remarked that his discovery of his three categories – Firstness, Secondness and Thirdness – represented his 'one contribution to philosophy', 'the gift I make to the world' following 'three years of almost insanely concentrated thought hardly interrupted by sleep'.[13] But notice that so far his suggestions have not been worshipped and glorified as 'great' in the way fashionable philosophers or ideas may be required to be in the modern world: they have been made a subject of scrutiny, and indeed modified to face the requirements generated by a different kind of cultural problematic.

NOTES

[1]'Dreamt-Pool', 5 September 1980.
[2]*Chamber's Large Type English Dictionary*, ed. by Rev. Thomas Davidson, Edinburgh: Chamber Press, p.108.
[3] Terms which are preceded by an asterisk appear in this book's glossary of terms: Appendix I.
[4]Quoted in the *North Kent Marshes Study* Applied Env. Research Centre Ltd., consultant's final report, presented and published by Kent County Council UK, May 1992; cf. 'I have always been passionately fond of the water. The sight of it throws me into a delicious dream, although often about no definite subject'. Jean-Jacques Rousseau, *The Confessions*, pp.592–593.
[5]*The Collected Papers of Charles S Peirce*, vol. I, para. 43.
[6]Joseph Brent, *C S Peirce: A Life*, specifically chapt. 6, 'The Wasp in the Bottle'.
[7]I owe the awareness of this negativity aspect to Jennifer Smith.
[8]*Ethics and Education*, cf. R S Peters, Chapt. V, 'Worthwhile Activities' and 'The Justification of Education' in *The Philosophy of Education*.
[9]T McCarthy, *The Critical Theory of Jurgen Habermas*, p.335. The lay person would simply use the word apathetic to cover all three of these negative attitudes. But there are three kinds of apathy: aesthetic apathy, a passive awareness awaiting distraction which can only be shocked to create attention; moral apathy – 'What can I do about these things?' – leading to an absorption in activities to divert the self away from moral responsibility; cognitive apathy – 'What am I to say about these matters seeing that there are so many conflicting views about it?' – labelled here as the enured.
[10]This fourth category – 'not being aware' – which includes the distractive, distracted and the enured, has been handled in a Peircian fashion. How it is to be explained, however, is a further problem within Peircian scholarship. Peirce blocks an explanation in terms of a weakened Firstness (Peirce: vol. 5, para. 68). It might be explained by pointing to the over-domination of instrumental concerns in our culture. Thirdness becomes undermined: Firstness is emptied (cf. Joseph Ransdell's 'Semiotic Activity' in *Frontiers in Semiotics*, pp.236–254). An alternative explanation might lie in what I call, in chapter V and Appendix IV Distorted Thirdness: an obsession with unexamined ideas. Peirce cites, for example, the proposition that 'Greed is good' or the claim that 'All humans only act selfishly'. (Peirce: vol. 6, para. 284; vol. 5, para. 382).
[11].Ludwig Wittgenstein, *Philosophical Investigations*, Part 1, **114** p.48e.
[12]cf. Simone Weil's, 'Philosophy' in her *Formative Writings*, (1929–1941), pp.283–288.
[13]Introduction to Writings of Charles S. Peirce, A Chronological edition, (1857–1866), vol. I, ed. Max H. Fisch et al., Bloomington, Indiana University Press 1982, p.xxvi

Chapter II

APPEAL AGAINST REJECTION

For s/he whose irregular life has as its fixed points the regulated repetition of set programmes on TV channels, the precarious beauty of a natural landscape, like the nature of s/he who might appreciate it, remains locked up within itself. In a world subject to the TV message that no thing is done for nothing, those, whose nature can yet extend beyond themselves in an appreciation of others or the natural environment itself, hardly remain immune from the chill of the charge that even the slightest gesture remains privately interested, inclined, 'as it must be', towards ends determined by some already preconceived project. Indeed the darkness of intent, cast across what might appear to be the most spontaneous of actions, hardly disappears immediately, even in the presence of someone filling the consciousness of s/he – who might be more natural – almost completely. Rather, such self interestedness is chased from a subject's sensibility only by the broader expanse of the other's actuality realised as that is in larger breaks in their attempts to conceal who they are. Hence, a form of consciousness can arise, made possible by the other's joy, in finding what it is that subject sought to communicate to s/he who can share the delight of mutual discovery. But the human energy so issuing remains as finite, yet as precious, as the sparkle of delicate raindrops clinging to long grass stems in an open field stretched out under an autumn sun.[1]

'Say No to Tiptree Tip' posters reappeared once more a year later. The village magazine announced that two public meetings, one for the benefit of the Lower Halstow and Upchurch residents and the other for those from Newington and Iwade, would take place to provide information on the Tiptree Site development. The former meeting took place in the Recreational Hall at Lower Halstow on Friday, 13 September 1996. The audience of over two hundred people were reminded by TAG's chairman that the site planned for clay extraction and subsequent waste infill would be larger than the villages of Lower Halstow, Iwade and Newington combined, that is to say at least one hundred and seventy-nine acres. Evidently the application by South East Minerals was officially rejected by Kent County Council on 13 October 1995 but the chairman of TAG had been informed that an appeal had been placed with the Department of the Environment against KCC's decision by the applicant. That appeal would be considered at a Public Inquiry to be held over ten days beginning on Tuesday 12 November 1996 in Iwade Village Hall.

The local Conservative Member of Parliament provided a polished speech to explain why the proposal, described as an 'abomination', ran completely against plans to maintain and improve the local environment. He assured the meeting that he personally was trying to persuade the Secretary of State for the Environment to decide the matter when the inquiry had been completed. He emphasised the fact that because KCC, Swale Borough Council, the five local parishes and TAG were so strongly united in opposition to this application, there were grounds for hoping for success. Without referring to a former Conservative Prime Minister's phrase, 'the unacceptable face of capitalism', the MP said that, despite the large sums of money being spent on the appeal by South East Minerals, unity provided the ground for optimism. As the chairman of Lower Halstow Parish Council was to put it later; 'We mustn't lose heart. We are a winning team!'

A co-chairperson of KCC's Planning Sub-Committee explained to the audience the various plans against which this application offended. In fact nine grounds of refusal were given against the application. Firstly, that the case had not been established for clay requirements (the Clay Need argument) nor was there a clear demonstration that harm to important acknowledged interests, for example alteration in the present land-form, would not be caused by this development. Hence the site was not judged to be an appropriate location for the pursuit of the proposed activities. Secondly, a Countryside Protection argument could be advanced protecting the countryside for its own sake in relation, for example, to the conservation of some trees – already subject to a tree preservation order – and the protection of other countryside amenities from the affects of increased traffic. The third ground advanced the second in developing both the Landscape and the Wild Habitat arguments. The fourth ground developed in its turn the third, in that this development would affect the historical, scientific and scenic value of this part of the coastline, and the estuary with its adjoining countryside. This can be called the Consequential argument because it dealt with the affects on other precious environments outside the area of the proposed site.

Fifthly, the character of rural lanes, which are important in relation to landscape features, nature conservation and so on, would be damaged (the Rural argument). Sixthly, such routes do not need to be improved on present safety grounds, as heavy goods vehicles are already discouraged from using these lanes and the proposed development was not well connected to the rest of the road network (the Roads argument). Seventhly, the proposed structure plans for Kent were designed to conserve and enhance the quality of Kent's environment and make good any damaging affects used by previous land development and changes in its use. These are undermined by the application (the Environment argument). Another policy, the Thames Gateway Planning framework provided the eighth ground. Its provisions were challenged by the application because that would cause a self-reinforcing blight on the local environment. Finally, the last, the ninth ground concerned the undoing of the Kent Local Plan for Waste Disposal (the Waste Disposal argument), which sets out the need for establishing new patterns of waste management other than by landfilling.

The biggest round of applause during the evening was reserved for Lower Halstow Parish Council's chairman when, on being introduced, the audience was told he was not only fighting against this application but also, personally, against cancer. After he joked about feeling so much better after taking a course of chemotherapy and how this battle against the application assisted his fight against this disease, he refuted the applicant company's claims about clay need. He argued that the value of the site to this company lay in its possible use as a major landfill site, even though present government policy advocated moving away from this way of dispensing with waste.

Questions were then put to the panel of presenters at the meeting. When a prospective candidate at the next general election, from a sitting position, prefaced his question by saying he was standing for Labour, the audience demanded to know why he was not standing up for New Labour now, so as to put his question; it pressed for more details as to the public bodies which may have been alerted with regard to the application. Had the shadow Minister for the Environment, for example, been contacted? The proposal by the applicant to dodge the Roads argument by proposing to build a new route from the Tiptree Site to run parallel to Stickfast Lane before joining the A249, Sheppey Way, at an improved Stickfast lane junction, was also discussed.

Towards the end of the meeting, the local vicar pointed out that St Margaret's Church would be open from 10 o'clock on each morning of the appeal. After all, prayers had worked before: KCC had rejected the application! Finally at the meeting's end TAG's chairman announced that seven thousand pounds would be needed for the appointment of a planning expert as representative at the inquiry. At least fifteen pounds from each household would be money well spent to prevent devastating effects on the locality and to prevent residents' homes from devaluing. In addition the members of the audience were reminded that they could attend the inquiry, since the more interest taken in it by the residents, the more influence could be brought to bear on its outcome. It was important, however, that no matter for what length of time a resident attended a particular day of the inquiry, s/he needed to sign the inquiry's attendance book! Each member of the present audience could place, too, a bright gold-yellow poster in their front window. The more the inspector saw, the better. One member of the audience said he would lose pay, if necessary, in order to attend the inquiry so as to display his opposition to the application. Others spoke of their disapproval that only attendance at the inquiry during the day would be possible!

Before considering the issues at this meeting an introduction to some terminology is once more required if we are to take up the philosophic issues raised in the last chapter. The writer, then, pleads for patience from his reader if the following terminological examination seems for the moment to be a slight diversion from what s/he has just read.

It was the psychologist Jean Piaget who gave credence to the idea of *egocentricity.[2] A child is described as egocentric when s/he is unable to appreciate things from a different point of view. One Piagetian test shows that if, after indicating the arrangement of objects on a table from their own point of view, children are then asked to do the same for someone standing on the other side of the table, those below a certain age are only able to indicate their position in the same way as they did initially. There is then no appreciation of the fact that the two perspectives on the objects, located upon the table's surface, would be different. Consequently it is claimed that the only points of view to which the child can relate are those of his own which then not only come to typify the child's speech and his or her activities but also serve to direct his or her plans for the future.

The meaning of the term 'egocentric', as found in the dictionary, extends the meaning of the term 'egocentric' slightly from its application to children so as to include older human beings as well since it defines an egocentric disposition as one which regards everything in relation to the self, offering the meaning as self-centred. Perhaps you may know someone whose discourse can be regarded as a running commentary on their own situation; in what they say they never seem to be able to escape from their own narrow concerns. Of course, you might notice yourself too in times of stress, for example in knowing that you are late for a key appointment, muttering to yourself as you attempt to leave the house 'Now I must not forget the keys', 'Check the backdoor and windows' or 'Must turn off the heating'. These locutions illustrate how such discursive commentary can serve to regulate ways of acting and direct activities to some end, in this case getting out of the house.

The term *anthropocentric, as a descriptive term is defined so as to cover cases where the human being is regarded as the central factor in the universe, just as *heliocentric is in relation to the sun as the centre, *geocentric as the earth and *theocentric as having God at its centre where 'it' refers to the universe. The term 'anthropocentric', however, is not usually opposed to any of these other three senses of 'centricity' but rather to the term *ecocentric which can be defined as a descriptive term to cover the ecological or the ecospheric! But, you may ask, why should the anthropocentric be opposed to the notion of the ecocentric in the way this writer has cast these two terms? A simple example can make a reply clearer.

In 1933, Aldo Leopold, famous, along with Charles Elton,[3] as a founder of ecological studies, wrote an article entitled 'The Conservation Ethic'. He drew attention to the fact that Odysseus, on returning from the battles for Troy, executed, by hanging, a number of slave girls he suspected of irregular behaviour whilst he was away. After all, they were his property and questions of right and wrong certainly did not apply to 'human chattels'. Hence the first step in ethical development, if it can be called that, interpreted as a narrow form of anthropocentrism, covered only the relationship between individuals. Leopold regards this view as embedded in the Old Testament. The second step he thought could be found in the New Testament. This newer ethic sought to relate the individual to society by extending ethical considerations 'to many fields of conduct', differentiating the social from the anti-social, enabling co-operation between human beings to take place. Leopold comments on this anthropocentric point of view:

> There is as yet no ethic dealing with man's relationship to the land and to the non-human animals and plants which grow upon it. Land, like Odysseus' slave-girls, is still property. The land-relation is still strictly economic entailing privileges but not obligations.[4]

Leopold then speculates as to the possibility of a third step in the development of human *ethics. He refers to the Old Testament prophets Ezekial and Isaiah as having claimed that it is not only inexpedient to despoil the land but wrong too! Yet society, at least in his own time, even if a conservation movement had developed within it, had not really underpinned their view. Indeed it is Leopold himself who is credited with formulating the land ethic: a 'thing is right only when it tends to preserve the integrity, stability and beauty of the community, and the community includes the soil, waters, fauna and flora, as well as people'.[5] Clearly Leopold was more anthropocentrically minded in this earlier formation of his land ethic, in that human society was regarded as an organicism, whereas in his writings from 1939 onwards the centre of his thought moves to the land which he comes to regard as part of a greater whole to *include* individual human activity in society.[6] In relation to his final position, it might be said that Leopold had become more ecocentric in his own thinking:

> A thing is right when it tends to preserve the integrity, stability and beauty of the biotic community. It is wrong when it tends otherwise.[7]

This description of Leopold's case might then serve to illustrate the passage from anthropocentricity to ecocentricity. However, that is a matter of some controversy.[8] What is important is that in his case we gain a glimpse of what it might be to exercise a more *non-anthropocentric point of view of the land, even if today's farmer has to spray it with chemicals to satisfy the demands of supermarket customers who may be blind to this ecocentric dimension.

Now consider the issues raised at the village meeting. In our first chapter Landscape and Wild Habitats arguments were labelled as aesthetic, the Need for Clay and Waste Disposal arguments as instrumental and the Agricultural land, Traffic and Pollution arguments, by extension, as instrumental too! The scientific argument, as that appeared as part of the Consequential argument above, does not seem to relate to either of the first two ways of classification. Moreover, some further arguments have been raised. The Thames Gateway Planning Framework argument appears to be instrumental too, as does the Rural argument in that it pertains to rural activities such as horse-riding, walking and other recreational occupations. But what about that Countryside Protection argument? Can the attitude of preserving the countryside for its own sake and its connection with the idea of tree protection, as covered by tree preservation orders, be subsumed under the aesthetic or does it belong to the scientific as a matter of course?

You may be gaining the impression, then, that our classification of attitudes into the aesthetic, the instrumental, the scientific and the non-aware is beginning to creak under the sheer number of arguments this classification is meant to cover. Before that issue is faced, however, another more difficult problem looms. Aren't *all* these arguments anthropocentric in character, thereby illustrating Wittgenstein's contention that we are imprisoned by the use of a 'picture' placing humans as a species at the universe's centre to yield an anthropocentrism. Isn't that anthropocentric view of nature, that anthropocentric picture continually reiterated to us in the language we use, just as egocentric speech – referred to earlier – directs the young child's thoughts and actions? After all, what was said at the Lower Halstow meeting concerned the affects of the landfill on those present in the hall and only those affects were considered. A disciple of Wittgenstein renders this idea in its clearest form: 'Our idea of what belongs to the realm of reality is given in the language that we use. The concepts we have settle for us the form of experience we have of the world'.[9] Here it is being suggested that our use of language creates a picture through which we construct the world in thought.

If all the arguments at the inquiry are anthropocentric, you might ask 'But surely the four types of attitude through which they are articulated must be anthropocentric too, mustn't they?' In the way Peirce defines the instrumental attitude, it must be anthropocentrically orientated; that goes for the not being aware since that too concerns human endeavours. But what of the aesthetic? Peirce casts this as if it can be ascertained by seeing nature, say, in terms of a painting which involves the use of a person's artistic imagination (Peirce: vol. 1, para. 43 and para. 48). Elsewhere, in 1902. Peirce clearly relates aesthetic enjoyment to art appreciation (Peirce: vol. 1, para. 281). Indeed, a year later he specifically tied aesthetic considerations to 'aesthetic enjoyment' released when the human subject attends 'especially to the total resultant Quality of Feeling presented in the work of art we are contemplating' (Peirce: vol. 5, para. 113). So when in 1893 he spoke of artists, for whom 'nature is a picture', it is not unreasonable to assume that our manner of creating or appreciating art – as he claims we do – can be projected upon nature in the way we take delight in contemplating it. In other words Peirce's use of the term 'picture' here does not refer merely to what is constructed by our use of language in thought, but by the way nature is seen as if it was a human creation. If that is the way the aesthetic is to be defined, then this aesthetic dimension hardly escapes the charge of being anthropocentric either!

We are left then with the scientific. Here, at least, one commentator – Hermann Deuseur[10] – has remarked that Peirce escapes anthropocentricity – even if not into some form of ecocentricity, then at least into a kind of theocentricity. The relevant quotation from Peirce's writings is given at some length not only for the sake of the beauty of its expression, but because there are a number of ideas present within it crucial to its interpretation. However, the reader is advised to be careful here since this passage was written in 1863 well before Peirce stumbled on his three categories – Firstness, Secondness and Thirdness – and before he had articulated to his own satisfaction the true nature of scientific inquiry!

> When the conclusion of our age comes, and scepticism and materialism have done their perfect work, we shall have a far greater faith than ever before. For then man will see God's wisdom and mercy, not only in every event of his own life, but in that of the gorilla, the lion, the fish, the polyp, the tree, the crystal, the grain of dust, the atom. He will see that each one of these has an inward existence of its own, for which God loves it, and that He has given to it a nature of endless perfectibility. He will see the folly of saying that nature was created for his use. He will see that God has no other creation than his children. So the poet in our days, – and the true poet is the true prophet – personifies everything, not rhetorically but in his own feeling. – He tells us that he feels an affinity for nature, and loves the stone or the drop of water. But the time is coming when there shall be no more poetry, for that which was poetically divined shall be scientifically known. It is true that the progress of science may die away, but then its essence will have been extracted. This cessation itself will give us time to see that cosmos, the aesthetic view of science which Humboldt prematurely conceived. Physics will

> have made us familiar with the body of all things, and the unity of the body of all; natural history will have shown us the soul of all things in their infinite and amiable idiosyncrasies. Philosophy will have taught us that it is this *all* which constitutes the church. Ah! what a heavenly harmony will that be when all the sciences, one as viol, another as flute, another as trump, shall peal forth in that majestic symphony of which the noble organ of astronomy forever sounds the theme.[11]

It does seem, at first sight, as if anthropocentrism has been overcome in Peirce's rejection of the idea that nature was created for man's instrumental use. But what makes it so delightful for the reader of Peirce's writings is the way he makes use of irony. Notice his use of the idea of materialism – the theory that everything which exists in nature can be reduced to matter rather than mind – as having done its 'perfect work' despite the fact that Peirce opposed this philosophy throughout his life! Bearing this reservation in mind, it is not clear whether Peirce does break out of an anthropocentrism for in his lauding of the poet as 'the true prophet' he credits him with *personifying everything. Examination of any dictionary renders the meaning of this term as attributing a human character to other things – even a personal nature – symbolising them in such a way as to regard them as persons. So our earlier considerations, in relation to his subsuming of the aesthetic perspective under art appreciation, seem confirmed. But then this anthropocentrism is overcome by a theocentrism, at least at a naïve glance, since humans and other things, living or inert, are regarded as God's creations. This view, he then goes on to argue, is displaced in its turn by a scientific 'centricity'. In other words, 'the religion of science' replaces his theocentrism. Here, at least, Peirce is quite certain that all knowledge can be identified with science – scientism – which is his way of representing science's belief in itself.[12] But can the activity of science prevent human beings from taking an anthropocentric standpoint on the universe?

One notion, the most important, which Peirce sustained throughout his life, was that the activity of *inquiry itself, for its own sake, was of supreme value. Only by a commitment to inquiry could any advocacy, on behalf of some private-interest theorising, be overcome. Only through inquiry can our obsession with our own subjective concerns be displaced. Your present writer follows in Peirce's footsteps here, even though that support might not extend as far as Peirce's personification which he expressed in 1903:

> The universe as an argument is necessarily a great work of art, a great poem – for every fine argument is a poem and a symphony – just as every true poem is a sound judgement. (Peirce: vol. 5, para 119)

You may want to know why exactly a concern for scientific inquiry either is or is not anthropocentrically centred. Surely science is a dispassionate form of inquiry which can displace man's narrow concerns with his own private welfare. If someone is to inquire about the right way to live and if the truth is to be really ascertained in these matters, then other human beings, equally committed to such an inquiry, are important in relation to seeing how the concerns of a mere egocentricity can be overcome. For only in the *activity of critical argumentation can the falsity of the ideas of any of the inquirers be made clear to them. As Peirce put it in 1878 in his 'How to Make our Ideas Clear', any human being would be 'delighted to have been "overcome in argument" for in the "real spirit of Socrates" that human being would have learned something by it' (Peirce: vol. 5, para. 406). Not only is this Socratic requirement indispensable for any inquiry whatsoever,[13] and not only informing standards of communicative understanding within scientific inquiry itself, but is certainly indispensable in finding out whether we do have grounds for ascribing moral considerations to other entities in the world besides humans. That Socratic requirement is sustained by an inquirer, one 'who recognises the logical necessity of complete self-identification of one's own interests with those of the community and its potential existence in man' (Peirce: vol. 5, para. 356), as Peirce put it as early as 1868. This claim, to be found in 1878 in his article 'On the Doctrine of Chances', provides the basis for his *'*logical socialism*'[14] to use Apel's phrase, that 'logic is rooted in the social principle' (Peirce: vol. 2, para. 654). Let us explore this issue a little further.

Because their mental development is so short in view of so brief a life, there is no possibility that one single human being alone can discover the truth either about him/herself or anything else. Peirce puts that point as follows:

> The individual man, since his separate existence is manifested only by ignorance and error, so far as he is anything apart from his fellows and from what he and they are to be is only a negation. This is man:
>
> > '...proud man
> > Most ignorant of what he's most aware,
> > His glassy essence.'
>
> (Peirce: vol. 5, para. 317)[15]

Only as a member of such fellows – those who make up a body of like-minded scientific investigators – can a human being begin to acquire virtue through the pursuit of truth. And the latter means being committed not to something individually personal, but to a greater aim which can be found within the activities of a larger group of persons: '...logic rigidly requires, before all else, that no determinate fact, nothing which can happen to a man's self, should be of more consequence to him than everything else' (Peirce: vol. 5, para. 354). In this way scientific cognizing must presuppose communicative understanding of other inquirers since they are co-subjects in such cognizing too and such communicative understanding in itself presupposes ethical relations.[16] So the activity of inquiring presupposes ethical understanding: 'For he who recognises the logical necessity of complete self-identification of one's own interests with those of the community, and its potential existence in man, even if he has it not himself, will perceive that only the inferences of that man who has it are logical' (Peirce: vol. 5, para. 356). In addition, this claim does not hold merely for scientific inquiry. Rather the possibility of reaching consensual agreements, at the level of communicative understanding, is presupposed by any kind of interaction between human beings who, in their argumentative discourse, are concerned to establish the truth about some matter as best they can.

This final claim represents the case for Peirce's logical socialism: human beings are to be held responsible for the way the world develops historically through the institutions they create and actions which they take in relation to their development, as these emerge out of their communicative interactions with each other. After all, only by being concerned for the interests of other human beings is it possible to realise the ideal that everyone can take part in such authentic communication acts with each other. At the same time, only by attempting to realise such an ideal can human beings seek to make any ethical sense at all of trying to preserve the members and the interests of such a communicating *community.[17]

For Peirce, the only way of sustaining that 'conceived identification of one's interests with those of an unlimited community' is by holding to three sentiments which constitute the rational enterprise, 'namely, interest in an indefinite community, recognition of the possibility of this interest being made supreme, and hope in the unlimited continuance of intellectual activity'. These three are to be regarded not only 'as indispensable requirements of logic' but also as 'pretty much the same as that famous trio of Charity, Faith and Hope' which St Paul estimated to be 'the finest and greatest of spiritual gifts' (Peirce, vol. 2, para. 654–655).

Now in seeking to ascertain whether a case can be established for ascribing moral considerations to entities in the world other than human beings – the heart of the ecocentric point of view – we do find that we can ground an ethic at the level of world-citizenship at least.[18] This is because Peirce's three regulatives do apply to a community extending 'to all races of beings with whom we can come into immediate or mediate intellectual relation' so as to reach 'however vaguely, beyond the geological epoch, beyond all bounds' (Peirce: vol. 2, para. 654). The anthropocentric concern then is not overcome unless Peirce – like Leopold, whose case we examined earlier – can extend the notion of community beyond the range of human enquirers or potential human enquirers to other entities in nature. 'But can he?' you may ask.

It is true that on at least one occasion Peirce suggests that the community 'may be wider than man', (Peirce: vol. 5, para. 57) but the passage is set within a discussion 'of private logicality'. In addition, this idea is tied to the issue as to whether humans have any other kind of existence save as individuals, that is to say whether 'men really have anything in common'. Can the idea of a community be 'considered as an end-in-itself'? (Peirce: vol. 8, para. 38) Given, too, that the significance of this idea arises out of the importance for Peirce of the activity of inquiry, of finding out, it is hardly surprising that he ties his notion of the community to what is 'capable of an indefinite increase of knowledge'.[19] 'That point is re-emphasised when Peirce discusses what will count in terms of the future existence of a particular thought. That thought will only sustain in the light of what is thought subsequently, 'so that it has only a potential existence, dependent on the future thought of the community' (Peirce: vol. 5, para. 316). Moreover it is quite clear, on the basis of this, that his notion of community can only be based on a conception of human beings, the cognitive requirement for his sense of community, as we can refer to it.

You may want to argue at this point that the charge of anthropocentrism only works against the claims of science because it is Peirce's account of science with which we are concerned but not scientific activity in general. But, at least within our culture, the accepted view of science is that it is a form of knowledge which is made up of statements or systems of such, which are capable of clashing against possible or conceivable factual observations, a point of view defended in the well-known writings of Karl Popper.[20] But Peirce has the edge on Popper's account for at least three reasons. First of all, the idea that only those statements, which survive our past attempts to prove them wrong, is anticipated by Peirce.[21] In discussing in 1896 what he calls the uncertainty of scientific results, after he points out that 'any scientific proposition whatever is always liable to be refuted and dropped at short notice', he says the 'best hypothesis, in the sense of the one most recommending itself to the inquirer, is the one which can be the most readily refuted if it is false'. That fact far outweighs the advantage it has of its being merely likely since it is only going to be likely because it falls in with our previous ideas. These more often than not are false. So if an *hypothesis* such as one the mind suggests 'can quickly and easily be cleared away so as to go toward leaving the field free for the main struggle, this is an immense advantage' (Peirce: vol. 1, para. 120).

Peirce's actual practice provides grounds for two further reasons. Peirce in 1905 identified his own way of thinking as that of the experimentalist type whose '...disposition is to think of everything just as everything is thought of in the laboratory, that is, as a question of experimentation' (Peirce: vol. 5, para. 411). During the 1870s Peirce became something of a scientific authority in relation to his pendulum experiments, where he sought to measure gravity so as to determine its variations in different places, and his photometric researches. Indeed, on his visit to Europe in 1875 he became the first American participant to play his part in the committee meeting of an international scientific association, namely the Permanent Commission of the International Geodetic Association at which he reported his findings on his own pendulum experiments in September of that year.[22] If his paper 'created considerable opposition and heated discussion' at the time but two years later his results were approved,[23] then clearly it is one thing for a scientist to perform experiments and test something to his own satisfaction, quite another to get other scientists subsequently to approve of his findings arising from them. So it is not as Popper would have it that science is made up of statements which are capable of being refuted by possible or conceivable observations, but rather that it is made up of statements which are capable of being refuted *in the eyes of a contemporary community of scientists* by possible or conceivable observations.[24]

The third reason is provided by the fact that he did not simply swing his pendulum at the Commission's meeting in Paris; he had to report to this meeting his findings of his pendulum experiments conducted, as these had been, at the Observatory in Geneva.[25] So Popper's account of science has to be modified a second time: science is a form of knowledge which is made up of statements or at least systems of statements which are capable of being refuted, *in the eyes of a contemporary community of scientists*, by possible or conceivable *observational statements*, that is the kind of statements which would appear in a report or a speech, and, it would seem, in Peirce's case at least, that these observational statements took two years to be accepted by that scientific community![26]

You may want to reply by pointing out that what you have just been told about science – whether or not this is an acceptable account – has not been concerned with 'The Logic of Scientific Discovery', as Popper once entitled one of his earliest works, but rather with the logic of testing the results of human creations within scientific activity itself. Surely something has to be said about how discoveries first occur to scientific inquirers. Popper treats this merely as a psychological matter. Why should we be concerned with where our ideas come from – last night's dream, mother-in-law's wisdom or a knock on the head? What matters is the theory arising from such an accident and how it can be put to the test! How then, according to Peirce, do possible explanations, as to how processes in nature occur, arise?

In facing this question it might appear, at first sight, as if Peirce was breaking out of his anthropocentrism in an article he wrote in 1908 'A Neglected Argument for the Reality of God'. In this article he defends what he calls the idea of *Il Lume Naturale*, an idea he credits to Galileo, of a natural light, the idea 'that man has, in some degree, a divinatory power' for discovering hypotheses. Without such '...a natural bent in accordance with nature's, he has no chance of understanding nature at all' (Peirce: vol. 6, para. 477). In order to illustrate this idea imagine the activity of a water-diviner. This person has a special stick of hazel twigs stretched out before him which can be used to locate water underground. Now think of this divining rod as an analogy for the human mind. According to Galileo the human mind has a divine faculty for discovering a range of likely hypotheses to explain some phenomenon.

In studying Peirce, however, we have to be ready to read his reservations in relation to this suggestion. Galileo's idea is, for Peirce, simply a way of explaining why only one, two, three or so hypotheses are suggested to the scientist in response to some phenomenon rather than one hundred and one, two, three and so on! Secondly he always holds this suggestion of Galileo at arm's length, in order to dispel what he calls 'rash cocksuredness': 'If, however, the maxim is correct in Galileo's sense...' (Peirce: vol. 6, para. 477) Examination of the processes through which this natural light works should also put us on our guard.

First of all, Peirce does not focus on direct experience to elucidate this capacity; rather he turns to 'a certain agreeable occupation' of mind, when for example we are out for a stroll say at dawn or at dusk. Unlike for the water-diviner, then, scientific hypotheses do not arise when we are consciously concentrating on forming them. Rather they occur to a special, relaxed 'frame of mind' which takes part in what he calls Pure Play. Now, Play, we all know, is a lively exercise of one's powers. Pure Play has no rules, except this very law of liberty. It bloweth where it listeth. It has no purpose, unless recreation' (Peirce: vol.6, para. 458). But if such creative activity originates in humans, regarded as perceiving subjects, then in exercising this capacity for Pure Play, the human subject can only deal with what s/he has created. Such creations may arise on purpose or by accident in what Schiller once called 'the insubstantial realm of the imagination'.[27] Secondly, this mind-dependent account is developed when Peirce speaks in terms which might remind you of Wordsworth's 'Prelude'[28] as he enjoins his reader to enter this 'frame of mind' as if it were his or her 'Skiff of Musement'. **Musement* he regards as reflective speculation on what is produced by Pure Play (Peirce: vol. 6, para. 458): to 'leave the breath of heaven to swell your sail. With your eyes open, awake to what is about or within you, and open conversation with yourself'. But this skiff is located on 'the lake of thought' realised through inner 'meditations' not on a Wordsworthian lake at all!

In these last paragraphs, we have dealt with Peirce's account of how scientific discoveries can be made. Unlike Popper, he claims that it is important to investigate how hypotheses 'pop' into the mind as a result of trying to explain how some natural phenomenon occurs. Peirce thinks that scientists may have a divine faculty for originating such hypotheses once they can enjoy a special 'frame of mind', where the mind can enjoy Pure Play with Possibilities. Reflective speculation on this Pure Play – Musement – may assist the sudden discovery of a likely hypothesis to explain that phenomenon. But notice that all this free activity is still very mind-dependent.

The issues, explored in this chapter, can be brought together by drawing attention to a work important in considering the relevance of philosophy to ecological concerns. That book was published in 1974 and corrected with a new Preface and Appendix in 1980: John Passmore's *Man's Responsibility for Nature*.[29] In it he defended the thesis that ecological problems arise from man's dealings with nature and their practical consequences. His list refers to four of these: Pollution, the Depletion of Natural Resources, Over-Population and the issue of Species Extinction. The ecological problem, of most concern to Americans, that wilderness areas should be preserved and their surrounding areas considered in the light of this imperative, is subsumed under the fourth category.

His work is referred to since he treats these matters anthropocentrically: the first three problems, for example, are particularly significant in relation to the future quality of life human beings are likely to experience. The last is lumped under the heading 'Conservation' and focuses upon the issue of what future humans are likely to enjoy or can seek if the present generation is prepared to undergo sacrifices for posterity. If the arguments in relation to the Tiptree development site are considered, Passmore's Pollution category encompasses the Traffic and Pollution arguments directly and the Rural argument partially. His Depletion of Natural Resources category refers to the case for the Need for Clay and the Waste Disposal arguments. The Human Over-Population argument does not fully surface save for the issue of pollution in that the Thames Gateway Planning Framework allows for housing development in the area which may be affected by the proposed Tiptree development. The Landscape, Wild Habitats argument, the Rural argument and the Countryside Protection concerns would then be covered by the Conservation category.

In relation to this chapter, a Peircian distinction between four different kinds of attitude towards Nature was evoked and support was sustained for the view that these too were anthropocentrically centred, justifying initially the idea that the arguments raised at the Tiptree meeting focused upon an anthropocentric view of nature. That view is sustained in Passmore's work. Reticence was expressed, however, about the scientific attitude and that of the aesthetic. In Peirce's case the Scientific argument is cashed in terms of an anthropocentricity: first because it emerges from a contrast with egocentricity and secondly because, unlike Leopold, his conception of community does not extend to the ecological! Where he might be taken as arguing otherwise, he claims that those entities in nature other than man have a value because of their place in a possibly theocentric world!

Unlike Popper, however, Peirce seeks to show how the activity of science is rooted in the aesthetic dimension; the origination of scientific theories originates in an aesthetic reverie made up of Pure Play and Musement This means that if his account of the aesthetic is rendered anthropocentrically too, then neither the aesthetic nor the scientific attitude will ground non-anthropocentric points of view. That his view of the aesthetic is anthropocentric was established by showing how it is, following the philosophers Schiller and Kant, quite mind-dependent. Whether elements within Peirce's own writing can sustain a more non-anthropocentric view of these matters, is an issue which remains to be explored in subsequent chapters. Now consider, as a disciple of Wittgenstein might claim, that it is not possible to say *what there is* but only to *say* what there is independently of human concerns. If this be so, then it may be that an approach to these matters which is narrowly concerned with pure *epistemological matters – matters pertaining to theories of knowledge – will forever imprison inquiries within the cognitive realm. That approach would thereby remain, as Passmore shows, quite anthropocentric. In addition, whilst we stay with the way Peirce describes his three categories it might appear that we have to remain within an entirely anthropocentric position. How far that last claim is true is also something which is the subject of later chapters. Having examined his treatment of the scientific category we now need to say something further about the others, particularly the way they may be related to each other.

NOTES

[1]'Only Connect', 25–30 November 1995, Ilford Fields, Avon.

[2]*The Moral Judgement of the Child*, Jean Piaget.

[3]Charles Elton, *Animal Ecology and Evolution*.

[4]'The Conservation Ethic' in *The River of the Mother of God and other Essays* by Aldo Leopold, p.182.

[5]Charles Elton, op. cit., p.345. Leopold's land ethic is discussed in N E Boulting's 'The Emergence of the Land Ethic: Aldo Leopold's Idea of Ultimately Reality and Meaning', *Ultimate Reality and Meaning*, vol.19, no.3, pp.68–188.

[6]Curt Meine, *Aldo Leopold: His Life and Work*, p.395.

[7]Aldo Leopold, *A Sand County Almanac and Sketches Here and There*, 1949, Special Commem. Ed., Oxford UP, 1987, p.225.

[8]cf. N E Boulting, 'Between Anthropocentrism and Ecocentrism', *Philosophy in the Contemporary World*, vol. 2, no. 2, pp.1–18.

[9]Peter Winch, *The Idea of a Social Science*, p.15.

[10]Hermann Deuseur, 'Charles S Peirce's contribution to Cosmology and Religion, *Religious Experience and Ecological Responsibility*, pp.159–162.

[11]C S Peirce, 'The Place of our Age in the History of Civilization', 21 November 1863, *Writings of Charles S Peirce – A Chronological Edition*, vol. 1, p.114.

[12]Jurgen Habermas, *Knowledge and Human Interests*, p.4.

[13]Karl-Otto Apel, 'The Common Presuppositions of Hermeneutics and Ethics: Types of Rationality Beyond Science and Technology', in *Research in Phenomenology*, vol. IX, pp.35–53, p.**48**.

[14]Karl-Otto Apel, *Charles S. Peirce, From Pragmatism to Pragmaticism*, p.213, n.107.

[15]Peirce quotes from William Shakespeare's *Measure for Measure*, Act II, Scene 2, lines 117–120.

[16]Karl–Otto Apel, 'The Common Presuppositions of Hermeneutics and Ethics: Types of Rationality Beyond Science and Technology' in *Research in Phenomenology*, p.47.

[17]Karl-Otto Apel, 'Types of Rationality Today: the Continuum of Reason between Science and Ethics' in *Rationality Today*, pp.307–350, p.**338**.

[18]Karl-Otto Apel, *Towards a Transformation of Philosophy*, pp.226–227.

[19]*Writings of Charles S Peirce*, p.239.

[20]Karl Popper, *Conjectures and Refutations*, p.39.

[21]See N E Boulting, Charles S Peirce (1839–1914), 'Peirce's Idea of Ultimate Reality and Meaning Related to Humanity's Ultimate Future as seen through Scientific Inquiry', *American Philosopher's Ideas of Uram*, pp.28–45

[22]Introduction to *Writings of Charles S. Peirce*, vol. 3, p.xxv.

[23]Murray C Murphey, *The Development of Peirce's Philosophy*', p.102.

[24]This criticism is valid if Popper is seen as a dogmatic falsificationist; scientific theories are refuted by hard facts. If interpreted as methodological falsifications, then the rejection of a scientific theory can be separated from its refutation, the

former subject to a scientist's decision and the latter by a concern for observational statements. (see Imre Lakatos 'Falsification and the Methodology of Scientific Research Programmes' in *Criticism and the Growth of Knowledge*, Imre Lakatos and Alan Musgrave, sect. 2.

[25]Introduction to *Writings of Charles S. Peirce*, vol. 5, p.xxv.

[26]Joseph Brent, *C S Peirce: A Life*, p.120.

[27]See N E Boulting, 'Grounding the Notion of Ecological Responsibility', *Religious Experience & Ecological Responsibility*, p.129.

[28]For a relevant treatment of Wordsworth's poetry see 'The Romantic Reaction' especially paragraphs 35 and 44ff, A N Whitehead's *Science and the Modern World*.

[29]John Passmore, *Man's Responsibility for Nature*.

Chapter III

THE FIRST MORNING: CONFUSION

Is it true that the mere striving for existence is what keeps living things both in motion and fully occupied? Surely, just as seagulls beat up against a stiff Westerly wind only to fall steeply downwards before being carried effortlessly into an easy flight sustained by a sudden smooth passage of air, so the thought of what ought to exist struggles to gain more than a mere reality before relaxing into the recognition that presence to consciousness is neither necessary nor sufficient to render it an existence. But if 'What still might be so' is thereby rendered fictional, is it essentially or inessentially so? For just as the seagull's steady beating of its wings, when the wind's buffeting momentarily ceases, allows some progress to be made, can't the thought of something inessentially fictional originating in possibilities drawn from past experience, now essentially 'fictional' yield a passage through what is both essentially and inessentially non-fiction, namely life itself?[1]

It was a cold, grey drizzling morning on Tuesday, 12 November 1996. This was the day of the inquiry at Iwade into the appeal by South Eastern Minerals against the rejection by Kent County Council (KCC) of their application to develop the Tiptree Site, Inside the village's community hall the heating was still noisily operating. In a chair on the stage was a Lancashire man who introduced himself as the inspector. He held a BSc. and was a companion to the Royal Institute. It is hardly likely that this was the Royal Institute of Philosophy. That last remark indicates one difficulty in reporting this first morning's activities: his soft voice and the noise from the heating made it difficult to catch what he said. Such difficulties were not overcome for the laypersons attending, since the room was orchestrated so that all remarks were directed to the stage and non-participants sat between the contestants and swinging doors which, as soon as the heating stopped, began to creak as cold air pressed into the hall.

The contestants introduced themselves. The representative for the Local Planning Authority (LPA) was a dark-haired, young man – soon to be labelled enthusiastic – representing KCC, sitting on the left-hand side of the room before the stage so that he appeared side-on to his audience. With his back to the creaking door on the left side of the speakers parallel to the stage sat the moustached representative for the Environment Agency (EA), formerly the National Rivers Authority (NRA). Next, directly in front of the stage, was the bespectacled, balding representative of the Swale Borough Council (SBC). His name was not discovered since his back blocked the view of the stage where the inspector sat. Next, on the right, were the representatives of TAG (Tiptree Action Group) who represented the local residents from such villages as Lower Halstow, Bobbing, Iwade, Newington and Upchurch. Their spokesperson was a small bespectacled woman with excellent diction, referred to with some difficulty by the inspector, as Ms rather than Mrs or Miss. South Eastern Minerals (SEM) were appealing against KCC's decision not to allow an extension to what was once a borrow pit, now called a clay quarry, and the restoration of this site by infilling. The appellant's representative was a clear-speaking, moustached gentleman, a Queen's Council (QC), whose thick arrangement of hair made him look, sideways-on, like the one-time chancellor of the exchequer Nigel Lawson in his former, tubbier appearance at least. Throughout the hearing, besides the inspector, he appeared to be the one most concerned that the members of the audience should hear what he had to say.

All other participants were then required to introduce themselves and an argument continued as to whether audibility conditions were really satisfactory, though the members of the audience were not really consulted as to whether they were so. There seemed to be agreement between those sitting at the tables that microphones were not necessary. What was clearly audible was a point made by the inspector: at this inquiry conditions would have to be agreed upon which would be imposed on the appellant if their appeal was won. The stating of this point, he said, in no way was to be taken by anyone in the room as grounds for thinking that the result of this appeal was a foregone conclusion!

The QC, the appellant's representative, then introduced his case for the appeal by claiming that the amounts of clay to be extracted were to be changed, adding that these changes were only 'refinements' of the application as originally presented to KCC. Whilst vehicle movements would remain the same, the size of the extension and its depth had been altered of which some months' notice had been given to KCC. The inspector required clarification in relation to what really was part of this present appeal. Was the haul road proposal, outside the site but connected to it, part of this appeal or rather a matter to be agreed upon later? The appellant's representative said he needed to take instructions.

During this pause, the audience was given the impression of a man somehow distracted, as though he hadn't, quote, 'got his act together'! This report's writer felt some sympathy for him, since his own head was filled with the Nick Ross Radio 4 debate about the European directive on working hours in Britain. Ian Lang, a government spokesman, had spoken about fighting for Britain, the *Sun* newspaper backed him by reporting on 'The Battle of Britain'. Most people who rang Nick Ross, in the first half of the programme at least, were overwhelmingly in favour of that newspaper's position, namely, that what hours were worked by people per week in Britain should be decided by them so that they would not enjoy necessarily a shorter working week, as would be imposed by the directive. Perhaps the QC too had been listening to the debate on his way to this inquiry. In addition it was getting colder, those doors wouldn't stop creaking and the contestants' voices in this inquiry did seem to keep dropping, something about which, in relation to himself, the QC kept apologising during the proceedings.

The KCC representative had already made it clear that KCC was not really given much notice of these latest changes, claiming that the appellant was being 'disingenuous'. Loudly and more clearly, this enthusiast, who had an uncanny skill for making his point succinctly and then sitting down, challenged – was it the inspector or the QC – to say how far it was possible to continue to make amendments to documents altering the nature of the appeal. A quite wounding debate then followed, the final outcome of which was hardly audible. It appeared that certain papers regarding changes to original documents had been forwarded to KCC by the appellant in either June or August but had had the word 'confidential' attached to them. So what was their exact status? In addition an argument broke out between the inspector, the QC and the KCC representative as to how much clay was to be extracted in total and over what time span.

KCC's representative, appearing more and more like the actor Hugh Grant, rose to the occasion. He emphasised that the extension period was to be raised from thirteen to nineteen years. The infill rate was misleading in that it appeared the amount of waste to be deposited was to be increased over an even longer period of years. For him, the whole nature of the application had been changed from one of clay extraction to one concerning waste disposal, since the amount of clay to be extracted was now halved, yet there was to be a fifty per cent increase in waste. What he wanted to point out to the inspector was that the whole inquiry should be adjourned since the amendments now made to the appeal were no longer relevant to the nature of the original application. For the first time ever it now appeared that toxic waste was to be deposited in the quarry, discharges from which might affect the estuary itself. No assessment of environmental consequences of these proposed changes to the original application had been provided. The KCC representative also pointed out something as to the road changes being put in late, although this point was not clearly audible in the meeting. Because the scheme had changed so dramatically and because the presentation of the appellant's case was so 'shambolic' the inspector had no choice: either he could throw out this application or at least adjourn the proceedings. According to the KCC representative, what was really relevant was the question as to who would pay the costs of this appeal; what was the exact corporate identity of the companies involved in this application?

A black-moustached man spoke on behalf of the Environment Agency (EA). He made it very clear from the start that he was *not* opposed to the original scheme. Since November, however, the agency had been opposed. He agreed too that the inquiry should be adjourned since the conditions of the present appeal would now have to be explored, just as the environmental consequences of the amendments to the original appeal would have to be assessed, because of the substantial changes these amendments now introduced. He seriously doubted whether the agency could play a proper role in this inquiry since the full details of the new proposals were not known. They could only advise on the basis of the appeal as it was originally forwarded, that is with the appeal based on the extraction of clay as opposed to waste disposal and it was thanks to KCC that the agency had been alerted to the manifest change in the conditions of the application for an appeal. The original depth of the quarry was to be 120 metres (approximately 400 feet). Consequently if less clay was to be removed and if more waste was to go into the site instead this could affect the activities of the nearby brick company. Did they know the details of the new proposals?

The QC intervened to defend the appellant. He pointed out that less clay would be coming out but its exact amount could only be given when further work had been done in ascertaining the nature of present geological levels. As to the 'actual' bottom of the excavation, that would have a liner of not less than one metre's thickness.

The black-moustached man continued, however, to press his case and his presentation wasn't always clearly audible, whilst the door continued to creak even after it was oiled. What part of his case was he defending? Permeability questions, in relation to the clay becoming sandy and silty at particular levels, were mixed up with questions concerning the changing of the kind of waste and issues relating to the estuary discharges. Even the inspector seemed to have difficulty following his train of thought when he was speaking about gravel being *above* levels of the London Clay. Perhaps he was objecting to the idea of excavating through a permeable area, particularly where what he called fissure flows were in fact fast flows. He spoke of a two-fold effect, but he couldn't be heard to name them clearly. Presumably he was referring first of all to effects on the brickworks borehole reading taken to ascertain the depth of clay and which also indicated something as to the level of water at this point and, secondly, effects upon the domestic water supply as well as ground water entering the excavation. Liners have to stay where they are put but this liner would shift if water was still coming out of the clay. So how deep was the newly proposed excavation? What is the geological nature of the land? Did we have proper details of local variations? What kind of liner was proposed? The agency was concerned with all the risks now involved in the proposal, but real objections could not be expressed until more details about it were known. Such issues would take months to resolve and these would have knock-on effects in relation to other parts of the inquiry. He was as concerned with inconsistencies in different parts of the amendments to the appeal as he was about local variations of up to ten metres or so between the sample borings the appellant seemed to have made. These indicated that the materials discovered by such borings were out of phase with what was recorded on the geological map for the area. A thorough geological exploration was needed. Only in the light of this could sense be made of the appellant's present proposal. The agency could not fulfil a proper role at the inquiry without some clarity being brought to bear on these issues.

The spokesperson on behalf of Swale District Council rose to say, much to the exasperation of the KCC representative who expressed his displeasure facially, that though there were severe problems with this application, the borough council would be reluctant to see this inquiry adjourned. He believed that there were other aspects of the appeal which could be explored besides those concerned with hydro-geology! For example, it would be the responsibility of the appellant to satisfy the Secretary of State for the Environment that there would be no damage to the adjacent international Ramsar site – a site with worldwide recognition as important for waders, waterfowl and other waterbirds – nor to the local environment as a result of the proposed new form of waste to be disposed.

The KCC representative then pressed his case. Even with the amendments to the original application, the appellant was still proposing a substantial extraction of clay. Such amounts of clay were not needed. Moreover, fundamental consequences would follow, subsequent to the changes to the depth predicated, so how could this appeal be forwarded? Once it was ascertained clearly what the appellant was proposing, it could then be possible to assess the impact of nature conservation issues as well as the consequences for international sights of ecological interest. Proper questions about the appeal couldn't be put to witnesses if it was not known how far these amendments would be allowed to go. Finally, he quoted a case of a Wessex company versus Salisbury District Council taken from the Journal of Planning where an inquiry had had to be adjourned because in the appeal a sudden alteration had taken place in the number of houses it was proposed would be built!

The QC then called for an early adjournment of half an hour before lunch at 12.30 since he needed to take advice from the appellant. As to who stood behind the company South Eastern Minerals Ltd., the KCC representative would simply have to speculate on the matter and find out for himself. That issue was not relevant to this inquiry!

Some thirty years after Peirce distinguished the three kinds of attitude, considered in chapter I, the German philosopher Martin Heidegger wrote as follows:

> 'Nature' is not to be understood as that which is just present-at-hand, nor as the *power of nature*. The wood is a forest of timber, the mountain a quarry of rock; the river is water-power, the wind is wind 'in the sails'. As the 'environment' is discovered, the 'Nature' thus discovered is encountered too. If its kind of Being as ready-to-hand is disregarded, this 'Nature' itself can be discovered and defined simply in its pure presence-at-hand. But when this happens, the Nature which 'stirs and strives', which assails us and enthrals us as landscape, remains hidden. The botanist's plants are not the flowers of the hedgerow; the 'source' which the geographer establishes for a river is not the 'springhead in the dale'.[2]

Notice that the same three attitudes towards nature are displayed. There is the aesthetic perspective where we can be enthralled by landscape, where we *feel* that nature appears to stir and strive so that we might delight in a 'springhead in the dale'. The instrumental is captured by his use of the descriptive phrase the *ready-to-hand. When a hammer is used it is not regarded aesthetically, as if some craftsman had produced it in an earlier historical age. Our actions put it to good use in such a way that, in skilled activity, its existence as a tool is hardly noticed as a nail is driven 'home'. So a river's current or a wind's power in a yacht's sail may give a vessel passage whilst its passengers can view a surrounding countryside from its deck. The third dimension, a more theoretical one, is referred to in such phrases as 'the botanist's plants' or the geographer's mapping of 'the "source"' of a river. For Heidegger this dimension indicates the *present-at-hand meaning what can be *thought about in relation to something's *existence, an existence which is at a distance from an observer thinking about its properties in what is sometimes called a *scientific, more *objective way. After all, most people would agree that a large wood could be called a forest of timber without necessarily agreeing it possessed any special aesthetic features. Again, someone might regard a hill as suitable for quarrying because of its minerals, without perceiving it as having any special landscape qualities!

Now why has your attention been drawn to this passage? Notice that in the last sentence of the quotation the aesthetic attitude is hidden by the theoretical or more scientific way of regarding nature; as Heidegger identifies it – 'Nature'. The latter dimension only appears when the practical or instrumental standpoint is 'disregarded'. You may have noticed, for example, at this inquiry's first session, how the KCC representative was concerned with instrumental considerations connected to the use of clay, whereas the QC's interest was much more distanced in seeing Tiptree Hill simply in terms of a resource to provide clay for sale and a deposit for waste. Moreover, since we live in a culture completely dominated by the influence of technological advance, the aesthetic perspective is demoted to *What still might be so*, as this chapter's opening *aestival refers to it. But you might object, why does Heidegger contrast the scientific and the aesthetic attitudes in this way?

Peirce regarded imaginative activity as central in scientific inquiry, as will be demonstrated in a subsequent chapter: 'The scientific imagination dreams of explanations and laws' (Peirce:

vol. 1, para. 48). In saying that, we can see Peirce defining science in terms of a process of discovery conceiving it thus 'as a living historic entity' not as a body of 'organised knowledge' (Peirce: vol. 1, para. 44). Heidegger, however, seems to view science in term of the products, which emerge from such scientific inquiries, employed for modern technological purposes; knowledge about a forest of timber or a quarry of rock. Consider the objects which are identified in nature as timber, whose wood science has discovered possesses certain properties, or rock, which science shows has crucial features; all may be regarded as thing-like. They can then be employed for man's purposes. Whereas for Peirce it is quite a separate matter how the results of scientific activity are used for technological purposes, for Heidegger the interest in particular scientific inquiries is in fact driven by the state of man's technological needs and purposes. The interests of science and technology can no longer be separated in the way a scientist may have conceived them in an earlier century.

Another way of expressing this difference would be to say that a technological attitude emerges from the relationship of the scientific to the instrumental attitude. But how is that to be conceived? Is it, as Peirce thought, that a technological perspective emerges from a scientific attitude guided by instrumental concerns? Or is it, rather with Heidegger, that the technological perspective issues from the instrumental guided by the scientific? That debate is a separate issue in itself, but you may still be wondering why Heidegger's work has been referred to in relation to the proceedings of the first morning of this enquiry.

The reason is that the original distinction between the three basic attitudes, as outlined in chapter I, has once more come under pressure. Despite both thinkers agreeing about the importance of distinguishing them, clearly the way they are characterised and related to each other is different for each thinker. Heidegger does not speak of seeing nature as a picture nor does Peirce regard experiencing nature aesthetically as hidden by the scientific in the way that Heidegger does. But the real reason for that pressure on the relationship between the three basic attitudes is because, unlike Peirce, who reflected on these kinds of attitudes in a more philosophic moment, in relation to this inquiry we are concerned with how these attitudes relate to or react upon each other in practice. Notice that, despite taking the theoretical attitude in speaking about the supposed results of hydro-geological findings and their threat upon what the appellant was proposing with respect to the quarrying of clay, the environmental agency spokesperson seemed quite opposed to the appellant's case. This spokesperson's opposition was sustained even though the QC appeared to be adopting the scientific attitude too in presenting details of the discoveries made by the company's borings and despite the QC's attempt to reassure him in regard to the pit's liner and so on.

Pressure on the distinctions between these three basic attitudes, however, comes from another direction. You will have noticed that one party called the other 'disingenuous'. To accuse someone of not being free from deception or of not being sincere is to make a ⋆moral accusation about a person's character or the actions s/he is forwarding. Regarding something, someone or indeed nature itself in a moral way doesn't seem covered by our four-fold classification either. So how does a moral attitude relate to this classification between the aesthetic, the instrumental, the scientific and the not being aware, the latter being noncognitive unlike the other three? Of course we have already referred to the moral attitude in chapter II. There it was indicated how Peirce showed the necessity for certain moral principles to make genuine scientific inquiry possible.

Peirce listed three: 'interest in an indefinite community' referred to as charity; 'recognition of the possibility' of this former interest 'being made supreme' – faith – and, finally, hope identified in terms of the limitless 'continuance of intellectual activity' (Peirce: vol. 2, para. 655). The first – charity – inspires, minimally at least, the moral ⋆precept that a person should make him or herself as clear as possible to another so that some kind of understanding between them can be achieved, a claim that is to say for ⋆intelligibility. The second – faith – seeks to make the former interest supreme thereby underpinning a claim for ⋆fairness and ⋆impartiality in recognising the members of a ⋆community of inquirers as bearers of a moral point of view. That recognition is presumed in others too, who seek to understand that point of view. The third – hope – explicated as the continuance of intellectual activity, focuses upon the issue of truth. Such an unlimited continuance of intellectual activity can only be sustained if the parties, engaging in such activity, try to be both *truthful – in intending to speak what is true so that knowledge can be both shared and advanced – and *sincere in that others engaging in this activity can believe an attempt is being made to tell the truth. In this way mutual trust between the different parties can come to exist.[3] If someone is accused of being disingenuous presumably s/he is being accused at least of being insincere and, in addition, may be failing to be truthful even if s/he seeks to be intelligible as well as fair and impartial!

In chapter II one obvious difficulty with grounding morality in this way was raised: this kind of defence for moral principles only appeared to be available to language users, those who could articulate such moral precepts. Such a defence entailed an anthropocentric point of view. Secondly, and more seriously, it might be pointed out that such moral precepts were binding only on those capable of sustaining a 'continuance of intellectual activity'. Thirdly the demand to sustain such a morality only holds whilst such intellectual activity continues. Outside a devoted and passionate concern for the truth in scientific matters, for example, Peirce displayed an ignoble character![4] Perhaps, reflecting on this ignobility – his womanising, the squalor of his money-making ventures which made him bankrupt and his wife-abuse – such reflections caused Peirce, later in life, to ground value claims in an entirely different way. Let us examine that attempt.

The way in 1903 that he tried to do this proves relevant for our four-fold classificatory system. We can start by considering his observation that a righteous man is he 'who controls his passions, and makes them conform to such ends as he is prepared deliberately to adopt as *ultimate*' (Peirce: vol. 5, para. 130). The question then arises, what is the nature of that ultimacy, that 'admirable ideal' (Peirce: vol. 5, para. 130) to be adopted? We are dealing with *action – *Secondness – and we have focused upon the instrumental, in the sense of pursuing means to achieve an appropriate end. But what end? Peirce replies that the end 'must be something, good or admirable, regardless of any ulterior reason. Such an end can only be the aesthetically good. But what is aesthetically good?' (Peirce: vol. 5, para. 594; vol. 5, para. 36) is his next question.

Elsewhere he points out that for him 'aesthetics considers those things whose ends are to embody qualities of feelings', (Peirce: vol. 5, para. 129). We can see that what is important for his analysis is the way *Firstness can inform Secondness, the way feelings can relate to the way we act. In answer to his question, 'What is aesthetically good?' he considers what he calls 'the mood of aesthetic enjoyment', that is 'the mood of simply contemplating the embodiment of the quality' or qualities of feelings. He provides an example. He considers the aesthetic contemplation of the Alps and notes how people in previous times 'when the state of civilisation was such that an impression of great power was inseparably associated with lively apprehension and terror' were thereby unable to enjoy 'a calm aesthetic contemplation' of such mountains (Peirce: vol. 5, para. 132). They clearly did not regard such aesthetic awarenesses as good!

Having got this far, where can he move next? He seems to doubt whether his suggestions have much worth. Instead of attempting to consider further an examination of the nature of the object under *contemplation, or critically examine the status of cultural ideas which might prevent an exercise of the aesthetic dimension,[5] he suddenly concludes that what he has said about aesthetic contemplation show' that there is 'no such thing as aesthetic goodness'. Rather 'there are innumerable varieties of aesthetic quality, but no purely aesthetic grade of excellence' (Peirce: vol. 5, para. 132). In view of his own admission that he has failed to 'go into details' in relation to his argument, we can leave for the moment an examination of the adequacy of his own conclusions. What is clear is that for Peirce, after 1903, a moral attitude is to be based somehow on the aesthetic dimension.

You may wish to make two objections at this point. The first might be, why should any contemplation of examination of nature have anything to do with holding a moral attitude? The second, why should aesthetics have anything to do with morality?

Peirce would not be the first philosopher to think that considerations concerning nature might have something to do with human conduct. A thinker about whom Peirce was contemptuous (Peirce: vol. 1, para. 5), Herbert Spencer, directly recommended such an approach to morals. Incidentally, the term ★morals is connected with the idea of something being 'a moral matter' concerning something which ought to be done whereas the term ★ethics is used in the context of a theory as to how a moral position is to be justified; it 'is reasoning out an explanation of morality' (Peirce: vol. 1, para. 666). Herbert Spencer (1820–1903), the philosopher who invented the term 'the survival of the fittest', believed that 'Nature takes it into her own hands' to provide the child, for example, with knowledge which will 'secure direct self-preservation' for that child. In other words, in order to find out what ought to be done in relation to human affairs we should turn to nature to see what it can teach us about such matters. Such an ethical theory is termed ★ethical naturalism, and in the case of Spencer's naturalism bringing up a child was not a matter of mere whim but rather a case of recognising not only a physical continuity between the way animals and humans live but also, at the instinctual level at least, a mental affinity too! So, for example, the training of the child in powers of observation through nature study would serve 'to habituate the mind from the beginning to that practice of self-help which it must ultimately follow'.[6]

Of course, there are many kinds of naturalism. One of the first philosophers to claim that 'nothing which is contrary to nature is good' was Aristotle. He regarded nature as a purposive system; that is to say he adopted a ★teleological view of nature. This means that each living species has its own specific goal, end-state or telos to which it strives during its development. Hence, nature displays ★rational principles in that each species develops its own natural and specific capacities which ought to be exercised if a normal or natural development is to take place. Nature, then, makes everything for a single use and Aristotle's form of naturalism is sometimes called ★functionalism, in that his model is generated by the idea that 'every instrument is best when intended for one and not many uses.'[7] The latter remark might make some sense by the way if we are considering a career that an individual has chosen to develop – a farmer, soldier, teacher and so on – but even then arguments might break out in terms of evaluating the role performances which may be involved in farming, soldiering or teaching as activities. None the less, it is to be hoped that in such evaluations the reasons why someone chose to act in one way rather than another within such role performances are to be connected to certain interests s/he had in performing appropriately. Where anyone else shared such interests they too might be expected to have reasons and feel urged to perform in a similar fashion. But what sense can be made of the claim that human beings as such have a specific role, function or mission to perform in life itself? Indeed, even those of us who might feel on occasions that we are on a mission might like to know perhaps what exactly that mission was! We need not explore such arguments here. It is sufficient to claim that much of the theorising which has occurred in the discipline known as ethics has been in relation to forwarding or rejecting some form of ethical naturalism.

In this way, then, an attempt has been made to deal with your first question. An ethical naturalist sustains the thesis that we must look to nature to provide moral guidance for human action. Much writing in ethical theory has addressed that proposal either by defending the naturalist's thesis or by attacking it.[8]

Turning to the writings of Simone Weil may assist us in dealing with your second objection. In *The Notebooks* she asks us to consider a child choosing a dainty cake from others on a plate. The reason for the child's hesitation may not be because s/he can't choose. It may be that the children enjoy looking at the dainty of their choice, 'and though they finally end up eating it, they have the feeling that in doing so they are lowering themselves a little.' In case you are tempted to say that this explanation appears somewhat fanciful, do bear in mind the ordinary everyday expression 'You can't have your cake and eat it!' On the basis of such reflections she suggests the child may be touched by beauty already: 'for a delicious flavour which is present to the sensibility without one experiencing or wanting to experience it through taste, is in a sense comparable to beauty'. Such considerations enable her to conclude: 'Beauty is a sensual attraction that maintains one at a certain distance and implies a renunciation –

including the most intimate form of renunciation, that of the imagination. One wants to devour all other desirable objects. Beauty is something that one desires without wanting to devour it. We simply desire that it should be.'[9] That last wish could well apply in relation to objects – such as art objects – we admire, human beings whom we love or indeed nature itself which we enjoy contemplating. If so, since it is not logical to seek to possess or alter such an object, person or landscape – otherwise it would no longer be what it was – then it must be morally wrong too, a view Fleda Vetch for example takes in Henry James's novel *The Spoils of Poynton*[10] and a stance which is expressed by Simone Weil when she comments 'Purification is the separation of good from covetousness'.[11] True, such an object, person or landscape is subject to processes of change or is in flux, but in seeking possession or bending them to our will we ignore the end-states which could be achieved if that object, person or landscape were left to its own devices. To ignore these considerations in our actions would be to act immorally.

Perhaps this argument is the first one we have discovered which strikes in favour of an ecocentric concern. But an old joke returns us to the anthropocentric point of view. A vicar passes someone working in his garden; 'What a beautiful garden!' he exclaims. The gardener replies: ' You ought to have seen the mess when God had charge of it!' Yet this implied objection only sustains at most a stewardship claim in relation to the end-states of things in nature. It does not sustain the domination view where nature is considered only in terms of its resources for human use!

The argument sustaining this more ecocentric view can be put in a Peircian way. Peirce's line is that an aesthetic experience is characterised by its intrinsic value. Now if a sense of goodness is tied, as Charles Hartshorne puts it, to 'the disinterested will to enhance the value of future experiences' then ethics must presuppose aesthetics. Where Peirce stumbled was in focusing upon a particular individual's enhancement of the value of his own future experiences for that person – the Dandy, Peirce – himself, in order 'to be a great man and a saint *by one's own standards*' as Baudelaire once put it![12] From a moral rather than an aesthetic standpoint then it is not the value of experiences in themselves nor indeed for an individual that matters. It is, rather, their instrumental value in providing guidance 'so as to increase the intrinsic value of future experiences',[13] particularly for others, which determines their goodness or ethical worth. Again, however, we have the problem as to what that term 'others' refers to here: simply to human beings or is it to be more widely understood? If the former, then we are still entrapped by some form of anthropocentrism!

Having tried to defend the claim that the moral attitude can be derived from the aesthetic inspiring the instrumental attitude let us return to the attitudes struck by the Environmental Agency spokesperson and the QC for the appellant. The latter, though invoking what might be regarded as a scientific attitude in referring to certain hydro-geological findings, seemed to escape from that attitude in referring to their possible effect in handicapping quarrying at the Tiptree Site. Really, these findings can be regarded as Heidegger described a wood as timber or a mountain as a quarry of rock, that is to say as carrying the attitude of seeing Tiptree Hill simply as a site for exploiting minerals. Now can that attitude be really described in terms of one of the four attitudes so far delineated – the aesthetic, the instrumental, the scientific or the non-being aware – or even in terms of our derived ones, the moral and the technological? The nearest seems to be the instrumental, the use-value of the clay for human purposes. But is the appellant's case really to be spelt out in terms of this clay's use or does the reason for quarrying it lie elsewhere. Consider the following quotation in this context:

> Nature becomes for the first time simply an object for mankind, purely a matter of utility; it ceases to be recognised as a power in its own right; and the theoretical knowledge of its independent laws appears only as a stratagem designed to subdue it to human requirements, whether as the object of consumption or as the means of production.[14]

In using the phrase 'for the first time', Marx is referring to the development of capitalism: a form of production which is based on private property controlling the means of production for a market, now a world-market, giving a 'cosmopolitan character to production and consumption in every country'. Within international capitalism new forms of production are introduced to generate 'a life and death question for all civilised nations' where industries 'no longer work up indigenous raw material, but raw material drawn from the remotest zones; industries whose products are consumed, not only at home, but in every quarter of the globe'.[15] Tiptree Hill then no longer satisfies an instrumental view merely embodied in using it say as a high ground from which messages can be conveyed,[16] on which a gun can be placed to control the estuary or from which clay can be dug for local emergency sea-defence purposes. Rather its clay can be produced to be consumed in a wider market and the void space produced in its turn consumed too for waste disposal purposes. What connects that production to consumption is money so that the use-value of the clay is converted into exchange value, since the clay as well as the void space can be regarded as a commodity to be bought and sold in a market place! A third reference to Marx's writings makes these distinctions clearer:

> A thing can be a use-value, without having value. This is the case whenever its utility to man is not due to labour. Such are air, virgin soil, natural meadows etc. A thing can be useful and the product of human labour, without being a commodity. Whoever directly satisfies his wants with the produce of his own labour creates, indeed, use-values, but not commodities. In order to produce the latter, he must not only produce use-values, but use-values for others, social use-values.[17]

Of course, in accepting the validity of these claims we are not thereby bound by his conclusion: 'Lastly nothing can have value without being an object of utility' since that would rule out both aesthetic and scientific as having independent value, but for the moment we are not interested in pursuing a critique of Marx's position! What has been established is that just as moral attitudes can be regarded as generated at the interface between the instrumental and the aesthetic and the technological between the scientific and the instrumental, so someone sustaining the ⋆commodific or commercial attitude would see nature simply as resources to be bought and sold. Peirce, of course, employed to some degree this kind of reasoning. After all, for him the instrumental attitude was captured in seeing nature as 'an opportunity' to realise 'some ulterior purpose' and he lists three: 'to make money, or to amend his life, or to benefit his fellows' (Peirce: vol. 1, paras. 43–45). These are left undescribed, but it does not seem unreasonable to regard the first as the commodific or the commercial and the last as possibly involving the moral, but what about doing something 'to amend his life'?

Two interpretations of the latter are possible, depending on what could amend a person's life as seen by others or as seen by him or herself. In the former case such a person might purchase a car, introduce central heating into a home or pay a city a visit. We are returned to the instrumental. But consider the second, the prospect say of purchasing one of Princess Diana's dresses; refusing to part with a 1964 Queen's Theatre programme for a performance of Chekhov's *The Seagull* directed by Tony Richardson;[18] buying an antique at a sale. Though we might agree with Marx that the dress, programme or antique are 'in the first place' objects 'outside us' and were once 'products of labour', we may have to reject his constant talk of commodities only in such terms.[19] Human beings can take an entirely *emblematic attitude towards them.[20] The dress, programme or antique serve no use-value at all nor does their value depend on the price they may currently possess. Rather their owner may wish to be associated with Princess Diana's world in some way, be reminded of a rather special theatre performance or be pleased by the grin issuing from an antique Toby jug! Such commodities no longer signify the thing for which they were once created but have become emblematic for the imagination of their possessor whether that be in terms of an association with Princess Diana, a memory of the company of someone at the theatre or the enjoyment of some past event associated in childhood with a Toby jug grin.[21]

In this chapter we have explored different ways in which the four-fold classification of attitudes might be put under pressure to yield further attitudes which human beings can take in viewing the world: the moral, the technological, the commodific – and the emblematic. It has been argued that following the ethical doctrine of naturalism – captured in Aristotle's claim that 'we must always look for nature's own norms in things whose condition is according to nature'[22] – a moral perspective can issue from an instrumental standpoint inspired by a concern with the aesthetic, as Peirce suspected, just as a technological concern emerges from the relationship of the scientific attitude to the instrumental. A return was then made to a further examination of the instrumental attitude. Despite his obsession with *instrumentalism* – the doctrine that something, can only possess value if it has a use – Marx's insights were used to show how, in the contemporary world, instrumental value can be transformed into commercial value so that use-value is converted into exchange value and objects become subject to commodification. That last term, meaning subject to the commercial, can be labelled in relation to one of two possible attitudes. The first, the commodific refers to the attitude of seeing everything in terms of exchange value. Nature is seen as offering materials or resources which can be sold. The second, the emblematic, projects onto things a subjective value displacing its use or exchange value. So Tiptree Hill could be converted into some theme park: besides the cafeteria and souvenir shop, areas could be allocated for viewing the estuary; examining a model of the old Admiralty Telegraph system;[23] looking through an eye glass used to monitor a reconstructed First World War gun; learning about the life of the local owls – if there are any left – or exploring stages in the development of the horse seeing as it emerged during the Eocene period at the time when this hill was formed. Is this the way historical value could resonate in the future, through the emblematic – an attitude emerging from the commodified's contact with the aesthetic?

NOTES

[1]*Existence or Mere Reality*, Westwood, Wilts., 1 February 1993, cf. 'What keeps all things occupied and in motion is the striving for existence,' T Adorno *Minima Moralia* p.175.

[2]Martin Heidegger, *Being and Time*, p.100.

[3]Jurgen Habermas, *Communication and the Evolution of Society*, pp.2–3.

[4]Joseph Brent, *Charles Sanders Peirce: A Life*, pp.336–337.
[5]N E Boulting, 'Grounding the Notion of Ecological Responsibility: Peircian Perspectives', *Religious Experience and Ecological Responsibility*, sect. VI.
[6]H Spencer, *On Education* (1854–1859), pp.16, 93, cf. Rousseau's Naturalism: 'Observe nature and follow the path it maps out for you. It exercises children constantly; it hardens their temperament by tests of all sorts; it teaches them early what effort and pain are.' *Emile or On Education*, Jean-Jacques Rousseau, p.47.
[7]Aristotle, *The Politics*, sects. 1325, b9–10 and 1252 b4, Everson p.p.161–162.
[8]Cf. For example its critique in R M Hare's *The Language of Morals*, chapt. 5 and its defence in the writings of Philippa Foot in her Introduction and chapt. 5 of *Theories of Ethics*.
[9]Simone Weil, *The Notebooks*, vol. 2, pp.335, 416.
[10]Henry James, *The Spoils of Poynton*.
[11]Simone Weil, *Gravity and Grace*, p.20. cf. N E Boulting, 'Necessity, Transparency and Fragility', Simone Weil's Conception of Ultimate Reality and Meaning', *URAM*, vol. 22, no.3, pp223–246.
[12]Joseph Brent, op. cit. p.23, n.38.
[13]Charles Hartshorne, 'The Aesthetic Matrix of Value', chapt. XVI of *Creative Synthesis and Philosophic Method*.
[14]Karl Marx, 'The Rise and Downfall of Capitalism' in 'Grundrisse' included in *Karl Marx: Selected Writings*, pp.363–364.
[15]Karl Marx, 'The Communist Manifesto', op. cit., p.224.
[16]An Admiralty Telegraph system was set up on Callum Hill, adjacent to Tiptree Hill, Lower Halstow during the Napoleonic Wars. See *The Old Telegraph* by Jeffrey Wilson.
[17]Karl Marx, Chapt. 1, 'Commodities' in 'Capital', vol. 1, op. cit. p.425.
[18]Starring Peggy Ashcroft, Vanessa Redgrave, Peter Finch and Paul Rogers.
[19]Karl Marx, chapt. 1, 'Commodities' in 'Capital', vol. 1, op. cit. pp.421, 429, 436.
[20]It can be argued that Marx says this too when he speaks of 'The Fetishism of Commodities' where exchange value is converted into 'a social hieroglyphic' but his obsession with his instrumentalism prevents the full development of that insight. Karl Marx, Chapt. 1, 'Commodities' in 'Capital', vol. 1, op. cit. pp.435–443. For a contrary view see Susan Buck-Morss, *The Dialectics of Seeing*, Chapt. 6, sect. 4.
[21]'We dream of again meeting with the particular circumstances which made for our now lost happiness, the signs in which the latter may be read. The imagination, filler up of the void, attaches itself to the signs but not to their significance, which is not an object of the imagination. Freedom as such is not an object of reverie.' *The Notebooks* of Simone Weil vol. 1, p.160.
[22]Aristotle, 'The Politics', Everson, p.6
[23]As stated earlier, an Admiralty Telegraph system was set up on Callum Hill, adjacent to Tiptree Hill, Lower Halstow during the Napoleonic Wars. See *The Old Telegraph* by Jeffrey Wilson.

Chapter IV

THE AFTERNOON: SOME SEMBLANCE OF RATIONALITY

Hobbes's words point to the true nature of rationalisations: 'For – thoughts are to desires, as scouts, and spies, to range abroad and find the way to the things desired'.[1] Even if Hobbes does indicate the source of rationalisations themselves – their affixation to desires – his analogy fails to explain why such thoughts do not appal their thinker, since they obviously exploit that age-old cleavage between a man's genuine thinking and the form of life which makes that thinking possible. Rationalisations, as thoughts which charm just because they offer no clue to close the gap between what one feels ought to be done and a life that generates them, are the very ones which consciousness allows to be separated from itself since in the alienated life humans are forced to live, they must cause no pain. This is true even of those thoughts which, like the living feeling that fate will usher in a worse tomorrow, conspire with their thinker to ensure the undoing of a present commitment to the conscious life. In rejecting the general seal of approval society awards to rationalisations, however, genuine thoughts, which that same consciousness will have created, may yet still hover over its surface defying its attempts – simply because they are not rationalisations – to reclaim and so reform them in an ongoing process which represents the true nature of consciousness itself. But in that case a new problem arises, not one concerning the actual content of thoughts themselves, but rather with how to catch those very non-intentionalities which consciousness seeks to absorb within its own activity.[2]

The Environment Agency's representative identified Biffa Waste Services as a controlling interest in the latest submission by South Eastern Minerals Ltd. Two issues disturbed him: that the results of the borehole tests – one must assume he meant those done by the appellant – were only received by himself on 11 November and the information now that industrial special waste was to be put into the quarry, which raised permeability issues. But the nature of that issue was to be determined not by the permeability of the clays involved but by the nature of the disposed material. What was planned in the proposal? Given the new nature of the waste to be disposed, perhaps forty to fifty metres of clay should be left in this pit which would, thereby, make a material change to its depth. Two options were available to the inspector. Firstly, matters could be ended now by adjourning or even ending the inquiry to enable a proper environmental assessment to be made to ascertain how deep the quarry could be excavated, the nature of its extension, the viability of the site, transport implications and so on in view of the new proposals. Secondly, adjourn part of the inquiry, in relation to the nature of the permitted waste. He inclined against the latter proposal.

The spokesperson on behalf of Swale Borough Council then rose to correct a wrong impression he may have caused before lunch. Clearly the inquiry could not continue on the basis of entirely fresh amendments but he was instructed that there was no reason why the inquiry could not be pursued on the basis of what still stood in the original application. 'We are in your hands', he said to the inspector, 'in that you will have to decide, sir, whether the nature of the amendments was so far reaching as to justify an adjournment'.

The appellant's QC said emphatically 'We shall not withdraw!' Issues surrounding waste licensing should not be convened at this inquiry. 'Has the application changed?' he asked. In reply to his own question he re-examined the original application in order to show what remained the same: plan 1: the location; plan 2: the landscape setting; plan 3: the context for the original site; plan 4: the proposed development; plan 5: stage three extraction; plans 6 and 7: with cross-sections north/south, east/west and three subsequent phases of development. Figure 8, the restoration plan and plans 9A and 9B: the planning application for the road. He admitted that there was what he called 'the missing link' in all this: the time issue. That was because estimating time actually can't be part of an application like this. Nor was there a full story actually about quantities; that was why only hectarage indications could be given, but the basis of the appeal was still grounded actually on clay extraction. As you will see the term 'actually' was a word he could not stop himself using! In addition, noises of people moving in the room and the creaking door prevented him from being heard clearly but he seemed to give the impression that waste disposal lorries would treat the waste off-site! Approval for the extension of the site to unpack an extra million tons of clay would be pursued for stage four. Approval, 'actually', for that part of the application was not being sought now. Admittedly the actual size of the present proposal had changed from 90.5 to 72.5 hectares (a hectare being about 2.5 acres).

After being pressed as to whether the road was or was not part of the present proposal, he stood up quite upright and delivered the following memorable phrase to his audience: 'Let us now', he said, 'put a semblance of rationality on all this!' He continued by explaining that the area would be exactly the same. It was just that the site had been remeasured. The depth of the extraction 'actually', had been reduced. The restoration levels remained the same. Eight hundred thousand tons of clay would be removed during the first five years and replaced by two hundred thousand tons of waste. This fact might reduce the time involved, though that was not stated in the application, and thereby traffic use 'actually' would be lessened if less clay was removed. So eight thousand cubic metres of clay would be replaced by two thousand cubic metres of waste. There would be 'actually' no change in the landscape, no change in the access points and no change in the workings so that ecological landscape potentialities would not be changed either. Only the shape of the site would have to be slightly altered. During stage one the creation of protection measures were still covered by the scheme, using some of the clay extracted. Since the planting of trees to give visual protection to the site would start as soon as work at the quarry began, so, again, there were no real or substantial changes 'actually' to the original scheme.

The QC then referred to his experiences since the 1970s. Such applications as this one simply have to be progressive. He referred to a Planning Policy Guidance (PPG) note from the Department of the Environment, PPG 23, as justifying a distinction between the planning stage of an inquiry and pollution matters. The latter issue raises far more wide-ranging issues but there would be no excavation 'actually' below the level of the London Clay. He spoke of his company having twenty-two sites, they knew what they were doing and because they were experts in eco-management they were given a European award recently. The company was answerable to the Waste Dispensation Authority. Because of such responsibilities it had undertaken new borings on its own initiative. As a result the company had decided to amend the base of the quarry. They had been too honest! They could have kept all this information to themselves. These results were known in October and he admitted, in forwarding them to the Environment Agency, he may have caused some problems.

Now there was a pause. KCC's representative was required to move his car outside in the car park. On resumption of proceedings the QC offered a long exposition about the nature of the clay. He admitted that questions were raised about the different levels of clay measured but noted ground water was not recorded in London Clay even if water was found in the Woolwich clay beds. The lining for the quarry would require some engineering to satisfy a Waste Licence Application and to meet criteria nine times the European standard, but London Clay is actually the most researched clay in Britain. It has wholesale use 'actually' because of its low permeability. A reader has only to read the documents then with some intelligence to see that the company was as concerned to provide a proper containment system as was the Environment Agency. This should be possible with London Clay because it has 'actually' an integrity of a very high order. The QC seemed to suggest that if ten metres of such clay needed to be provided in the base with the liner, this would be done. He stressed once more the evolving nature of the application and that the application did specify the details of actual waste disposal in the supplementary statement.

Finally, he dismissed the reference to the Salisbury versus Wessex Company case because the action of the company was judged to have acted unreasonably in preparing their application. The notion of unreasonable action was the issue. In the present case the issue was one of a planning judgement and the inquiry should go forward on the basis of the present application. The public interest lies in efficiency and that is part and parcel of the government's own position too! Perhaps it might be reasonable to adjourn the inquiry to the new year after the geological matters had been further investigated in terms of the waste deposits. But what was the benefit of a new application? He admitted to the inspector that it might well be a circumstance he had not met before, but the issue concerning the kind of waste, which could be deposited on this site, would be settled by the Special Waste Regulations. In this regard, with respect to a special government document issued on 30 September 1996, many things could come under the term *special waste*. The nature of this evolving definition would be settled as a licensing matter. Such issues need not be raised now but rather related to the supporting material for the application. In this way he supported the position of the borough council: the inquiry should continue.

During the last half-hour of this session many of these arguments were reiterated Two issues were stressed, particularly by KCC's representative: the nature of the application had been altered to a waste-led application from a mineral-led one and the time scale of the application, taken as a whole, had obviously increased. The inspector said that during the first hour the next day he would review all that had been said and, at exactly eleven o'clock, the inquiry would start again with his decision as to whether it would or would not be adjourned.

The attention of someone with an interest in philosophy would be held by the idea of 'putting a semblance of rationality' on something. What could the QC have meant? In what followed this remark he attempted to show how the changes to the application were reasonable in the light of the original proposed scheme for Tiptree. But in that case why not say simply that he sought to draw out the rationality of what was now proposed? But saying that one is putting a semblance of rationality on something – say on an application – this move seems to imply that there is nothing in the application which displays rationality in itself. So the application is made to appear to be something which it is not!

In the article where Peirce distinguished his 'three classes of men', he draws attention to the danger of looking 'upon reasoning as mainly decorative': 'sham reasoning'. Peirce does this because he is very suspicious of the effects of 'mixing speculative inquiry' (Peirce: vol. 1, para. 58), such as philosophy and science, with 'questions of conduct' (Peirce: vol.1, para. 56). The answers to the latter take time to evolve and such matters are conservative by nature, since they refuse to submit their absolute distinctions between right and wrong to experiment (Peirce: vol. 1, para. 55). But those who employ 'sham reasoning' continue to tell themselves they regulate their conduct by reason; but they learn to look forward and see what conclusions a given method will lead to before they give their adhesion to it. In short, it is no longer the reasoning which determines what the reasoning shall be' (Peirce: vol. 1, para. 57). But this description of sham reasoning doesn't seem to cover quite what the QC said! Rather than use reasonings so as to appear to be doing so without having a fixed pre-planned conclusion already in mind – despite that conclusion already predetermining the kind of reasoning employed – the QC, in what he said subsequently, tried to show that the appellant was acting in accord with reason. A spectator, on the other hand, might receive the direct impression that these were nothing but, 'excuses which unconscious instinct invents to satisfy the teasing "whys" of the ego'. According to Peirce, such an activity renders 'philosophical rationalism a farce' (Peirce: vol. 1, para. 631).

In these few lines, however, the possibility of making a number of distinctions has already been suggested. We have the cases of sham reasoning – reasoning given a 'spin' by a conclusion it is meant to lead to – and those of rationalisation where 'reasons' are invented to legitimate or make what has been said or done appear reasonable. But that leaves us with two further terms to distinguish: valid reasoning and that of rationality itself. Perhaps recalling once more Peirce's categories Firstness, Secondness and Thirdness might throw some light on these distinctions. Indeed, the idea of a conclusion already pre-eminently fixed upon recalls Secondness, since it appears to be something which can 'resist our will' (Peirce: vol. 1, para. 419), we are constrained by it (Peirce: vol. 1, para. 325) as it forces 'its way to recognition as something *other* than the mind's creation' (Peirce: vol. 1, para. 325) predominating as 'what *has been* done' (Peirce: vol. 1, para. 343). Both *sham reasoning and *rationalisation then appear to be coverings for the already decided: a conclusion in the first case or the legitimation of an argument in the second.

Firstness, you will recall, tied to immediacy, to a sense of freshness, to the delight in a possibility (Peirce: vol. 1 para. 25), and characterised 'that which has not another behind it' (Peirce: vol. 1, para. 302). The 'positive internal character' (Peirce: vol. 5, para. 469) of *reasoning is that it is pursued for its own sake alone 'without any sort of axe to grind' (Peirce: vol. 1, para. 44). As Peirce puts it in 1877 in his *The Fixation of Belief*: 'The object of reasoning is to find out, from the consideration of what we already know, something else which we do not know. Consequently, reasoning is good if it be such as to give a true conclusion from true premises, amid not otherwise' (Peirce: vol. 5, para. 365).

For Peirce, the category Thirdness ties not to the past nor to the present so much as to the future, in relation to our capacity for 'making some kind of prediction' (Peirce: vol. 1, para. 126) since we consider that phenomena are rule-governed or subject to some kind of law (Peirce: vol. 1, para. 420). Peirce specifically ties the issue of *rationality to future concerns: 'the rationality of thought lies in its reference to a possible future' (Peirce: vol. 7, para. 361). He also directly connects it to Thirdness: rationality is 'conformity to a widely general principle' (Peirce: vol. 8, para. 152). The 'essence of rationality lies' in 'being governed by final causes'[3] (Peirce: vol. 2, para. 66). But now at least three questions arise: 'Aren't all attempts to be rational, rationalisations?'; 'How can we be rational?' and 'Why be rational? What grounds the activity of being rational itself!'

One way of establishing the distinctions we have sustained so far is to examine the claim that all reasons are rationalisations. That claim would make the QC's position quite reasonable! To deal with this issue we can begin by referring to Plato's philosophy. Plato founded his Academy in Athens in 386 BC and his *Republic* can be regarded as 'a statement of the aims to which the Academy set itself to achieve'.[4] In it we can see Plato advocating the cultivation of reason so that the vocation of the philosopher and the ruler of his ideal state would be one. Reason then, manifested in the life of the philosopher, provides the basis for the exercise of *authority on the part of the ruler. His disciple, Aristotle, however, was not nearly so optimistic as his teacher: the offices of philosopher and ruler are separated. Rather, the true philosopher should advise by the ruler in order that his reign can display good words and deeds.[5] Rumour has it that, for example, Churchill enjoyed the advice of Isaiah rather than Irving Berlin at the end of the Second World War![6]

Although Thomas Hobbes, born in 1588, was an admirer of Plato, this doctrine that the due application of reason underpins authority, was stood on its head.[7] According to Hobbes reasoning between individuals can't even begin without presupposing a sovereign or at least some kind of imposed authority to provide the necessary stability in which such an activity can take place. So instead of reason providing the basis for the exercise of authority, Hobbes's *Leviathan* makes manifest the idea that the activity of reasoning itself is made possible by being grounded on an antecedently agreed-upon constituted authority which adjudicates between disputants.[8]

You may wonder why Hobbes took this position, why he was so pessimistic about the individual's pursuit of reason. His case is that with respect to any subject of reasoning '...the ablest, most attentive, and most practised men may deceive themselves and infer false conclusions' so that '...where there is a controversy in an account, the parties must by their own accord set up, for right reason, the reason of some arbitrator or judge, to whose sentence they will both stand, or their controversy either must come to blows, or be undecided, for want of right reason constituted by nature; so is it also in all debates of what kind so ever' (Hobbes: *Leviathan*: chapt. V, para. 3). Indeed even before 'the names of just and unjust can have place, there must be some coercive power' (Hobbes: *Leviathan*: chapt. XV, para. 3) so as to make men perform according to their agreements through fear of some punishment which only the creation of the civil realm can inaugurate. The necessity for some arbitrator or sovereign does not, then, lie in the need to coerce those in the civil realm to become civilians but in the need to provide incentives for them to keep to their covenants with each other and to punish renegades.

Sufficient has already been said for you to see that Hobbes regards human beings as completely monadic, completed entities unto themselves and, even if at times gregarious, hardly related to each other as social persons at all. This conception follows from his view that human beings are to be regarded merely as physical existents. So in *De Cive* he writes: 'For every man is desirous of what is good for him and shuns that which is evil for him, but chiefly the chiefest of natural evils, which is death; and this he does by a certain impulsion of nature, no less than that whereby a stone moves downwards.'[9] Other human beings are seen as adversaries, threatening objects to one's own health and safety and with whom one would be constantly at war but for the imposed order the civil realm can supply. Without the civil realm, Hobbes regards human beings as something quite other than Marx's members of a 'species being'[10] and sees human beings as dissociated from one another to the extent that '...men can have no pleasure, but on the contrary a great deal of grief, in keeping company, where there is no power able to overawe them all' (Hobbes: *Leviathan*: chapt. XIII, para. 5).

Interestingly, again, that sense of dissociation comes out more in *De Cive* where Hobbes speaks of men 'as if but even now sprung out of the earth, and suddenly (like mushrooms) come to full maturity without all kinds of engagement to each other'.[11] Even in the civil realm the only reason for one individual to hold to his covenant with another is out of self-preservation for he '...therefore that breaketh his covenant, and consequently declareth that he thinks he may with reason do so, cannot be received into any society, that unite themselves for peace and defence, but by the error of them that receive him' (Hobbes: *Leviathan*: chapt. XV, para. 4). Again, such a conclusion is hardly surprising in view of what he tells us about human beings in their natural state since there '...everyone is governed by his own use of reason' (Hobbes: *Leviathan*: chapt. XIV, para. 4), that is to say by his own thinking. No grounds exist for believing that the institution of the civil realm basically changes men's nature in this regard so that his concern in the 'Of Commonwealth' section of the *Leviathan* is not to show how man's nature can be transformed by the civil realm – it can't be – but rather with how it can be controlled. That is because man is a creature subject to his passions which can 'tend to be evil', (Hobbes: *Leviathan*: chapt. VIII, para. 20), but he makes it clear that '...desires and other passions of man are in themselves no sin. No more are the actions that proceed from those premises, till they know a law that forbids them' (Hobbes: *Leviathan*: chapt. XIII, para. 10). We arrive now at the remark opening this chapter relating reason to passion: 'For the thoughts are to the desires, as scouts, and spies to range abroad, and find the way to the things desired' (Hobbes: *Leviathan*: chapt. XIII, para. 16).

So far you have seen how Hobbes's conclusion, that all attempts to be rational amount to generating rationalisations, follows from his conception of human beings as being self-sufficient and atomistic, subject to passions and desires which drive the activity of reasoning itself. But his conclusion is sustained too by the way he regards language. He points out that the '...general use of speech, is to transfer our mental discourse into verbal; or the train of thought into a train of words;' (Hobbes: *Leviathan*: chapt. IV, para. 3) the former called 'mental discourse' as opposed to discourse in general. (Hobbes: *Leviathan*: chapt. III, para. 1).

Just concentrate for a moment on the first part of that claim. Hobbes is saying that the use of the communicator's discourse can always be manipulative since that discourse can always stand outside, be separated from, the thoughts and desires of the communicator, which gave rise to that communication in the first place.[12] As Peirce says of sham reasoning, Hobbes shows us how the communicator's thoughts, desires and decisions can be brought into existence prior to and without even implicitly acknowledging particular rules of discussion or argument prevalent in the way others communicate or discuss with him or her.

Now for some technical jargon: in philosophy the term *solipsism is used to describe the idea that a human being only knows the condition of his or her own private states and that only these are real compared to what is taken to exist in the world. For the solipsist the world is a personal dream! Hobbes could, however, be regarded as advocating a *methodological solipsism:[13] he adopts solipsism in such a way as to generate views about language, action and the way human beings can relate to each other so as to show that all human behaviour is a matter of strategy in striving for a set of private ends!

Not only does Hobbes adopt this 'transfer-thesis' – the idea that thoughts are translated into utterances for self-interested purposes when humans address each other – to account for how language is used, but he plays down the significance of *language use* more broadly by saying that no '...man can know by discourse, that this or that is, has been, or will be' (Hobbes: *Leviathan*: chapt. VII, para. 3). Ordinary language usage, infected by mere opinion, even lacks the status of what he calls conditional knowledge which at least, he considers, scientific discourse enjoys. This is because descriptive terms, what he calls 'names', 'can never be true grounds of any ratiocination' (Hobbes: *Leviathan*: chapt. IV, last para.) any more than can metaphors or 'tropes of speech', but at least these latter two, unlike the former, make obvious what he calls their 'inconstancy', and inconstancy arises because men use all their descriptive terms subjectively. Indeed descriptive terms or names '...are ever used with relation to the person that useth them; there being nothing simply and absolutely so' (Hobbes: *Leviathan*: chapt. VI, para. 6). Consequently Hobbes sees the kind of discourse employing such terms as contrived in the sense of being 'regulated by both desire and design' (Hobbes: *Leviathan*: chapt. III, para. 4). Small wonder then that a modern commentator – Richard Peters – was able to use the image of the wrestling bout where the point is '...to throw the other fellow and glory in his discomfiture'[14] to characterise Hobbes's conception of language use between people.

So far you have been presented with a philosophical defence of the claim that outside scientific discourse, all attempts at trying to be rational are really exercises in generating rationalisations. That particular defence has been largely ignored by philosophers not so much because it invokes non-testable or *metaphysical assumptions, but because these seem largely counter-intuitive or too sweeping; 'All thinking is self-interested'; 'All human behaviour is strategic'; 'All human beings are self-sufficient, atomistic individuals rather than social creatures' and that 'All uses of language involve attempts to manipulate the addressee into thinking or doing what the communicator intends him or her to think or do'. Such claims may well typify situations where conflict physically is taking place and so much broadcasting involves propaganda, or where a psychological war is occurring say between politicians at election time or between sellers of products competing for business in the market place. But whilst these activities keep to their own domains or at least outside the institutions where philosophising takes place, not much attention need be paid to Hobbes's outlandish ideas!

Yet though some doubt has been cast on the thesis that all attempts to be rational produce only rationalisations, little has been said positively so far to show how this thesis can be overcome. This is where Peirce's famous article *The Fixation of Belief* comes in, for in it he considers four 'synoptic' methods[15] in which opinions about matters can be settled; those of tenacity, authority, the *a priori* method, and the scientific. If you are aware of Plato's Divided Line as illustrating four stages of knowing (Plato: *Republic*: sect. 509d–511e) or you have read his *Cave Allegory* (Plato: *Republic*: sect. 514a–521b),[16] your awareness may prove useful in appreciating something as to the status of Peirce's methods. In *Republic*, those tied by chains so that their heads are fixed to see shadow on a wall in front of them, like watching a TV screen, enjoy the lowest form of cognition, as this deals with images and likenesses, taking things at their face value without questioning. Similarly in relation to the *method of tenacity, we find Peirce stressing that, in order to settle belief at his level, any answer to a question which may be fancied is ripe for adoption. Like the unreleased prisoners in Plato's Cave allegory, men's '...instinctive dislike of an undecided state of mind, exaggerated into a vague dread of doubt, makes men cling spasmodically to the views they already take' (Peirce: vol. 5, para. 377).

The next stage in Plato's Divided Line concerns right or correct beliefs without knowledge. They are what everyone shares at the level of common sense beliefs for which no rationale can be given by believers, yet which can successfully guide action.

Similarly Peirce's *method of authority is concerned with the way in which it is possible '...to keep correct doctrines before the attention of the people, to reiterate them perpetually and to teach them to the young' (Peirce: vol. 5, para. 379). In this condition people are '...kept ignorant, lest they should learn of some reason to think otherwise than they do.' And just as Plato remarked that common sense beliefs at this level are more genuine (Plato: *Republic*: sect. 510a) than images or likenesses because at least these can now be seen for what they are, once the prisoner is released from his chains to appreciate how the fire at the back of the cave makes such shadows possible, so Peirce remarks that the method of authority possesses '...immeasurable mental and moral superiority to the method of tenacity' (Peirce: vol. 5, para. 380). He adds: 'Its success is proportionally greater; and, in fact, it has over and over again worked the most majestic results'.

The third level of Plato's Divided Line marks a superior advance, since there is a shift in analysis from the physical to the intelligible realm. The prisoner is forced out of the cave to experience a greater reality beyond, even if only, at first, shadows of things or their reflections because of the brightness of the sun outside. We can compare this standpoint to Peirce's **a priori* method. The term **a priori* is usually connected to statements or propositions which are regarded as necessary and can't be conceived to be false: consider, for example, 'What is extended is coloured' or 'The angles of a triangle are 180 degrees'. Peirce's third *a priori* method is considered to be far more intellectual and respectable from the standpoint of reason. (Peirce: vol. 5, para. 383) At this level, Socrates, Plato's fictional character, speaks of basing inquiries on *assumptions – claims that may be true but need testing, or claims merely proposed for the sake of argument – from which other conclusions can then be deduced. Odd and even numbers, geometrical figures and three kinds of angle are cited as adopted assumptions which, because they are so intuitively obvious, need no further explanation (Plato: *Republic*: sect. 510c–d). Peirce speaks, in relation to this third method, of propositions being adopted because they seem agreeable to reason since the beliefs deriving from them are developed 'in harmony with natural causes'. Indeed Peirce cites Plato who 'for example, finds it agreeable to reason that the distances of the celestial spheres from one another should be proportional to the different lengths of strings which produce harmonious chords'. Peirce also cites Hobbes's view that the actions of men are purely self-interested as another example of this method and concludes: 'This rests on no fact in the world, but it has had a wide acceptance as being the only reasonable theory' (Peirce: vol. 5, para. 382).

Finally, in Plato's Divided Line, at the fourth and final level, there is a move from an assumption to a non-hypothetical principle which is assumption-free (Plato: *Republic*: sect. 510b), indeed to a permanency offered by the ultimately real (Plato: *Republic*: sect. 511c) pictured as the free cave dweller able to look at things – trees, boulders or clouds – directly, without the aid of shadows or reflections. Correspondingly in Peirce we read about the necessity for finding a method which is determined '...by nothing human but by some external permanency – by something upon which our thinking has no effect'. And in his 1903 footnote he adds '...but which, on the other hand, increasingly tends to influence thought; or, in other words, by something Real' (Peirce: vol. 5, para. 384).

Now, of course, you may have spotted Peirce's possible implicit use of his categories which, regarded singly, may have generated the first three rejected methods. If so, then the first three can hardly be regarded as 'straw methods'[17] since the method of tenacity could be regarded as based on feelings alone, that is to say Firstness: the believer 'feels that, if he only holds to his belief without wavering, it will be entirely satisfactory'. This can yield 'great peace of mind' (Peirce: vol. 5, para. 377). The method of authority can be seen to be associated with Secondness alone. It is introduced in relation to 'the will of the state' acting, rather than out of the felt whim of the individual, to create something 'to keep correct doctrines before the attention of the people' (Peirce: vol. 5, para. 379). Finally, if Thirdness is considered singly, we arrive at the *a priori* method, since it is introduced by considering the possibility of human beings thinking about and reflecting on the other two methods (Peirce: vol. 5, para. 381–382) and is legitimated by their 'conversing together and regarding matters in different lights' so that beliefs 'in harmony with natural causes can be developed' (Peirce: vol. 5, para. 382). Plato regarded the former stages of the Divided Line as significant in making the next level possible and sent his released man back into the cave to battle with the images and illusions that the still imprisoned cave dwellers suffered. Similarly, Peirce, after establishing the *scientific method as 'the only one of the four methods which presents any distinction of a right and a wrong way' for settling opinions by criticising the others, then passes down his hierarchy in considering the particular merits of each in turn. That is hardly surprising in view of the fact that in explaining what his scientific methodology amounts to, he has to make use of these categories in what he says about the Logic of Abduction or Indagation – 'things that are not governed by rational considerations very seldom have three elements' (Peirce: vol. 7, para. 197).

That logic will be explicated in chapter VI. For the moment we need only note that whereas the other three methods are concerned with belief in something as a way of settling opinions about some matter, Peirce's fourth, the scientific method, focuses upon doubt. And even in this article, as Max Fisch has noted, the categories are not far away: 'Doubt is an uneasy and dissatisfied state (Firstness) from which we struggle to free ourselves (Secondness) and pass into the state of belief (Thirdness)' (Peirce: vol. 5, para. 372; Max Fisch's bracketed interpolations).[18] Elsewhere, Peirce claims that if we are to choose what he calls 'his favoured method', 'that of reasoning' (Peirce: vol. 7, para. 324), we ought to do so 'on rational grounds' (Peirce: vol. 7, para. 325). Having then dealt with our first two questions – 'Aren't all attempts to be rational, rationalisations?' and 'How can we be rational?' – and if the answer to the second question is that we should adopt the scientific method or method of reasoning, we now face our third: 'What grounds the activity of being rational itself?'

Peirce gives a straightforward answer to this last question. He hooks rationality to generality or 'law to be true' (Peirce: vol. 7, para. 186) regarding the finding of this as defining the scientific interest. If the latter is identified with processes of investigation then its life is that of 'living doubt'; 'when doubt is set at rest inquiry must stop' (Peirce: vol. 7, para. 315). So 'the only rational ground for preferring the method of reasoning to other methods is that it fixes belief more surely' (Peirce: vol. 7, para. 325). Lines from William C Bryant's 'The Battle-Field' express 'the philosophy of the matter perfectly': *Truth crushed to earth shall rise again; the eternal years of God are hers; But Error wounded, writhes with pain, And dies among his worshippers.*[19]

Now for a bit more jargon! Peirce in advocating the value of inquiry, states his doctrine of *credibilism. Since we can't at one fell swoop doubt everything at any one time, there must be *propositions which are free from genuine doubt if we are to doubt others. For you, standing at a railway station, to doubt that the train will be on time, there must be other assumptions you can take for granted to make sense of such doubtings: that trains do run on these tracks, that there is a railway station and that it is correctly named and so on. For an inquiry to lead to a demonstration a start has to be made then with 'propositions free from all actual doubt' (Peirce: vol. 5, para. 376). Yet such *indubitables* may not be regarded as absolutely certain in the light of a theoretical or reflective point of view taken subsequently. So you may indeed realise that this line has trains which are subject to cancellation, that this railway station has been closed or that there are two stations serving this same town for example. Hence a proposition can be regarded as indubitable at one time, say in practical affairs, or when we are concerned with testing some other proposition(s), but not at another – when we are subjecting that apparent indubitable, for example, to doubt or criticism at some other time. The thesis that all our ideas or beliefs can be doubted constitutes Peirce's doctrine of *fallibilism and is central to his idea of what can be called *procedural rationality. Opinions, conjectures or theories are proposed which can be criticised and tested to decide the validity of accepting them provisionally. Because science satisfies the requirements of this procedural approach it is legitimated, rather than justified, and hence regarded as a highly rational activity.[20]

We can now formulate our third question more precisely: 'Why be Rational?' For a Procedural Rationalist, such as Peirce, the scientific, or what he regards as the rational attitude of considering seriously what we learn from experience or through argumentation is simply adopted. Insofar as this attitude is so adopted, a commitment to the use of this procedural sense of rationality cannot be justified[21] since it hangs merely on a 'faith in reason'. Accordingly, since no grounding can be provided for engaging in the rational enterprise itself, we either choose to adopt such procedural rationalism or we are left with what Popper calls 'comprehensive irrationalism'[22] and brands of this we have now considered under the methods of tenacity, authority and the *a priori* method.

You were initiated into philosophical ruminations in this chapter by considering the QC's position at the inquiry when he said that what was important in his presentation of the appellant's case was to put 'a semblance of rationality' on the application. The question was raised whether, outside scientific activity, reason is ever used in a disinterested sense to ascertain the truth or whether such attempts fall into the pit of rationalisations. A strong defence of the claim that such attempts do indeed fall into that pit was made by considering Hobbes's philosophy in these matters. But his position is undermined by his use of far-reaching metaphysical claims he needs to sustain it: that all men are atomistic and self-centred; all human action is strategic and that all human discourse employs language manipulatively. Once doubt was cast upon his thesis, an attempt was made to show why three methods – those of tenacity, authority and the *a priori* method – of settling opinion were open to the charge of legitimating the creation of rationalisations rather than the giving of reasons for beliefs. That attempt forwarded the idea that all three such methods are concerned with proving or justifying an already accepted assumption which rationality had to serve. None of these methods is concerned with criticising or doubting a claim which the pursuit of rationality in the hunt for truth would either reject or fix more securely.

Peirce's 1877 article, then, *The Fixation of Belief* shows what he considers the rational attitude ought to be and this attitude of seriously considering what we learn from scientific inquiry or what he calls later in 1893 'the experiential method' (Peirce: vol. 5, para. 406) is simply adopted unquestioningly, so that a commitment to this method is irrational in that it hangs merely on a 'faith in reason'. No logical grounding can be given for doing this since we choose either to adopt such a procedural rationality or plump for one of the three other methods. He illustrates the arbitrary nature of this adoption in the closing paragraph of this article:

> The genius of a man's logical method should be loved and reverenced as his bride, whom he has chosen from all the world. He need not contemn the others; on the contrary, he may honour them more deeply, and in doing so he only honours her the more. But she is the one that he has chosen, and he knows that he was right in making that choice, And having made it, he will work and fight for her, and will not complain that there are blows to take, hoping that there may be as many and as hard to give, and will strive to be the worthy knight and champion of her from the blaze of whose splendours he draws his inspiration and his courage. (Peirce: vol. 5, para. 387)

But 'in making that choice' of a bride, it is merely a *contingent* matter as to whether people choose, or even employ without necessarily intending to, any of his four methods, even if it happens to be the method of critical argument or rational discussion which characterises his, the scientific philosopher's mode of consciousness. Put like that, however, and it appears simply a matter of choice, a contingent choice, whether we commit ourselves to the pursuit of rationality or merely 'the semblance of rationality', as the QC put it, by adopting one of the three other methods. Of course, when Peirce came to see what is involved in pursuing scientific investigation, when he saw that such values as 'an interest in an indefinite community', 'the recognition of this possibility' 'being made supreme' and 'the continuance of intellectual activity' were underpinned by his commitment to his fourth method, then he had a different rationale for choosing it. But that view, examined elsewhere, did not come until his later article *On the Doctrine of Chances*. Clearly, if a different sense of rationality[23] dawned in his mind so that the idea of an arbitrary adoption of a method was overcome by his later 'logical socialism', which was discussed in chapter II, then he would no longer need the above 'remarkable simile in an eloquent passage'.[24] So we should not be surprised that he deleted this final passage to his 1877 article in its reprint in 1893! Will the QC display any similar revision in relation to his 'semblance of rationality' attitude?

NOTES

[1]Thomas Hobbes, *Leviathan*, chapt. VIII, para 16.

[2]This piece arose out of my objection to 'Ride Down St Morgan' stated to Arthur Miller at a question and answer session at the National Theatre. He remarked 'The man who believes in his own sincerity can achieve anything'. This piece, 'Rationalizations' represents my reply, 6 August 1994.

[3]I am particularly indebted to a remark of Wendy Probyn here: 'Rationalisations are always backward looking, whereas reasoning involves looking forwards' made at the Cyril Hodgkinson Philosophical Society Meeting, Leigh-on-Sea, 3 April 1988.

[4]Desmond Lee's *Introduction to Plato: The Republic*, p.19.

[5]See L Nannery, 'The Problem of Two Lives in Aristotle's Ethics' in *International Philosophy, Quarterly*, September 1981, vol. 21, no. 3, pp.277–293, especially fn. 21 on p.283.

[6]He thought Berlin might be an expert on American political life (cf. R Jahanbegloo, *Conversations with Isaiah Berlin*).

[7]Much of what now follows is taken from N E Boulting's 'Conceptions of Human Action and the Justification of Value Claims' in *Inquiries into Values*, pp.173–193, particularly pp.186ff.

[8]The capacity for thinking, for using one's wits is, for Hobbes, intrinsic to what it is to be a human being, even if some are better at it than others! 'For such is the nature of men, that howsoever they may acknowledge many others to be more witty, or more eloquent, or more learned; yet they will hardly believe there be many so wise as themselves; for they see their own wit at hand, and other men's at a distance. But this proveth rather that men are in that point equal, than unequal. For there is not ordinarily a greater sign of the equal distribution of anything, than that every man is contented with his share'. *Leviathan*, chapt. XIII, para. 2; cf. Botwinick, A, (1983*) Hobbes and Modernity*, p.32.

[9]Thomas Hobbes, *De Cive: English version* – a critical ed., chapt. 1, sect. 7, 'Philosophical Rudiments concerning Government and Society'.

[10]Karl Marx, *Economic and Philosophic Manuscripts of 1844*, pp.67, 89.

[11]Thomas Hobbes, *De Cive*, op. cit. chapt. 3, sect. 7.

[12]I Hungerland and G R Vick, 'Hobbes's Theory of Significant' *Journal of the History of Philosophy*, October 1973, vol. XI, no.4, pp.459–482

[13]Karl-Otto Apel, *Towards a Transformation of Philosophy*, p.269.

[14]R S Peters, *Hobbes* (1956), p.164.

[15]Morris Grossman, 'Interpreting Peirce', *Trans of C S Peirce, Society*, vol. XXI, no.1, p.114.

[16]The numbers refer to the sections of Plato's *Republic*, not to page numbers; see *The Republic of Plato* trans. with an Introduction by F M Cornford, Oxford: Clarendon Press 1961, pp.216–221 and 222–230.

[17]A term used by Murphey to dismiss them! (Murphey, M C, *The Development of Peirce's Philosophy*, p.164. He says there is 'no historical just justification' for Peirce's listing of the three alternative methods to the scientific 'and it can hardly be maintained that these are all the possible methods. Rather Peirce seems to have drawn up this list more or less arbitrary so as to have some alternatives against which the superiority of the method of science could be shown.'

[18]Max Fisch Peirce, *Semiotic and Pragmatism*, p.96.

[19]*Writings of Charles S Peirce – A Chronological edition*, vol. 3, pp.16, 500.

[20]J W N Watkins, 'Decision and Belief' in *Decision Making*.

[21]W W Bartley III, *The Retreat to Commitment*, chapt. V.

[22]Karl Popper, *Open Society and its Enemies*, vol.2, p.230.

[23]N E Boulting, 'Hobbes and the Problem of Rationality' in *Hobbes: War Among Nations*, pp.173–193, especially p.183ff.

[24]Morris Grossman, op. cit., p.109.

Chapter V
CLARIFICATION?

Why does a cloudless dark blue sky perceived in a wind, conditions whose chill harbours no compromise, yield such excitement? Is it that it provides an image of something so absolute the more vividly because of the conditioned nature of the consciousness perceiving it, forced, by this very act of perception, out of its more usual disposition of inner emigration necessary for most of our more normal behaviour? Is it that this very act offers the possibility in experience – hinted at in moments of pure contemplation where what is contemplated can never be possessed – of a transparent consciousness without anything to distract it from a complete release from itself? Or is it rather that, like looking down through a once muddy pool, now finally still and clear, the hope is released that something authentic will be revealed to transcend our more usual reflections, masquerading, as these do, as something other than what they really purport to be?[1]

It was a brilliantly sunny morning without a cloud in the sky. Everyone gathered in the hall, which seemed on that Wednesday morning to be so much warmer, to hear the inspector's decision. His view was that the supplementary and supporting statements did change the nature of the appeal in that 'a series of refinements' altered its proposals. Consequently there was a difference between the appeal KCC was supposed to be considering and the one now under examination at the inquiry. Indeed, doubts about parts of the application, so transformed, did not make it possible to consider adequately the basis of the application itself. However, it was clear that at least 1.83 million tons of clay was still to be extracted. The inspector also noted that the term 'approval' not 'permission' was to be sought in relation to phase four of the application, as though this was part of a further application. Ten years seemed to be the time-scale for the clay extraction; a further three years at least seemed conjectured for continuation afterwards. The claim for clay extraction was part of the application and did still seem more important than space for landfilling. The indication of the time-scale for the rate of the landfilling was not given, nor as to when the site would be restored. So the question remained: 'What is the extent of the landfilling?'

As to the issue of toxic waste, it could not be merely a matter for the Licensing Stage nor could consideration of pollution matters be left solely to the concerns of the Special Waste Regulations. To cite just one issue: the nature of the waste would have severe transporting implications for the movement of materials finding their way into the excavation. Note too must be taken of the Environment Agency's present rejection of this application based largely on all kinds of considerations now raised in relation to this new kind of waste.

In summary then, the inspector concluded there were many parts of the original application still retained but there were insufficient details generally about what was proposed; that the possibility of toxic waste raised other implications; that it was not clear whether the application had always been concerned with the waste disposal issue; that environment assessment questions needed to be fully explored – a process which could take months – and the time-scale changes now involved had to be clarified. Finally, it was important to deal with the issue as to what arrangements were specifically proposed in relation to the matter of toxic waste. However, his final decision seemed to be that the present inquiry should proceed except for matters concerning clay need and environmental issues.

In response to all this, the QC, after consultation with the appellant's representatives, thought that the inquiry should continue too, putting on one side the question of clay need on the one hand and the question concerning the implications of toxic waste on the other. KCC's representative called for the whole inquiry to be adjourned. How could he even begin to question witnesses until the new parameters of the application had been fully explicated? If there were to be two inquiries then the two main issues referred to by the QC would have to be adjourned to some later date, when everything, including investigations proposed now, would have to be explored for a second time. The speaker for the Environment Agency supported this point of view since he thought it would be some six months before he could report back properly on the two separate issues, the one relating to the consequences of removing a specific amount of clay and the toxic waste disposal issue.

The representative for Swale Borough Council (SBC) argued that, despite having to submit to the inspector's ruling, he still had misgivings about suspending the issue of the need for clay. The inspector retorted that he had settled upon his decided course of action because of the Swale Borough Council's representation the previous day. TAG's representative quickly brought the proceedings round to focus on the main point: 'If there was no need for clay, then the application falls at the first fence. Clay need is the prior issue. If there is no need for clay then there is no need for the hole. If there is no need for the hole then there will be no need for a space for the waste disposal and that is the end of the matter!'

Immediately following her intervention the SBC's legal representative rose to say that he had been given fresh instructions: if the issue of the need for clay were not to be addressed at this present inquiry, it would be better adjourned! The QC pointed out that the issue of the need or the supposed lack of a need for clay in Kent, in fact dominated the rationale for the rejection of the original application by KCC! The present application was forwarded today in the light of the wider concerns for the need for clay and for land refilling. Forecasting for a clay need was difficult to ascertain 'actually'. There is a real need, however, he argued for landfilling. As for KCC plans, they must *not* be regarded as prescriptive; rather they are simply suggestions to be examined as such! The real interests, 'actually', are those of the country taken as a whole. 'Actually' there are only two real individual interests that his company needed to satisfy: those of the brick company in relation to its own water borehole and those of the local residents, as opposed to the irrelevant issues raised by KCC!

The KCC's representative focused on the clay need. This inquiry was a meaningless exercise until this issue was settled because it would in the end determine the extraction rate too. That was quite a separate issue from the concerns of the Environment Agency even if it did have implications for the presentation of their case. In addition, without knowing the full details of the new proposals, it would be impossible to evaluate the application in any general attempt to weigh planning priorities. At a continued inquiry, witnesses would be cross-examined on details of a set of issues whose nature had not been fully declared.

The QC responded by raising the case of a liner used at Shelford: that had been shown to be highly satisfactory. Moreover, the KCC spokesperson demonstrated more enthusiasm 'actually' rather than good sense in the things he had been saying. However, there was a public responsibility for a fair hearing so his side was prepared for an adjournment. As to the case for the need for clay, such a national issue could hardly be settled at an inquiry of this kind, since that was a matter for the Secretary of State himself. The inspector then adjourned the inquiry for fifteen minutes in order to consider these latest representations.

The inspector's second decision of the morning was to adjourn the whole inquiry in the light of the unsettled issue of the need for clay. Without settling that issue it would be impossible to consider the discrete effect: or indeed avoid cross-issues arising from it. He then questioned the QC on the alternative access road proposal. The QC wanted to know in return whether this was a county or a district matter. An argument then broke out between the inspector, the QC and KCC's representative as to what kind of application this was, whether KCC was trying to make the application appear fraught with technicalities and whether only a proper road for a mineral site was being proposed by the appellant.

KCC's representative argued that following the Crown versus Berkshire CC case, at second appeal, if anything in a submission refers to a county matter that makes it something over which the county has in fact jurisdiction. The QC reiterated his claim that it was KCC who were trying to enmesh the issue in technicalities. Argument again broke out as to whether the inquiry was dealing only with an application in relation to a mineral site, whether Swale was the determining authority over the proposed road and why no environmental assessment for the new proposals had been provided.

In response to such issues as to whether the environmental assessment could be tagged on as a formal amendment, what the nature of the application now was, and whether that was now determined by the nature of the supplementary material, the inspector ruled that all these features were part of the appeal scheme, as he put it. His concern, addressed to the QC, was that in adjourning this inquiry there still might be further amendments for the next. The QC quickly assured him that if new rules emerged to inform the base of the quarry, the interested parties would be told in good time for the next inquiry!

Ignoring this assurance, the inspector reiterated that by putting off the inquiry now, there was the danger that the appellant's application could be changed again. He then challenged the QC by saying that he hoped the appellant wasn't trying to avoid an environmental assessment for the proposed road in his scheme.

At this point the representative for Swale Borough Council rose to point out that the local district council had refused permission for the road to be built. The QC challenged the legitimacy of this decision; it had been taken by the wrong authority. This was a matter for the Secretary of State under provision 2861 of the 1990 Act which was now part of the appellant's appeal. The Swale Borough Council's representative continued unabated; this was not a county matter, and any environmental assessment was irrelevant to that issue. In these matters the district council had jurisdiction.

For the next inquiry, the inspector insisted that the parties should try to ascertain areas where they did agree – site descriptions, the listing of any relevant building plans, the emerging plan for the development of the site, the relevant historical position, the nature of the proposal, the arguments for clay need, sorting out the relevant issues in relation to the landfilling issues and what can or cannot be seen in terms of the view of the landscape. Without some kind of agreements between the parties in these areas, insisted the inspector, it would not be possible to ascertain their differences accurately.

Finally, an emotive issue closed this session of the inquiry. KCC's representative rose pointing out that because the appellant had not given the parties sufficient time to consider all the proposed amendments to this appeal for this inquiry, the question of costs for these three sessions rested with the appellant. TAG's representative passionately defended this point too saying that the costs of this inquiry and its preparation so far would bite well into the seven thousand pounds the local residents had already raised as part of their resistance to this appeal!

In defence, the QC repeated once more his argument that the appellant should not be punished for bringing evidence, which could well have been kept within the company, into the public arena. In relation to the geological discoveries the appellant had made, it was necessary that there was not unreasonable action taken in this regard. Actually, it wouldn't be in the public interest for the appellant to be punished for being so frank in this matter; just think of the effect this would have on other public companies who might have information it was essential to share with the public community at large!

Swale Borough Council's representative thought that the appellant had acted disgracefully; the geological questions could have been raised much earlier. The inspector ignored this last interjection and pointed out that it was not usual to pay costs before an inquiry had finished its business. He requested that the officers directly involved in the inquiry meet again at two o'clock in the afternoon, after lunch, so as to see if all could agree on the date of the next inquiry, 17 June 1997, and whether in these rather exceptional circumstances, TAG's representative could collect any kind of interim compensation or costs. As it turned out his pessimism in this regard proved well founded: all claims for costs would have to be heard at the end of the inquiry. His last command for the morning was that in preparation of the documents for the next inquiry. Word Perfect 5.1 would be the system to use and no system of any other kind. Someone shouted out 'Windows or MS Dos?' When he replied in favour of the latter a howl of laughter rose to the roof of the community hall!

The main point of this inquiry's last session was the inspector's demand for greater clarity about the application and nature of the objections. Similarly you may now demand greater clarity as to what the point of the philosophic content of what has been discussed so far really is. And just as the inspector was able to focus his demand by bringing together some of the issues raised in the inquiry thus far, so it ought to be possible to state what has been established by the use of Peirce's ideas applied to the different issues raised by this inquiry itself. Notice, however, that 'the semblance of rationality' issue still haunted this final morning's activities: firstly there was the sense that the appellant's real interest lay in waste disposal rather than in digging clay; secondly the apprehension, incited by the QC, that human decisions – consider the mineral plan drawn up by KCC – were merely arbitrary suggestions if they opposed the appellant's interests and, thirdly, the suspicion that the QC's claim about a certain frankness in publishing geological findings hid a more genuine intention to alter the original application on the sly. But notice that all these reservations are still held within an overriding anthropocentric concern. Moreover, we can still ask, with respect to the QC's discourse, whether it does display what Peirce called 'the rationality of reasoning' (Peirce: vol. 2, para. 159), or whether his manner of reasoning lies outside the scope of what can be called rational and thereby has to be given a 'semblance of rationality' so as to make it appear so?

Peirce defines 'the rationality of reasoning' in terms of 'that character of a reasoning at which the reasoner aims' (Peirce: vol. 2, para. 159). Chapter I of this book opened by characterising three kinds of reasoning at which a reasoner might aim, whether he be an artist, a practical businessman or a scientist. A fourth attitude towards reasoning was identified – the non-aware or unconcerned – whose rationale lay in not being concerned with anything cognitive, a perspective which does issue in different ways into the other three forms of reasoning. In chapter II the status of the scientific attitude was examined since, in the forms of reasonings science employs, it signifies the requirement of a procedural approach to rationality thereby legitimating it as a highly rational activity. Attention was then given, in chapter III, to the instrumental and aesthetic attitudes along with the kinds of reasoning which might be employed to ground them. It then appeared that different kinds of attitudes such as the moral, the commodific and the emblematic could be distinguished which might carry distinct forms of reasonings of their own. As the kinds of reasonings of each of these three were explored, the reader will have noticed that s/he was increasingly distanced from viewing nature in the way each was grounded. Initially it appeared that reasonings could provide guidance in determining the intrinsic value of future experiences of people or in nature. But reasonings could also be driven by exchange-value. Finally, reasonings might be concerned pre-eminently with cultural desires projected onto things, persons or objects found in nature, making the latter, as thing-like, popular or fashionable at a particular stage in a society's development. But to make reasoning dependent on a particular culture's fancies seems to undermine attempts to provide a rationale for thinking or doing anything. Is it simply the case that the pursuit of rationality dissolves into the search for rationalisations: that one person's packet of rationalisations becomes another's rationality?

That last issue was addressed in chapter IV, Peirce's *The Fixation of Belief* was invoked in order to see how such a challenge might be met. Interestingly, you will have noticed that a return was made to analysing the scientific attitude once more, since it can be regarded as providing a *paradigm – the best possible case or the ideal model – for what it is to be rational! Yet in that article Peirce leaves us to plump either for the scientific method of reasoning to lead us to truth or some other method which might yield 'comfortable conclusions', 'uniformity of opinion' or 'strength, simplicity and directness' as he says of the other three methods – the *a priori* method, and those of authority and tenacity – in turn (Peirce: vol. 5, para. 386). True in his later article *On The Doctrine of Chances* he sets out a much broader conception of scientific rationality than the merely individualistic one offered in *The Fixation of Belief*, a conception which underpins and is underpinned by his 'logical socialism'. But as we saw in chapter II, such a grounding only seems to hold for those who are inquirers. Indeed, it may not be sustained outside that form of intellectual activity, never mind for any entities which do not belong to the class of human beings as such, those capable that is – no matter for how long – of engaging in such activity! This line of argument, then, seems to keep being blocked by the anthropocentric character of the rational enterprise.

You might have some objections to this whole line of enquiry as it has been pursued so far. Firstly, its approach to philosophy may have been wrecked upon these difficulties because we started our journey under the guidance of a wrong authority: the philosopher Peirce. Secondly, every time the question of rationality has arisen the writer has turned to science or at least to the idea of *inquiry. This move seems to make philosophy appear scientific, as if this kind of activity would provide the grounds for what it is to be rational rather than say religion, clairvoyance or the authority of a 'proper' twentieth-century philosopher, whether a Ludwig Wittgenstein or a Richard Rorty! Thirdly, the writer is simply out of touch with what counts as philosophy in the contemporary world: there can be no grounds validating philosophy, science, aesthetic concerns, morality, making money or packaging oneself – or what remains of the self – to please the idiosyncratic desires of societal subjects in a given culture. These are just activities in which people engage – 'What has to be accepted, the given is – so one could say – *forms of life*'[2] – so that's the way things are and once realised we can see 'how little theoretical reflection is likely to help us with our current problems'![3]

One way of meeting the challenge – that we have been guided wrongly by choosing to explore aspects of Peirce's philosophy – can be met by examining its basis, so as to show that his three categories – Firstness, Secondness and Thirdness – are not arbitrary. First of all, in a Harvard lecture of 1865, he distinguishes three kinds of inference or forms of argument.[4] (When considering these forms of inference by the way, do concentrate on their form rather than their content since you may wish to reject the latter.) The most common one he rendered in the first two of his Lowell lectures[5] in 1866: **Deduction* the simplest form since it unpacks what can be *thought. Indeed

Peirce at first was tempted to use the term **a priori* for this form of inference! Consider, then:

All representatives speak the truth:	The assertion of a Rule
This man is a representative:	The assertion of a Case
∴ This man speaks the truth:	The assertion of a Result

Another commonly accepted form of inference was analysed in Peirce's next two Lowell lectures:[6] **induction*, sometimes called the logic of scientific inference.

This bird is a swan:	The assertion of a Case:	This QC is a man
This bird is white:	The assertion of a Result:	This QC is capable of truth-telling
∴ All swans are white:	The assertion of a Rule:	All men are capable of truth-telling

Here 'this bird' can be regarded as 'an *index' (Peirce: vol. 5, para. 559) of what is to be regarded as a 'swan', just as 'this QC' can be for what is to be a 'man'. Now you can see that, unlike the case of deduction the conclusion of such an argument can be false: swans discovered in Australia are black. That recognition draws attention to a problem with induction as an argument: we can't infer from some cases to all. This is regarded as a scandal in philosophy – the problem of induction – since scientific laws are meant to cover all cases, yet can never be justified for future practice. This problem is too large to be handled here! What you need to bear in mind is that induction is still a form of inference, meant to indicate or pertain to what exists.

A third form of argument Peirce defended is not nearly so widely accepted.

This man is familiar with Latin pronounced in the Italian manner	The assertion of a Result
All Catholic priests are familiar with Latin pronounced in the Italian manner:	The Assertion of a Rule
∴ This man is a Catholic priest:	The Assertion of a Case

Peirce refers to this form of inference as **retroduction* (Peirce: vol. 6, para. 525ff) which in his fifth Lowel Lecture,[7] he called **Hypothesis* to distinguish it from induction. Indeed Peirce defines 'making an hypothesis' as 'the inference of a case from a rule and result (Peirce: vol. 2, para 623).

These beans are white:	The assertion of a result
All the beans from this bag are white:	The assertion of a rule
∴These beans are from this bag:	The assertion of a case

Now again, you may be uneasy about both these inferences. They seem somehow too conjectural, too hypothetical and only merely possible. You might be happier with:

Few beans of this handful are white:	Denial of a result
Most beans in this bag are white:	The assertion of a rule
∴ Probably, these beans are from another bag:	Denial of a case

The point is that such 'an hypothetical inference' (Peirce: vol. 2, para. 627), whatever it be, is only suggestive, 'something *like* the conclusion is proved' (Peirce: vol. 1, para. 559) and so has to be subject to further inquiry.

In 1868 Peirce admits to creating his theory of signs to conform to the nature of these inferences: '…a sign has, as such, three references: first, it is a sign *to* some thought which interprets it, second, it is a sign *for* some object to which in that thought it is equivalent; third, it is a sign *in* some respect or quality, which brings it into connection with its object' (Peirce: vol. 5, para 283). And earlier in his *On a New List of Categories* of 1867, he identified, in the light of these 'fundamental senses of reasoning' and on the basis of 'differences of form among signs of all sorts', that there were but three 'elementary forms of predication or signification', (Peirce: vol. 1, para. 561) three basic ways of thinking about experience, three basic conceptions which could be applied to experience or what he called categories: 'Quality', 'Relation' and 'Representation' (Peirce: vol. 1, para. 555). Originally then, these fundamental categories were regarded as deriving from a thought's representative function. Later, in the development of his phenomenology, they were named Firstness, Secondness and Thirdness respectively.

Now, however, you are able to make your challenge against being introduced to Peirce's ideas more openly: this book is not called *To Be or Not to Be Phenomenological* but rather *To Be or Not to Be Philosophical*. But to say that my emphasis on Peirce's ideas makes him appear to be doing *phenomenology rather than *philosophy implies that we know what these terms mean.

Throughout this book we have focused upon an inquiry and have done so in order to deal with 'assumptions on which a great many normal beliefs rest – not the sort of things which people like to dig up, because people sometimes do not want their assumptions examined over much' as Sir Isaiah Berlin put it once in an interview.[8] All the investigations in this book have been carried forward in that spirit. As for the term 'phenomenology', that was used by Hegel to refer directly to what you, I or anyone else receives through what can be called sense-consciousness – what appears to the mind. That received material, by passing through a number of stages in *consciousness, can render understanding so that what we refer to as objects can be brought under concepts.[9]

For Peirce, Hegel's idea of phenomenology was too cerebral, over dominated by Thirdness! So although he did use that term to define the nature of his own philosophising, after 1904 he preferred the term *phaneroscopy to describe where philosophy must begin and this book has tended to follow his advice. Peirce used the word *phaneron 'to mean all that is present to the mind in any sense or in any way whatsoever, regardless of whether it be fact or figment' (Peirce: vol. 8, para. 213). And, in relation to someone wishing to exercise his capacities in this kind of activity, s/he must possess 'first and foremost' that rare faculty 'of seeing what stares one in the face, just as it presents itself unreplaced by any interpretation…' (Peirce: vol. 5, para. 42):

> …what we have to do, as students of phenomenology is simply to open our mental eyes and look well at the phenomenon and say what are the characteristics that are never wanting in it whether that phenomenon be something that outward experience forces upon our attention, or whether it be the wildest of dreams, or

> whether it be the most abstract and general of the conclusions of science. (Peirce: vol. 5, para. 41)

Having established his case once more for the categories – Secondness, Firstness and Thirdness as presented above (Peirce: vol. 5, para. 41) – let us move to the second objection which is made up of two parts: why should we always turn to science to settle the case for rationality? Again, why are the activities of science and philosophy spoken of in the same breath?

Peirce often contrasts what he calls living belief with the provisional acceptance of theories in science, the latter examined in *The Fixation of Belief*. He gives an example of what he thinks to be the superiority of the former: a ship's captain in a storm relies on what Peirce calls 'the stout belief of any common seaman' of what to do (Peirce: vol. 5, para. 60). Indeed, he openly advocates rule of thumb techniques based on *instinct in matters say '...of business, family, or other departments of ordinary life' (Peirce: vol. 2, para. 177; vol. 4, para. 450). And in chapter I of his *Minute Logic* he says, in relation to someone playing billiards or a carpenter building a pigsty, '...it would be most unreasonable to demand that the study of logic should supply an artificial method of doing the thinking that his regular business requires every man to do' (Peirce: vol. 2, para. 3).

Yet to leave one's actions unconscious or to the instinctual would be to act blindly: Firstness and Secondness without Thirdness.[10] Rather, a rational person exerts 'a measure of self-control over his future actions' to make action something conscious through '...a process of self-preparation', tending '...to impart to action one fixed character' (Peirce. vol. 5, para. 418). So, Peirce recommends not reflection, introspection or contemplation in relation to what you or I have done in the past. Rather inferential or counter-factual reasoning – of a kind we have been exploring above – is emphasised, which would look rather to the future where 'an experiment shall be an operation of thought' (Peirce: vol. 5, para. 420). Indeed he goes as far as to say: 'Whenever a man acts purposively, he acts under a belief in some experimental phenomena' (Peirce: vol. 5, para. 427). Finally, once the action decided upon 'is repeated again and again' it takes on

'...the perfection of that fixed character', 'marked by an entire absence of self-reproach'. The more this ideal is approached, the more it becomes automatic, unconscious once more, so that conscious self-control will be absent and thereby it can 'be marked by entire absence of self-reproach' (Peirce: vol. 5, para. 418).

What then is the reader to conclude in relation to this question as to whether an account of what it is to be rational is to be sought in the canons of scientific activity or elsewhere. At least three conclusions can be drawn. The first is that Peirce does endorse a faith in commonsensism, trusting in the instinctual, provided this keeps us in harmony with existing natural and social tendencies. That is why he can define common sense in 1905 as '...those ideas and beliefs that man's situation absolutely forces upon him' (Peirce: vol. 1, para. 129). As an illustration he remarks 'that it is better to recognise that some things are red and some others blue' rather than 'that some things are resonant to shorter ether waves and some to longer ones' (Peirce: vol. 1, para. 129), the latter deriving from the science of his own time. And by 'better' I take him to mean more convenient in the business of ordinary day living; being able to get done the things we seek to do!

Secondly, Peirce's real position is manifested in the following 1902 quotation:

> But *fortunately* (I say it advisedly) man is *not so happy* as to be provided with a full stock of instincts to meet all occasions, and so is forced upon the adventurous business of reasoning, where the many meet shipwreck and the few find, not old-fashioned happiness, but its splendid substitute success. When one's purpose lies in the line of novelty, invention, generalization, theory – in a word, improvement of the situation – by the side of which happiness appears a shabby old dud – instinct and the rule of thumb manifestly cease to be applicable. The best plan, then, on the whole, is to base our conduct as much as possible on instinct, but when we do reason to reason with severely scientific logic. It has seemed to me proper to say this in order that I might not be understood as promising for logic what she could not perform. Where reasoning of any difficulty is to be done concerning positive facts, that is to say, not mere mathematical

> deduction, the aid that logic affords is most important. (Peirce: vol. 2, para. 178)

Taking up 'the adventurous business of reasoning' then is likely to occur when doubt about a belief or an action arises, so that a person is no longer in harmony with existing natural or social tendencies: 'I do reason not for the sake of my delight in reasoning, but solely to avoid disappointment and surprise' (Peirce: vol. 2, para. 173).

Thirdly, Peirce is offering his reader a description or an explanation of how man's attempt to be rational arises; he is not recommending a particular account. So, in 1896, he pointed out that man's capacity to reason develops because his instincts become smothered in his highly individualistic and originatory nature (Peirce: vol. 1, para. 178), a result implied by the above quotation (Peirce: vol. 2, para. 178).[11] This claim, that he is merely describing the origins of rationality, could be rejected, however, by pointing out that Peirce seems to be saying that 'fortunately' man is forced into the adventurous business of reasoning. But in the substitution of 'rational purport' for sential qualities (Peirce: vol. 5, para. 428) in the pursuit of reasoning, where Thirdness substitutes for Firstness (Peirce: vol. 6, para. 132), Peirce takes you in the opposite direction. Who would want to lose happiness 'for its splendid substitute success' (Peirce: vol. 2, para. 178) which reasoning may bring? Perhaps Peirce desired to sustain a ⋆dualism: two claims at once! On the one hand the instinctual or commonsensism, which he defined as covering what was 'inherited' or 'due to infantile training and tradition' (Peirce: vol. 2, para. 170), whilst on the other, reasoning as employed in scientific enquiry (Peirce: vol. 2, para. 253–254). Yet the latter, he claims, can never surpass the instinctual in its degree of certainty (Peirce: vol. 1, para. 204; vol. 5, para. 522; vol. 5, para. 496).

We may now, however, be in a position to deal with the second part of your question. For his own time he was anxious not to be 'understood as promising for logic what she could not perform' (Peirce: vol. 2, para. 178). In the nineteenth century it was possible for him to say that it '...is the instincts, the sentiments, that make the substance of the soul. Cognition is only its surface, its locus of contact with what is external to it' (Peirce: vol. 1, para. 628). But if, since then, as a result of our awareness of Freud's insights, which the passage (Peirce: vol. 2, para. 178) above from *Why Study Logic?* itself anticipates, more and more of what is instinctual in man becomes rationalised, then it is not that Peirce is claiming that scientific inquiry is the model for philosophy. Rather he is to be interpreted as anticipating that philosophy would have to face the scientific challenge in relation to the tremendous intellectual developments he anticipated in the new century.[12] In 1905 he recognised that modern '...science, with its microscopes and telescopes, with its chemistry and electricity, and with its entirely new appliances of life has put us into quite another world; almost as much so, as if it had transported our race to another planet' (Peirce: vol. 5, para. 513). And even if he admitted to the disposition to 'think of everything just as everything is thought of in the laboratory' (Peirce: vol. 5, para. 411) yet he still rejected what he called the ideas of those practical men who 'look to science as a guide to conduct, that is, no longer as pure science but as an instrument for a practical end' (Peirce: vol. 1, para. 55).

As to your third claim it is indeed hard to see what the role of philosophy might be in life at the end of the twentieth century. Peirce can be called a philosopher in the *modernist tradition. His philosophical practice lies in the tradition of the enlightenment'[13] and he believed in the importance of establishing some kind of rational society in view of it.[14] He is also a modernist in emphasising the significance of brute fact, action to create what exists in the world. He would, then, be able to agree with a man of Kent who admired his writings, Alfred North Whitehead who wrote:

> Direct experience is infallible. What you have experienced you have experienced. But symbolism is very fallible, in the sense that it may induce actions, feelings, emotions and beliefs about things which are mere notions without that exemplification in the world which the symbolism leads us to presuppose.[15]

Symbolism, the way we are provided with 'the link between sign and signified' to enable us to read the world, as Simone Weil[16] puts it, can be fallible in at least two ways. Firstly, since we are dealing with what Peirce called Thirdness here, there will be implications in relation to what we say we know about the world because of the fallibility of our thoughts. For the moment we are concerned with how language relates to the world in the light of that fallibility. We can be concerned with the language we speak but symbolism, of course, can refer to films, TV broadcasts, books etc., but in each case we are concerned with how such symbolism serves us indexically, how it points to states of affairs in our experience. One way language, for example, can mislead us, is through the employment of distorted Thirdness. Peirce provides two examples: the two doctrines that 'man only acts selfishly' (Peirce: vol. 5, para. 382) and that 'greed is the great elevation of the human race' (Peirce: vol. 6, para. 290). We can add a third namely that the only purpose of education is to improve the economic prosperity of the nation. This kind of fallibility is more easily recognised in that it is the stock-in-trade of any politician seeking re-election and we have a name for that kind of distortion: propaganda.

The second kind of fallibility is much harder to elucidate. It is tempting to call it distorted Secondness but how can something existing be described as distorted unless an idea of what it ought to be is ready-to-hand with which it can be compared; a stunted tree, blind creature or deformed plant. This second kind is not, as in the first case, the arbitrary way symbolism relates to what takes place in the world, but the way objects can be stripped of their historical significance to be ascribed an entirely different meaning. The artistic movement surrealism paved the way for this development[17] in that objects might be 'discovered' as having a new contingent meaning in themselves. A toilet or a pile of bricks could be presented in an art exhibition as if these had an aesthetic meaning of their own. More topical is the representation of the Union Jack not as a celebration of the integration of three sets of values embodied in the lives of St George, St Andrew and St Patrick but as the definitive statement of some form of aggressive nationalism! This kind of problem has, in fact, already been raised at the end of chapter III when attitudes that were increasingly distanced from nature – the commodific and the emblematic – were considered.

Both these arbitrary ways of using symbolism have been displayed at the tribunal: notice the way evasion may have been used to hide real intentions in speaking of frankness and honesty in relation to changes in the application and also the constant talk of speaking of Tiptree only as a site of clay production and waste disposal. But given the increasingly arbitrary use of symbolism in this way, it becomes easier to see why philosophers would regard reflections on that language use as even more arbitrary! That is the position of the *postmodern philosopher. And s/he can then characterise the term 'modern', as outlined below:

> to designate any science that legitimates itself with reference to a metadiscourse of this kind (i.e. a discourse of legitimation with respect to its own status, a discourse called philosophy) making an explicit appeal to some ground narrative, such as the dialectics of spirit, the hermeneutics of meaning, the emancipation of the rational or working subject, or the creation of wealth.

This account allows its writer, Lyotard, to define 'postmodern' as incredulity 'toward meta-narratives'.[18]

If you haven't followed all the implications of what has just been said, focus on that last sentence! Peirce adopts a *meta-narrative in advocating his doctrine of categories – Firstness, Secondness and Thirdness – to enable you to make more sense of your experience than you might previously have done. Of course, other meta-narratives could have been used. One might be a sociological meta-narrative forwarding the idea that different ways people view the world is due to the social class within which they were reared. Or a psychological meta-narrative might be used: people view the world in the way they do because of the psychological traumas they may or may not have overcome in their development. Peirce's meta-narrative is more philosophical: it accounts for the differing world-perspectives on the basis of his categories – Firstness Secondness and Thirdness – categories which in their turn relate to possible ways human beings can reason about their experiences. Now has his meta-narrative helped you to make more sense of your own experience? If so, then you, like me will be able to express incredulity in relation to the phenomenon postmodernity. But we still have a problem with Peirce's own meta-narrative – his 'story about experience which hovers above and serves to interpret it': can it break out of the charge of being anthropocentric? Does it not suffer as postmodernism does from being entangled in the sophisticated maze offered by Thirdness despite its employment of real senses of Secondness and Firstness? The answer to that question carries you to the next two chapters; chapter VI in relation to Firstness and its place in scientific inquiry and chapter VII, the importance of Secondness in appreciating landscape!

NOTES

1An Image of Undistorted Nature? (Upchurch, 14:1:1993), 'The image of undistorted nature arises only in distortion, as its opposite.' T Adorno, *Minima Moralia*, p.95.
2L Wittgenstein, *Philosophical Investigations*, p.226.
3R Rorty, 'Philosophy as Science as Metaphor and as Politics' in *The Institution of Philosophy: A Discipline in Crisis?* p.28.
4*Writings of Charles Peirce*: vol. 1 (1857–66), p.186
5*Writings of Charles Peirce*: vol. 1 (1857–66), pp.358–392.
6*Writings of Charles Peirce*: vol. 1 (1857–66), pp.393–423.
7*Writings of Charles Peirce*: vol. 1 (1857–66). pp.423–40.
8Isaiah Berlin, 'On Man of Ideas and Children's Puzzles' in BBC 2's series *Men of Ideas: Creators of Modern Philosophy*, *The Listener* 26/1/1978 p.110.
9Para. 27 of the 'Preface: On Scientific Cognition' in Hegel's *Phenomenology of Spirit*, pp.15–16; para. 387 of 'Section One: Mind Subjective' in Hegel's *Philosophy of Mind*, p.27.
10I Stearns 'Firstness, Secondness and Thirdness' in *Studies in the Philosophy of Charles Sanders Peirce*, p.195.
11The instincts of the lower animals answer their purpose much more unerringly than a discursive understanding could do. But for man discourse of reason is requisite, because men are so instinctively individualistic and original that the instincts, which are racial ideas, become smothered in them. (Peirce: vol. 1, para. 178).
12cf. N E Boulting, 'Charles S Peirce (1839–1875): Peirce's Idea of Ultimate Reality and Meaning Related to Humanity's Ultimate Future as Seen through Scientific Inquiry' in *American Philosophers' ideas of Ultimate Reality and Meaning*, sects. 3, 4.
13S Buck-Morss, *The Dialectics of Seeing*, p.447, n.35.
14Richard J Bernstein, 'Towards a More Rational Community' in *Proceedings of the C S Peirce Bicentennial International Congress*, Graduate Studies of Texas Tech. Univ., pp.115–120.

15A N Whitehead, *Symbolism: Its Meaning and Effect*, Capricorn, 1959 p.6. Alfred North Whitehead was born 15 February 1861 at Ramsgate in the Isle of Thanet. His father was a parson from 1871 at St Peter's parish, Broadstairs where he guided the development of the inner architecture of that church. But just as 'Process Theology', which his son's thought inspired has little pride of price in this church, similarly, Whitehead's philosophical works have enjoyed little attention in his own country.
16Simone Weil, *The Notebooks*, vol.1, p.160.
17S Buck-Morss, op. cit., p.227.
18J F Lyotard 'The Postmodern Condition' (1984) quoted in *The Postmodern History Reader* p.36.

Chapter VI
THE BIRD HABITATS ARGUMENT

Why should human consciousness focus on the dunnock's song? Its twirling tones trill around a 'twseep' to create four stanzas persistently repeated. Their constant recurrence soon exhausts contemplation yet attention congeals on the possibility that meaning might be revealed in the apparent meaningless rush of such high-pitched sounds.

As the winter wind stiffens to shiver the soul, so this scattering of notes is the more clearly audible from a short beak raised like a small nose into the freezing fast-flowing air. This bird's fluffiness is puffed suddenly forwards, caught unawares by such merciless rufflings, yet its trills rise as from a desperate desire to transcend desperation.

At first its pipings recall the 'rash, fresh' skylark's carollings as that bird once rose high in a blue sky above golden corn swaying lazily in the sun. As the eye was carried upwards by this May-bird's calling, delight in its carousing prevented the immediacy of that experience withering. Such reflections now seem quite out of place since the sudden, sharp blasts from the north, seemingly designed to catch this present, grey-headed, streaky-brown plumed creature unawares, reminds the bird-watcher that the hunt for meaning requires pain from the meaningless in order to exist. The grey clouds continue to roll southwards above an inert landscape to provide no grounds for this bird's happy, rhythmic calls.

Using binoculars to make its contours clearer is to offend this bird's consistent desire for total inconspicuousness. It appears to defy such reification of its existence by moving from branch to branch, rearranging its position or by reshuffling its wings so that doubt may linger as to what it truly is. The reality then, about its natural state, can magically avoid the lie of suffering an exact description to provide comparisons with other birds.

Constantly under the pictured 'knowing' eye of an observing partner, its notes are accompanied, apparently innocuously, by this ever-watchful, highly nervous bird. Both keep their distance. Does such distance prevent either from sensing that its own movements in any way influence or condition a momentary feeling of what each is doing? And do their joint, scattered pipings, surely not entirely purposeless, offer the hope for the attentive human ear that two human beings can yet enjoy an undiminished purposelessness which only the warmth of a shared intimacy can kindle?[1]

During the time of my increasing interest in the Tiptree Inquiry, a leaflet was received from the Royal Society for the Protection of Birds (RSPB) – *Birds of Conservation Concern*. That enigmatic, inconspicuous, once abundant resident bird, the dunnock, has now been listed as a cause for medium conservation concern by this agency. According to the leaflet it is a bird in moderate decline (assigned priority amber); its breeding population has declined between twenty-five and forty-nine per cent over the last twenty-five years. Its existence can no longer be 'much taken for granted' as a book published by the Kent Ornithological Society put it some fifteen years or so ago.[2] The TAG member – referred to in chapter I – who won a Special Award for Environmentally Sensitive Tourism in 1994, points out in, his 'Barn Owl' guidance ornithological report in relation to the appeal,[3] that he saw a total of just eleven dunnocks on the site during five visits. There are estimated to be five million breeding pairs and twenty million such birds wintering in Britain and Ireland. But these figures derive from *Red Data Birds in Britain* published for the Nature Conservancy Council and the RSPB, which is still used to set the framework for the conservation debate despite being over ten years old.[4]

The case for the black-tailed godwit – an estuary bird with a long straight bill, white tailed base above a black tail end, grey brown on top, pale below in winter whilst in summer this bird develops rusty red plumage[5] – is much more serious. As a migrant, it visits Kent in small numbers but has been known to breed on the Medway. Indeed, twelve years ago it was possible to report that this species was wintering here more regularly and migrating in increasing numbers.[6] It is now listed as an historically declining breeding species (assigned priority red). Approximately fifty breeding pairs were registered some ten years ago with a figure of around four to five thousand as wintering in Britain, according to the *Red Data Birds in Britain* report. Moreover fifty per cent or more of their non-breeding population is found at ten or less sites in the United Kingdom and, in addition, it seems to be a species with an unfavourable conservation status in Europe. The writer of the 'Barn Owl' report claims to have seen flocks of this bird from his car, one mile north-east of the site on five occasions during January and February 1996.

The real reason why I refer to the state of the black-tailed godwit is not only because it has been assigned a priority red by the RSPB but because it is featured as one of the wader and wild fowl populations making the Medway Estuary of national and international importance. But the exact status of the bird seems disputed. According to the *North Kent Marshes Study* of 1992[7] over six per cent of the UK's national population of this bird appears in the Medway Estuary and just over four per cent in the Swale whereas in a *Nature Conservation Topic Paper*, produced to assist KCC's draft Estuary Management Plan for the summer of 1997, it is listed as internationally important![8]

I wish to stay for the moment, however, with the 1992 study. This is because it lists, in terms of their national and international significance, fourteen different species of birds. Two of these are interesting – the grey and ringed plover – interesting because nearly two per cent of the world's and at least twelve per cent of the UK population of the former is present on the Medway Estuary (similar figures for the Swale) and at least, in the case of the latter, just over two per cent of the world's and nearly six per cent of the UK population is present on the Medway Estuary. The RSPB assigns each species an amber priority rating, a still endangered species like the dunnock. In addition to the black-tailed godwit, a further ten species of bird – shelduck, wigeon, teal, pintail, shoveler, dunlin, curlew, redshank and turnstone – found on the Medway Estuary are given the same designation, whilst the case of two others remain exceptional. The great crested grebe is not listed red or amber and the brent goose is given an amber label though it is not clear whether that pertains to the dark-bellied kind from Siberia, of which farmers would say there are too many on the estuary, or the pale-bellied variety. The 'Barn Owl' report writer identified the brent goose and redshank one mile from the Tiptree Site and the curlew on ten occasions on the site itself.

In the later *Nature Conservation Topic* paper,[9] recorded on the Medway and Swale Estuaries as having nationally significant numbers: the little grebe, cormorant, white-fronted goose, gadwall, avocet golden plover, lapwing and bar-tailed godwit. The latter six are given an amber designation by the RSPB and our 'Barn Owl' reporter records significant numbers of golden plovers on the site during December 1995 as well as 792 Lapwings during ten visits from that time to August 1996.

I think the real significance, however, of the 'Barn Owl' guidance ornithological report lies in what it has to say about the reported sightings of six species of bird in addition to the black-tailed godwit which are designated a red priority by the RSPB: the grey partridge, turtle dove, skylark, song thrush, linnet and reed bunting. The latter three are registered as a rapidly declining breeding species whose breeding rate has withered by fifty per cent or more over the last twenty-five years. The case of the grey partridge, turtle dove and skylark is even more serious because they are not only a rapidly declining breeding species but also suffer from an unfavourable conservation status in Europe as a whole, according to the RSPB's *Birds of Conservation Concern* leaflet. Our 'Barn Owl' reporter recorded a sighting of a grey partridge on the 31 July 1996, six turtle doves during two visits in June and July, forty-two skylarks as a result of several visits, twelve song thrushes over six visits, ten linnets during two sightings and one reed bunting on the 9 January 1995. All these birds were seen on the site either from the vantage point provided by the Saxon Shore Way or from the byway on the north-east side of the site. Twelve short visits altogether were made to the site between December 1995 and August 1996 and five visits to the second location one mile from the site. All the visits occurred at high tide save one, when the tide was out.

In view of the 'Barn Owl' report and other considerations raised so far, especially those in relation to the limited number of visits made by our reporter and the restricted access there is to the site, since it can only be viewed from positions around its perimeter, seven arguments against developing the Tiptree Site would have to be considered at any inquiry. These seven arguments are based on an interest in birds alone. Additional ones might be put on behalf of the ecosystem more generally understood.

Firstly, I have not mentioned so far a concern for owls. A barn owl, along with short-eared owls – both amber listed species – has been reported near the site. The former breeds nearby therefore using the site for hunting, as is possible for the short-eared owl too. Long-eared owls have roosted in the vicinity during the winter; local farmers claim that a roost of long-eared owls is located near the Funton Brickworks in thick scrub.

Secondly, it is clear that waders, particularly in the case of the golden plover, use the fields adjacent to the Medway Estuary to include Tiptree Site fields at low tide. When wild fowling elsewhere disturbs birds, such as the golden plover, they leave the mudflats and descend on these adjacent fields. Indeed it can be argued that the Tiptree fields are particularly useful in this regard because other sites suitable for such birds as the golden plover, black-tailed godwits, lapwing and curlew are areas of disturbance from either game birding and wild fowling or from radio-controlled aircraft.

A third argument has to employ what can be described as the high tide haven implications more directly. At high tides more birds obviously leave the mud flats to occupy the fields. This argument is a particularly crucial one since it is by now universally recognised that sea-levels are rising and over the next fifty years are supposed to increase dramatically: an eight-centimetre sea level rise by the year 2025 for the Medway Estuary is predicted,[10] whilst erosion continues in relation to the saltings and islands within the estuary's Sites of Special Scientific Interest (SSSI). Birds displaced from these areas are under great pressure from human activities such as wild fowling, water sports, bait digging, not to speak of walking and birdwatching itself! So the significance of the fields, such as those at Tiptree, are very important and not just as feeding sites, especially if fields closer to Iwade are now to be part of a housing development.

A fourth argument focuses upon the kinds of birds which will be attracted to the area with the opening of a waste disposal site. At present the members of such bird species as the great black-backed gull, herring gull, carrion crow and magpie are not alarmingly high, although local reports confirm that the latter's numbers are rising! Such an increase in these birds would make the conditions for waders and terns more difficult – remember the red and amber priority assignations – not only in terms of breeding conditions but also in relation to food supply too. Non-breeding birds – referring, as our 'Barn Owl' reporter does, to larger gulls which do not breed before four years of age – would be present all the year. But in addition to this hazard from herring, lesser and greater black-backed gulls, it is well known that lesser black-backed and herring gulls often nest in very large colonies. These birds can range over large areas in search of food to the cost of protected species. To argue that such gulls would keep to their own routes of passage and their own present feeding sites would be to display a complete ornithological ignorance about such matters.

A fifth argument emerges from this latter one. In the Landscape Issues Proof of Evidence for the inquiry[11] it is pointed out that often trained hawks are used for deterring seagulls and many other scavenging birds attracted to waste disposal sites. But in that case the SSSIs adjacent to these Tiptree fields would then be affected.

A sixth argument emerges from the contents of this chapter as a whole. The majority of the birds concerned are 'amber' listed species and seven species – grey partridge, black-tailed godwit, turtle dove, skylark, song thrush, linnet and reed bunting – are red listed. Consequently one would expect the RSPB to be taking an active interest in possible developments at this Tiptree Site particularly if it is proposed that leachate – a noxious black syrupy liquid – is likely to enter the River Medway Estuary from the Tiptree Site if it is to be used, partially at least, for dumping toxic waste! But is the RSPB active in practice?

A seventh argument arises from considering an important international development in recent years, experience from America and the implications of a comparatively new law passed in England. Under a European Community Directive on the Conservation of Wild Birds (79/409/EEC) vast areas of the North Kent Marshes have been registered as of local, national and international significance for wildlife; 6,500 hectares of the Swale are now designated as special protection areas (SPAs). Interestingly, if it can be shown that the issue of wildbird conservation is not being fully addressed, the UK can be challenged in the European Court. Earlier, in 1971 in Iran, the UK, along with twenty-two other nations, became a signatory to the Convention of Wetlands of International Importance; the Ramsar Convention. This convention pledged its signatories to commit themselves to further the conservation of particular sites and to stop the destruction of wetlands whose existence is crucial for wildfowl, waders and other waterbirds. The United States signed up in 1987.[12] And in the United States there is a problem which throws light on this argument too. Much dispute has arisen as to whether national parks, such as Yellowstone, should be regarded as 'ecological islands'. That phrase usually pertains to the way they are managed determining a 'policy of taking whatever nature has to dish out'[13] rather than intervening to check over-population of some creatures or to encourage, say, small local fires from time to time to prevent the kind of large-scale conflagration of 1988. But it can also imply turning a blind eye to the kind of human activities taking place in forest areas surrounding these large parks, ecological buffer zones between man's utilitarian pursuits and the biodiversity of ecological niches to be found in these parks themselves. Now under the 'Wildlife and Countryside Act of 1991' English nature is required by statute not only to identify special areas because of their fauna, flora, physiographic or geological features as SSSIs but to act, if not in a proactive manner, then at least 'in a negative way'[14] to prevent adverse consequences upon those sites. It would be a great pity if it singularly failed to do so in the case of Tiptree. In addition, it would be morally reprehensible if it relied upon the developer's own accounts of the ecological significance of this site so as to avoid doing the appropriate research, to ascertain the potentially damaging nature of those developments, itself!

In chapter II you saw how hard it was for anyone concerned simply with the outcomes or products of human inquiry to overcome the anthropocentric charge. Will an examination of the processes of how we come to acquire knowledge – other than taking things as known from a screen – escape that charge? In other words, will a much more practical approach to these matters enable us to side-step the issues raised in that chapter?

Consider the following, what Peirce calls a 'thought-experiment', concerning an incident which may have happened to you. As you stand at the kitchen window looking out on your garden or backyard, if you are lucky enough to have one, you are watching birds trying to eat different kinds of seeds you have put out for them. You are now going to be asked to reflect on some thoughts you may be having as you watch them eating. This thought-experiment is going to be constructed then on something you have observed, 'an observed object' as Peirce refers to it. In his case in 1901 he was interested in whether the observed object, in his case a man was or was not a Catholic priest'.[15] In your case you are interested in whether the observed object, in your case a bird, is or is not a particular kind of bird.

Notice that you have not begun with any ideas. Your own observation, your own experience provides the starting point. Elsewhere Peirce asks but '...what is observation? What is experience?' He says it is 'the enforced element in the history of our lives', 'The act of observation is the deliberate yielding of ourselves to that *force majeure* – an early surrender at discretion, due to our foreseeing that we must, whatever we do, be borne down by that power, at last' (Peirce: vol. 5, para. 581). You will surrender to the possibility that that rather particularly excited female sparrow must have found something delightful to eat to be so fidgety! Your hypothesis is formed because that would help to explain the fact that it is feeding on the ground, it has no black bib under its beak and it seems rounded in its bearing. The forming of this suggestion as to what *may be* so Peirce calls *retroduction sometimes *abduction*, this 'first starting of a hypothesis and the entertaining of it whether as a simple interrogation or with any degree of confidence'.[16] Your 'inferential step' (Peirce: vol. 6, para. 525) can be related to the category *Firstness. You are appealing to something you feel about those perceived qualities: the bird's rounded bearing, its breast's markings and so on. Peirce remarks that this kind of 'suggestion comes to us like a flash. It is an act of *insight, although of extremely fallible insight' (Peirce: vol. 5, para. 181). If you have seen the statue of Sherlock Holmes, standing outside Baker Street underground station in London, you will have noticed how his stance is frozen – pipe held in his outstretched hand – because such an insight has struck him suddenly as to how a criminal act might have taken place!

Yet even though this bird appears to be like other sparrows you have seen, why does it appear to be so excited? This question now leads you to a second stage of inquiry: you are drawing out the consequences of your original hypothesis that you are seeing a female common sparrow before you. You are seeking a rule of the form 'if this were so, that would follow'. Your own inferences would be 'If that were a sparrow, it wouldn't be so excited and seen through binoculars it has a grey head with a grey forefront below its fine thin beak. Why does it keep flicking its tails and wings like that?' The drawing out of the consequences of your original idea or hypothesis Peirce called *deduction*. Whereas retroduction concerned what might be so, deduction helps you to claim what *must be* so from a logical point of view *if* this bird is a sparrow. Sparrows don't usually appear so agitated unless acting aggressively towards another. Sparrows, though rounded in appearance at a distance, have thick beaks and so on. It can't be a greenfinch, goldfinch or bullfinch because it has no discernible colouring. You still hold to your original hypothesis, but the ideas drawn from it don't seem quite to fit. Incidentally, the category *Thirdness* has now been evoked because you are concerned with ideas about what you have assumed to explain what you have just observed. And indeed you are troubled by your thoughts!

You now decide to test your deduction. It is most likely that this will involve you doing something. You might go to a guide-book on birds and check on what you have seen. You might ask someone else to come to the window to see what s/he thinks the bird is or you may move quietly to the open window to try to catch the 'brep' sound it chirps. Peirce called this stage *induction* 'because it is a test of the hypothesis by means of a prediction' (Peirce: vol. 6, para. 526). Your action in seeking to achieve a factual identification of the object, of this bird, in time and space, shows that the category *Secondness* is being invoked; induction can settle what *actually is the case*. Has your attempt to apply the rule(s) carried by your judgement 'It's a sparrow' been a valid one? Suddenly the bird flits to a nearby bush and you are sure you heard a high-pitched jingle around a 'twseep' to create a four part cycle. 'That must be a dunnock!' you say to yourself.

Now will you be satisfied at this point? You think this bird is a dunnock. Or will you wish to treat this stage as an attempt to enter a further cycle of inquiry – a new retroductive stage – drawing out further consequences of this new inference – deduction – testing again at a further inductive stage? Notice, however, that during this second cycle your experiencing will change. Whereas in the first case you simply observed something – a sparrow which appeared to be excited – unintentionally, by accident, your second observation will be much more purposively involved as part of a more systematic investigation.

At this point you might wish to claim that this 'road to inquiry' – what Peirce called 'indagation' (Peirce: vol. 6, para. 568)[17] – which dawned in his mind in 1868 at about the same time as he discovered his categories Firstness, Secondness and Thirdness, this road still leaves the inquirer held within an anthropocentric point of view. After all, s/he is still left with the ideas of a human being, derived or at least dependent upon the ideas taken as common parlance from within a society of which s/he is a member. Although your objection is valid, notice two possible reservations which might be made in reply. First, an attempt has been made to show how these ideas may fall or be corrected in view of your own observations. Now, of course, you will claim that these former ideas have been simply substituted by other ideas even if these are assumed in the light of experience. None the less there is a real relationship to experience acknowledged here but more importantly, consider that first experience which made retroduction, as a first stage of inquiry, possible. Notice the unintentionality involved in its initial creation and the idea of insight in response to it. Perhaps these two features of Peirce's indagation can be borne in mind in what follows below and in subsequent chapters. For the moment it is sufficient to point out that this 'thought-experiment' you have undergone has taken the three specific forms of inference, as unpacked in chapter V, to show you how they can be related to each other in the activity of inquiry.

Attention now, however, needs to be focused on another question you may have wanted to ask in relation to the contents of this chapter: 'Why do birds matter?' Why should we be so concerned if these varieties of our wildlife simply disappear? After all, little has been said so far about the rat, mole or toad creatures found to be so fascinating by one author that he wrote *The Wind in the Willows* about them? Over twenty-five years ago Charles Hartshorne, who has written about and originally edited – with Paul Weiss – the writings of Peirce, published an article 'Why Study Birds?' In fact this article draws to its reader's attention a number of reasons why we might value birds.

Like us, with the rat and the mole, they are warm-blooded but they are also the warmest, helping them to be quick and lending them a vivacity stimulating to the human eye. They migrate, exercising a capacity, which is only a recent acquisition for human beings, in relation to the speed we travel, even if this is so easily taken for granted by us today! As humankind is constituted by many races, so there is a vast variety of bird species – Hartshorne gave 9,000 in 1970. This is much greater than any mammal or reptile but not too many to prevent studying all their songs. He made his attempt in relation to 5,000 of them, pioneered in his book *Born to Sing*, whereas studying anything about beetles, for example, might well baffle comprehension since, even if several hundred have already been named, there are probably millions still waiting to be identified.[18]

Closer to home, it is clear that birds do not frighten us in the way smaller creatures like the mouse or rat can, despite Hitchcock's film to the contrary. They do not compete with us for survival and may assist our more modest agricultural purposes. Like us too, and unlike the rat, mole or toad, bird behaviour is guided first and foremost by sight and sound as opposed to smell, taste and how things feel. Sight and sound keep things at a distance: the latter three can overwhelm us! Unlike the rat or mouse they do indeed keep their distance from our homes too even if they live close by, and seem to celebrate a life akin to our own. Yet this staying at a distance from us may not always have been so. Consider Charles Darwin's discovery at the Galapagos Archipelago: 'It would appear that the birds of this archipelago, not having as yet learnt that man is a more dangerous animal than the tortoise or the amblyrhynchus, disregard him, in the same manner as in England shy birds, such as magpies, disregard the cows and horses grazing in our fields'.[19]

Be that as it may, reflect for a moment on the fact that they seem to tend to what we call monogamy and can be regarded as enjoying something akin to what we used to call 'family life'. Not only are their offspring cared for in a co-operative manner between male and female but, except in the case of egg laying, there is no real difference in function displayed by male or female.

So far, eight features of the existence of birds have been drawn to your attention which simultaneously make them 'wonderfully like and wonderfully unlike ourselves'.[20] These constitute certain brute facts about birds and ourselves – what Peirce referred to as Secondness. In acknowledging these features of the existence of birds, as these have emerged in evolutionary history, we acknowledge and recognise something about ourselves.

Moreover, if our interest in this list has been biassed in stressing what causally constitutes the kind of creature humans are, other features of the existence of birds might provide lessons for the kind of creature humans could become. Consider that last feature, the interchangeable role of male and female. Despite all the talk in the chattering classes, is it really the case in most of our homes that such flexible role models are provided for our children as the lives of birds can provide? Their example too, Hartshorne notes, supplies the additional moral lesson that individual 'self-protection is not the first rule of the universe' as the philosopher Hobbes may have thought. You only have to consider the behaviour of the female nighthawk as this American bird mimics having a broken wing to draw a predator away from her chicks or the female black throated diver which similarly spreads a wing as if broken to paddle around in circles, to be provided with examples of how a creature can risk destruction to protect the existence of future species members.[21]

Such features, showing how the example of birds might inform our conduct, illustrate the category Thirdness, ideas about the future behaviour of human beings. Birds are above all natural 'aviators' and it may be that humans too can find individual modes of air travel, akin to the bicycle, to enable them in the future to flit from place to place rather than just sit in huge car-jams. Again the bodily insulation provided by feathers may yet produce a revolution in the design of clothes in some future Paris fashion show! Consider their lightness yet their preservation of high body temperatures. But perhaps, as in many aspects of human culture, the Greeks anticipated these latter two rather bizarre possibilities. You only have to consider Aristophanes's play *The Birds* wherein it is shown that these creatures are worthy of some admiration. Indeed it is described as one of the most graceful of his plays and does emphasise the place 'where Wisdom, Grace and Love pervade the scene'.[22]

Yet, by contrast, in contemporary terms there is an overriding lesson birds can provide for human beings. That lesson is entirely an ecocentric one. The health of an ecosystem can be usefully and relevantly indicated by the presence or absence of birds.[23] Indeed, what was mere speculation for Aldo Leopold in his 1939 paper 'A Biotic View of the Land'[24] – that the activities of living species may be necessary to the maintenance of the biotic community as a whole – seems confirmed, even if in a negative way, by contemporary studies: the health or lack of it in an ecosystem is mirrored by the loss of various species.[25]

For most human beings, however, it is in their sheer presentness, rather than in their absence, that birds offer us something really important, which is, as Peirce once put it, a vivid qualitative something that is what it is 'quite regardless of anything else' (Peirce: vol. 5, para. 44), illustrating his category Firstness. But what is that something? As we have already learnt, such presentnesses catch us off our guard, they seem unintentional, involving an observation or observations gained quite by accident. An interesting case can be cited where a boy when he returned home from his place of study called his dog; 'Spitz! Spitz! Spitz!' After dashing downstairs the dog ran past a parrot perched in the hall who observed this ritual every day. Eventually the parrot called out 'Spitz', imitating the boy's voice, and then the parrot 'would jeer at it when it came running in, wagging its tail in welcome'. The boy was Charles's brother Jem (James Mill), who was a Harvard day student, and on the basis of these events, Peirce accepted the idea that animals and birds possessed reasoning powers akin to those found in humans.[26]

Peirce's own insight in these matters forming his inference about reasoning powers came by analogy, what he calls 'resemblance in form' (Peirce: vol. 7, para. 498). He defines analogy as 'the inference that a not very large collection of objects which agree in various respects may very likely agree in another respect. The example he cites is interesting: that the earth and Mars 'agree in so many respects' makes it 'not unlikely they may agree in being inhabited' (Peirce: vol. 1, para. 69). That particular hypothesis has fallen a century later, but the hypothesis that it is 'not unlikely they may agree in displaying life forms from the past' is very much with us. Peirce, however, pursued no investigations with respect to what he called these 'reasoning powers', in relation to birds. That was left to Charles Hartshorne who has pursued a form of enquiry more scientifically. His researches began because he says he has always enjoyed bird songs. His inference takes the form of a question: 'Do birds too?' At this first stage of his inquiry two possibilities can be forwarded: either musical sensibility 'is entirely unique to human life, or there are precedents or analogies in the older forms of animal life'.[27]

Drawing out the consequences of this hypothesis, that birds do have a musical sensibility and are thereby capable of enjoying other birds' songs, raises more questions. Consider initially the first alternative, that human beings are the only creatures with sufficient musical sensibility to appreciate the sounds birds make, other than warning calls, as beautiful. At least one philosopher has endorsed this possibility: 'The song of birds is judged beautiful by nearly everybody. No sensitive person of European background, for example, fails to be moved by the song of a robin after a shower of rain. All the same, there is something frightening lurking in the song of birds, which is not really a song but merely a response to natural necessity. The same frightful threat emanates from flocks of migratory birds; even today their formations bespeak the old practice of divination, forever presaging ill fortune'.[28] Whitehead draws out the implications of this point of view that birdsong is never beautiful in itself; rather its beauty lies in the eye of the beholder or at least within his or her inner ear: 'Thus nature gets credit which should in truth be reserved for ourselves: the rose for its scent: the nightingale for his song: and the sun for his radiance. The poets are entirely mistaken. They should address their lyrics to themselves, and should turn them into odes of self-congratulation on the excellency of the human mind'.[29]

Julian Huxley's response to this question was to say there need not be any incompatibility between these two ideas that birdsong can be regarded as a behavioural response favouring the success of natural reproduction and that it represented an expression of a creature's emotional state.[30] In the latter case it does not seem unreasonable that physical factors such as the presence of certain male hormones during the breeding season might not transform the feelings of a particular bird so as to make its tendency to sing more obvious.[31] Indeed, in this context, it is interesting to see that Charles Darwin noted that the severest rivalry between the males of many bird species in attracting females was expressed through singing. He concluded that he could 'see no good reason to doubt that female birds, by selecting, during thousands of generations, the most melodious or beautiful males, according to their standard of beauty might produce a marked effect' in favour of 'gorgeous plumage', the performance of 'strange antics' or a capacity to sing.[32]

To understand how a bird or a human being can appreciate something, even a song, as beautiful we need to focus on what it is that makes that something beautiful. Now, as Peirce pointed out, the 'beautiful is conceived to be relative to human taste' (Peirce: vol. 5, para. 128). But he avoids the claim that the beautiful is simply what gratifies desire or gives pleasure even if he does admit that in dealing with the question what makes something beautiful we are concerned with 'what it would be that independently of the effort, we should like to experience' (Peirce: vol. 2, para. 199). Rather he points to what he calls 'affinities to which the mind accords general approval' in somethings so as to pronounce them as beautiful (Peirce: vol. 1, para. 383). But if that latter remark pertains to the qualities of something which may make something beautiful, he suggests the idea of unity even in a complicated state of things (Peirce: vol. 1, para. 613).

It is this very suggestion 'unity in contrast' which Hartshorne uses to define beauty, enabling him to claim that in his research he found that bird 'songs are in varying degrees, objectively aesthetic in the sense of avoiding the most utmost extremes of mechanical regularity and mere chance diversity, as well as the ultra-simplicity of mere chirps or squeaks'. In his research, then we can see Peirce's indagation at work; first comes, at the retroductive stage, the possibility that birds enjoy bird songs – as Kierkegaard once said birds are living things 'who not only sing at their business, but whose business is to sing'. At a second stage – deduction – the consequences of this possibility are drawn out by considering whether there is any kind of pattern, order or harmony present in such songs that make them more than mere arbitrary sounds.[33]

We still remain haunted, however, by the question that began this very book you are now reading: in relation to birdsong, have we now overruled the utilitarian or territorial interpretation in favour of the aesthetic? An illustration might prove helpful here. Even in winter the ice-cream van patrols our cul-de-sac. Its arrival is announced by the playing of a tune – say 'the Harry Lime' theme. That tune not only may attract its hearers aesthetically, but it serves an instrumental use as well: it broadcasts the van's arrival. That may be the way to consider birdsong too!

Now having derived corollaries from the original retroductive inference based on an analogy, these can be tested in the light of scientific investigations, Peirce's inductive stage. According to these tests Hartshorne can argue that, like the ice-cream van's melody, bird songs seem similar to composed music acoustically. Secondly, like the ice-cream van's recording, birds can imitate songs, in their case other bird songs. Thirdly like the ice-cream van's tune, in that the driver can switch the sound off from time to time, birds seem to limit the monotony of their performances which makes such activities a bit different from more utilitarian ones such as flying and walking.

In three aspects, however, the analogy with the ice-cream van breaks down: the degree of development of singing skill seems to correlate with the biological need for singing. Yet singing performances, as Kierkegaard's remark implies, seem to outstrip any *pressing and immediate need*. Finally those birds which sing *more of the time* in any hour, day or year, seem to have more elaborately refined songs.[34]

Peirce once remarked: 'Analogies are never perfect for an analogy that should be perfect would be more than analogy' (Peirce: vol. 6, para. 325). The work of Hartshorne on his aesthetic analogy, however, seems at least to have weakened the smart, fashionable, anthropocentric view that musical sensibility is entirely unique to human life!

In this chapter you have faced two questions; 'Do birds have a value?' and 'Why study birds?' In relation to the last question Peirce's categories were connected with his logic of scientific investigation to show how it could be used to identify a particular bird. This raised the question as to why such an activity is worthwhile. That question enabled you to consider features of a bird's life which might have value because they enlighten our own human condition through considering our own evolutionary history, possible future human projects and the delights birds bring to us. It is that delight which might forward investigations into whether birds themselves have a musical sensibility akin to human beings and you saw how Peirce's logic of scientific investigation has been used in practice to throw light on this issue. Finally, as a result of holding to an aesthetic analogy, that birds may possess a kind of musical sensibility akin to humans and putting it to the test, we can see how a strict anthropocentrism in these matters can be undermined, even if not completely refuted.

NOTES

[1]'Dunnock Song', Upchurch, 22 February 1987.

[2]D W Taylor (et al.) eds. *The Birds of Kent: A Review of Their Status and Distribution*, p.2972.

[3]Ornithological Report re Tiptree Farm, Funton, Sittingbourne, Kent. Town and Country Planning Act, Appeal by South Eastern Minerals Ltd. Derek Tutt, Barn Owl Guidance for Birdwatchers and other Naturalists, 26 September 1996. I am very indebted to this report for much of the contents of this chapter.

[4]*Birds of Conservation Concern in the United Kingdom, Channel Islands and Isle of Man* designed and published by the RSPB, 1996 and *Red Data Birds in Britain*, Batten, L A (et al.) (eds.), a joint publication by the Nature Conservancy Council and RSPB, T and A D Poyser, London 1990. The position may be remedied by *British Birds*, Gibbons, D W in press. and *Birds Species of Conservation Concern in the United Kingdom, Channel Islands and Isle of Man: Revising the Red Data List*, RSPB Conservation Review in press at the time of writing this chapter.

[5]*The Mitchell Beazley Birdwatcher's Pocket Guide*, Peter Hayman, p.131; *Birds* M Woodcock and R Perry, p.94.

[6]D W Taylor, *The Birds of Kent*, op. cit. pp.176, 234.

7*North Kent Marshes Study* prepared by the Applied Environment Research Centre Ltd., Consultant's Final Report presented and published by KCC, May 1992, p.38.

[8]'Nature Conservation Topic Paper' prepared by members of the Nature Conservation Topic Group for the *North Kent Marshes Initiative*, Medway Estuary and Swale Management Plan; Topic Papers, Part 1, presented to assist the KCC's Draft Estuary Management Plan for Summer 1997, p.4.

[9]*North Kent Marshes Study*, op. cit. P.4.

[10]'Coastal Processes and Flood Defence Topic Paper' prepared by A Paley, Estuary Project Officer, for the North Kent Marshes Initiative, *The Birds of Kent*, op. cit., p.24.

[11]'Tiptree Farm, Funton Appeal by South Eastern Minerals Ltd., Proposed Extension to Clay Quarrying and Restoration by Infilling', 'Landscape Issues', Proof of Evidence by Geoffrey Beale, Kent County Council Professional Services, p.8.

[12]D W Taylor, *The Birds of Kent*, op. cit. p.39 and *Wildlife Policy and Management*, E G Bolen and W L Robinson, Prentice Hall (3rd ed.), 1995 pp.82–83.

[13]*Nature's Keepers: The New Science of Nature Management*, S Budiansky, p.133; pp.134, 137–138.

[14]*North Kent Marshes Study*, op. cit., p.4.

[15]C S Peirce, Hume on Miracles' (1901) in *Collected Papers of Charles Sanders Peirce*, vol. 6 paras. 522–547.

[16]Abduction refers to the tendency of humans to 'guess right' in forming scientific hypotheses, since there may be a correspondence between what is suggested to the human mind and what is found within the laws of nature (Peirce: vol. 1, para. 81) But this term, naming the process of reasoning involved in the formation of such hypotheses Peirce later called retroduction (Peirce: vol. 6, para. 525). He then spoke of the Logic of Abduction identifying this with pragmatism (Peirce: vol. 5, para. 196), a theory of meaning or inquiry 'capable of experimental verification'. (Peirce: vol. 5, para. 197).

[17]Boulting N E, 'Charles Sanders Peirce (1839–1914), Peirce's Idea of Ultimate Reality and Meaning Related to Humanity's Ultimate Future as Seen through Scientific inquiry in *American Philosophers' Ideas of Ultimate Reality and Meaning* ed. by A J Reck (et al.) Toronto UP 1994 sect. 2.4.1.

[18]S J Gould, *Life's Grandeur*, p.224.

[19]Charles Darwin, *The Voyage of the Beagle*, p.384.

---

[20]Charles Hartshorne, 'Why Study Birds?', *Virginia Quarterly Review*, pp.133–140, p.134.

[21]E O , *Sociobiology: The New Synthesis*, p.122.

[22]Introduction to Aristophanes *The Birds* in *The Birds and Other Plays*, 1978, pp.149–150.

[23]John L Craig draws attention to the work of Anderson – 'Pollination and Dispersal in New Zealand forests' (1997), 'Because the bellbird is a known pollinator and seed disperser of many forest tree and shrub species, the long-term sustainability of these forests appears partly dependent on this species...' 'Managing bird populations: For whom and at what cost' *Pacific Conservation Biology*, vol. 3, (1998) pp.172–182, p.177.

[24]Aldo Leopold 'A Biotic View of the Land' in *The River of the Mother of God & Other Essays*, p.271.

[25]J B Callicott and K Mumford (1997) 'Ecological Sustainability as Conservation Concept', *Cons. Biol.* vol. 11 pp.32–40.

[26]J Brent, *Charles Sanders Peirce: A Life*, pp.46–47.

[27]Charles Hartshorne, *Born to Sing*, p.xiii.

[28]T W Adorno, *Aesthetic Theory* tr. C. Lenhardt p. 99; tr R. Hullot-Kentor p.66

[29]A N Whitehead, *Science and the Modern World*, chapt. III, 7th para. from the end.

[30]Huxley, J, *Animal Language*, pp.17–18.

[31]Charles Hartshorne, op. cit. p.3.

[32]C Darwin, *The Origin of Species*, (1859), p.137.

[33]Charles Hartshorne, op. cit., chapt. 2 sects. B and C, pp.8,11 and xiii.

[34]Charles Hartshorne, op. cit., pp.12–13.

Chapter VII
THE LANDSCAPE QUESTION

Even for a non-Switzerland visitor, for whom the thick pine-scented inhalations bring consciousness back to an awareness of itself, Glacier Park's views, like standard National Geographic Magazine snapshots, hardly convey any sense of the beautiful as a direct perception above St Mary's Lake. Peirce's notion of Secondness – brute materiality – predominates, matter of fact solidity forces itself jaggedly upwards defying consciousness's attempts to issue delightful forms from this landscape. No sense of meaning is made available to resolve inner conflicts of a solitary self, anxious to discover its debt to what cannot be derived from its own solitariness. Even the shapes of the unmelted glacier sections are formed as if fixed in some Salvador Dalian painted construction, rather than charm the eye into believing they intertwine into some gentler Wordsworthian landscape naturally. As if especially cut out, the white rounded pieces of cardboard appear more naturalised as snowscaped coverings in what provides the mirror image of their surreality in the dark blue sheets of water below them.

Now, in relation to the environment outside Glacier Park, can an individual consciousness draw an analogical intuition of reality not yielded by the existence of what is actualised everyday, in and for itself alone?[1]

I was first made aware of landscape questions in what seemed to me to be a most striking document, called 'Landscape Issues', served as Proof of Evidence at the Tiptree Inquiry for November 1996. It is filled with delightful photographs and maps illustrating the historical development of the area. In section 3 – called 'designations' – the writer[2] draws attention to the fact that under the third review of the Kent Structure Plan, according to Environment Policy ENV4, the North Kent Marshes are designated as a Special Landscape Area (SLA) for which long-term protection will be provided.

...through local plans and development control and will normally give priority to the conservation and enhancement of natural beauty, including landscape, wildlife and geological features. The detailed boundaries of the SLAs will be defined through the local plan process'.[3]

We can now distinguish three kinds of designations: National Areas of Outstanding Natural Beauty (NLAs), e.g. the North Kent Downs Area; Special Landscape Areas such as the North Kent Marshes (SLAs) and Local Landscape Areas (LLAs) whose designation is the responsibility of a borough council. The Swale Borough local plan in a deposit draft identifies the hills between Lower Halstow, Newington and Iwade as worthy of a Local Landscape Area designation in its local plan yet to be adopted statutorily under its proposed policy E17. How did this idea arise?

To find the answer to this question I had to examine a document entitled the *North Kent Marshes Study* published in May 1992 since it was claimed in Swale's proposed Local Plan E17 that in this study 'a striking range of hills' had been identified 'in an otherwise flat marshland landscape' which presented 'impressive long-range views both inland and across the marshes'.[4] I have singularly failed to find in this *North Kent Marshes Study* any explicit mention of the hills around Tiptree accorded this description. Can the case be made on implicit grounds?

Consider the way the North Kent Marshes are defined. They are said to comprise:

> an extensive tract of largely undeveloped land. The landscape is very distinctive – broad vistas of mudflats and water, flanked by saltmarsh and grazing marsh with their creeks and ditches, punctuated by large industrial structures. The marshes have a wild, windswept and remote quality, in sharp contrast both to the generally softer and more wooded countryside which rises in some locations to the south, dotted with towns and villages, and to the major riverside industrial conurbations of the Medway Towns and Thames-side.

Now within this area the study identified a number of important landscape features, one of which was Ridgelines:

> Varying in height from 20–100m, ridgelines form an important backdrop to the open marshlands and provide visual relief. Ridgelines border the study area behind Gravesend, run along the Hoo Peninsula and rise behind Gillingham. Encroachment of development onto these ridgelines, particularly at Gravesend and Gillingham, has a damaging impact on the landscape.

In the recommendations of this study 'ridgelines' are once more emphasised as key landscape features and that, as important landscape elements, they should be positively managed 'through a North Kent Marshes Countryside Management project with priority given to landscape protection zones'.[5]

Curiously, in the *Landscape Issues* document there is no reference to ridgelines. So two questions arise: how was the case made in this latter document for a landscape interest and were the arguments used to make that case valid? Two moves can be made by this document's author in response to the first question; he distinguishes a section on what he calls 'Landscape Strategy' from one on 'designations', but nothing is said about a linch – another term for a ridge of land. What is also odd is that his document *Landscape Issues* contains extensive references to what is said about the Medway Marshes Character Area, as this is referred to in a latter publication *The Kent Thames Gateway Landscape*, a document published in 1994. So the following passage from that latter document is included:

> Small areas of marshland along the southern Medway have a predominantly rural character, mainly because they are visually isolated or distant from intrusive urban or industrial influences. Thin fringes of saltmarsh occur along the coast amidst the tidal mudflats of the estuarine system, physically and visually separated from adjacent land by the sea walls. These have a typically natural and unspoilt character, enhanced by the presence of wild birds and by the colour and charm of moored boats amongst the muddy creeks. The integrity of the marshland landscape is not significantly affected by distant views of industry on the opposite shores and is positively enhanced by boats and shipping which reinforce its coastal character.

Other passages besides this one refer to what can be seen from the hills, what surrounds the ridge, *not* to the ridge itself constituting the edge of the marshland. One passage, however, is quoted, one which describes the landscape type as occurring:

> ...just behind the marshland edge on slightly elevated ground to the north of Great Barksore and is typically open farmland which rolls down towards the marsh. Coastal exposure and the lack of tree or hedgerow cover contribute to a somewhat windswept and denuded character and significantly reduce the ability of the landscape to absorb change.

But the passage, which is significant in The *Kent Thames Gateway Landscape* study, is omitted from the *Landscape Issues* document. Under a sub-section headed 'Wooded hills and ridges' the following is written:

> The comparatively extensive area of Wardwell and Hawes Woods are located on a small but prominent ridgeline to the north of Newington. This unique landscape feature creates an attractive backdrop for views from surrounding roads and nearby properties as well as being a contrast to the flatter more open farmland in this area. Its visual prominence and woodland cover means that it is highly sensitive to development.[6]

Tiptree is part of an extension of this 'prominent ridgeline to the north of Newington' and thereby, according to this, *The Kent Thames Gateway Landscape* study, has as its positive attributes, a strong landform feature, a rural unspoilt character and an attractive backdrop to views' which the writer of *the Landscape Issues* document eulogises.

To put the matter another way, the 'Landscape Issues' document is concerned more with what might be called non-intrinsic features to be enjoyed in taking views from the hill rather than with its specific, intrinsic qualities as such. These latter are identified more in the section 'designations' where preservation of trees is referred to, along with natural beauty issues covered by the previously mentioned Policy ENV4 and Policies ENV7 and ENY10.

Policy ENV7 signifies that:

> It is policy to maintain tree cover and the hedgerow network in the County and enhance these where compatible with the character of the landscape. Local plans should include policies to protect, maintain, and enhance tree cover and woodlands.

Policy ENV10 insists that:

> Undeveloped coast and estuaries, except where allocated for port development and associated infrastructure, will be conserved, and enhanced. Development in such areas and in adjoining countryside will not normally be permitted if it materially detracts from the scenic, heritage or scientific value of these areas.[7]

But these claims indicate merely general policy statements which may or may not refer to an area waiting to be designated an LLA in a *de jure* rather than in a *de facto* sense.

Finally, the *Landscape Issues* document does refer to five other concerns which touch upon the landscape interest: the fact that the final landform will be altered in its final restoration; that the pleasure of those walking the Saxon Shore Way on the North West Side of the site would be diminished; that the visual impact to be enjoyed from sailing and anchorage viewpoints would be undermined; that rural lanes will have to be altered thereby undoing conservation and local landscape considerations and that nuisance issues are raised by unsightly fencing, smell, dust, litter, seagulls and noise.[8]

On the opposing view, it might be said – and indeed was said by Kent Property Services (Landscape) – that although the site did represent a noticeable topographical element in the landscape, any effects could be lessened by tree and bush planting as well as screen moulding. Moreover, the dispute as to whether the area should be assigned an NLA, SLA or LLA designation had not been settled in 1996. Finally, the timescale proposed in this site's development would ensure that any detrimental effects would be transient ones. Other than reservations about restoration proposals, there could be no real landscape objections.[9]

You might be interested to know how the term 'landscape' originated. At first it was used technically by artists; there is the

Dutch word *landschap* translated as 'landskip'. That the idea of landscape has painterly origins will prove important in what is now to be argued about this subject. My argument will show that the appreciation of a landscape may serve to recall in the viewer, either explicitly or implicitly, something of landscape pictures. Indeed such works of art might well provide a kind of pictorial ideal for viewing the countryside itself.[10] This is, of course, an additional point which can be returned to later; let us stay for the moment with the idea expressed by Peirce that for those who are moved by Firstness, by 'qualities of feelings', 'nature is a picture' (Peirce: vol. 1, para. 43).

Consider yourself standing before one of Constable's paintings of Salisbury Cathedral. You may see it representationally, especially if you are concerned, as you may be, that a motorway has been planned for the Meadows which Constable included in one of his views. In that case you are aware of the painting as a physical thing but suddenly in contemplating it there may come a moment when you 'lose the consciousness that it is not the thing, the distinction of the real and the copy disappears, and it is for the moment a pure dream – not any particular existence, and yet not general'. At that moment Peirce claims you 'are contemplating an *Icon'* (Peirce: vol. 3, para. 362). An icon refers to a possibility alone – with respect to what might exist – 'purely by virtue of its quality; and its object can only be an Icon' (Peirce: vol. 2, para. 276).

Consider your contemplation of a statue where you do not know whom it represents, an old photograph where nothing represented in it is known to you or a Blue Monochrome block by Yves Klein emanating the colour he invented: 1KB or International Klein Blue patent no. 63471, dated 19 May 1960! If you have had such an experience then, as a viewer, you may have:

> ...felt drawn into the depths of a blue that appeared to transmute the material substance of the painting support into an incorporeal quality, tranquil, serene. The emotion invested in the creation of such imagery was matched by the freedom of the viewer to see and feel whatever he was prompted to in its presence. Nowhere could the eye find a fixed point or centre of interest; the distinction between the beholder, or subject of vision, and its

> object began to blur. This, Klein believed, would lead to a state of heightened sensibility.[11]

This quotation renders a sense of what you could feel standing before Klein's canvas; it is what makes it an icon. As Peirce puts it in 1886, the icon has nothing 'to do with the sense of contact with the world, nor with the actual existence of its object. It is a mere dream'. But if you know Klein's work you might object on two grounds: firstly, there might be a case for saying that Klein attempted to capture the deep blue colour of the sky and secondly people viewing it might not 'get the point' if there is one! Again Peirce can indicate the right direction in relation to such objections:

> ...pictures, though their main mode of representation is iconical, yet depend very much upon convention. That is why new methods of painting are always unpopular; people cannot see, at first, that a picture resembles nature until they have become familiar with the conventions of the method.[12]

Your attention has been focused upon the painting rather than the landscape for the moment because what can be said about the former can, to a degree, be said of the latter too, as will be argued. Remaining with the painting, however, it can be said, in traditional terms at least, that it serves as a sign. As such, the picture can be regarded as a representation 'which stands to somebody for something in some respect or capacity'. The picture then addresses you, the viewer, in that it creates in your mind an equivalent or 'more developed sign, what Peirce called the *interpretant. In the case of a Klein painting you with your companion may want to argue about its interpretation and immediately you can see that whether or not Klein's Blue does represent the sky – whether it has something *symbolic about it – because thinking and reasonings are involved, the category *Thirdness proves relevant here. Moreover, if the picture does stand for something, say the blue sky in Klein's case, of Salisbury Cathedral in the case of Constable, then we are dealing with something which can be taken to exist. In that sense the picture can be regarded as an *index, in that it is pointing, like a weathercock or a signpost, to something other than itself. In referring to existence, the category *Secondness is invoked.

In addition, that existent is represented in relation to 'some respect or capacity'. You might gain a sense of warmth from the picture of Salisbury Cathedral or almost catch the sound of gurgling water depicted in the painting. Indeed, it may be that this picture always seems somehow to touch such resemblances from your own past experiences! In experiencing Klein's Blue Monochrome, you might compare your experience to your appreciation of a continuous tone in music in that 'the specific hue of the pigment engendered a visual sensation of complete immersion in the colour, without compelling the viewer to define its character'.[13] We have returned to the *icon* where *Firstness is manifested.

You are now able to see how the notion of landscape can be defined. It can't be said, then, that a landscape is merely 'a resultant of attitudes and actions'.[14] Instead, it is our everyday environment which is something created out of our attitude and actions, in what we do to it, just as that environment comes to create us. But a visual landscape is even more our creation than an environment is, since we view what we take the natural world to be offering to us from within a perspective. Indeed, it could be said that all nature viewing is perspectival in this sense: just as the painting is structured or fixed by its outer frame, so an expanse of what we consider to be natural scenery is constructed, fixed or framed by the eye of the viewer'.[15]

This approach is far from suggesting, however, as two geographers have done, that landscape can be defined merely as 'a cultural image, a pictorial way of representing or symbolising surroundings'.[16] Such a view, besides running the danger of individual subjectivism, might cut against landscapes not particularly memorable[17] – in words used by Aldo Leopold about wilderness – landscapes which are back washes from 'the River Progress' such as 'the wild roadless spot of a few acres 'adjacent to areas of timber growing. Otherwise there is the danger that beauty spots, satisfying 'more Thoreau-like aspirations'[18] become extra-special reified to serve as examples of 'a particular version of "high" or elite Western culture'[19] to the cost of other examples not rendered the same hype!

More seriously, emphasis on the iconic approach to landscape presents the natural world as fixed, as a picture, whereas what is seen is far from static. Consider the after-effects of the storm in Kent during the night of Thursday, 15 October 1987: branches of fruit trees torn from the place where they bore fruit; oaks ripped from water-sodden soil; scores of trees flattened in straight lines, leaving others upright or leaning north-eastwards in their woods, Any so-called 'normal' state of nature is hardly one of balance, fixity or repose; rather it can be described as a state 'recovering from the last disaster'.[20]

The iconic approach too, considered as the sole one, generalises the pictorial sense of landscape at the expense of cultural processes which have formed the landscape we see, just as Eric Hirsch claims, but again his account[21] of what makes up a landscape is deficient too in that the pictorial and socio-cultural aspects are only two of a three part analysis of what constitutes the idea of landscape. A return to a Peircian suggestion in relation to a tripartite analysis appears prudent. That tripartite analysis can be applied now that we are able to gather together some of the results established so far and at the same time do justice to the notion of something being permanent about a landscape.

A landscape can be defined as a human creation which stands to its creator as a perspective from some place upon a physical environment capable, in its infinite potentialities for profusion and diversity, of being appreciated panoramically as a whole. Here justice is done to the three elements which constitute landscape: a *cultural representation* referring to some *physicality* beyond ourselves which can allow the viewer to lose the distinction between both in an *iconic experience*. Of course, if such a physicality was an unbounded coastal area rather than ground, we would have a seascape. But in relation to either case Peirce offers us a warning about our tripartite analysis: in any instance where such a representation is applied 'there is one of the three characters which completely overshadows the others.'[22]

We have already seen the dangers of that overshadowing in relation to an emphasis placed upon iconicity: subjectivism, reification and the representation of nature as static as well as the ignoring of cultural processes informing our vision of the landscape. But in relation to this issue of iconicity, the danger in seeing a landscape merely pictorially lies particularly in missing certain cultural conventions – referring to Peirce's category Thirdness – evoked implicitly. These can be called the ideological implications of such viewing. So you will sometimes find in the literature a distinction drawn between 'outsiders' who may operate merely pictorially, taking an objective view of a part of the environment which can be consumed or possessed in a photograph, and 'insiders', those who live within this area regarded as having pictorial value.[23] The pictorial view then can emphasise 'a restrictive way of seeing' since it may undercut the perspective of those who either work on the land or who live in that environment. And even if this distinction 'savours of romanticism',[24] it may yet help to explain the difficulties an English person can experience in visiting national parks say in Montana USA. An American sense of wilderness hardly conveys any kind of vision of nature where man has a place at all, whereas 'a sense of rural belonging' is what is most taken for granted by an English consciousness, s/he as a visitor to Glacier Park, bears.[25] Moreover the latter sense of wilderness, experienced by 'insiders', makes it hard 'to abstract from what ecologically is'.[26] Alternatively, the viewer stays ensconced within his or her car ready to drive into pull-over sights where a view of the Rockies, for example, can be photographed or compared with an 'ideal' view provided by a purchased postcard. Once tramping begins in the forests away from the 'safer' camping sights, to be unaware of what ecologically is the case is to risk the danger of losing one's life, as the notices signifying the possible presence of grizzly bears make only too clear. Of course, if you are a perverse English person, you may actually enjoy the aesthetic experience of that special sense of rejection in finding yourself in such an environment and thereby come to regard the lectures of Montanians on the delights of this 'beautiful scenery' as somewhat strident attempts to force you into a kind of spectatorship[27] you would prefer not to enjoy!

By such considerations we are led to consider the sheer Secondness present in a landscape experience through which a sense of iconicity is mediated, that 'continuing, organic, ecological and evolutionary unity'[28] of the environment itself, which underlines the sense of an ongoing permanence a landscape has. At this juncture you may wish to raise an objection which you have held ever since this discussion began, that is to say, that there are no natural landscapes anyway. Your objection has two implications. Reference has already been made to Budiansky's rejection of any state of nature which is 'normal' as opposed to its being regarded as simply recovering from a previous disaster or upheaval; nature 'is ever changing, a creature of the endless geologic upheavals that are as old as the planet itself, of the instabilities and chaos that are the fate of all complex systems, and of chance'.[29] Even the Rockies are subject to that condition. But, secondly, your objection could be interpreted much more in a cultural context. Outside the great national parks[30] there are really no natural landscapes as such; rather, to coin a term of Theodor Adorno – though perhaps using that term in ways he did not anticipate – we should speak of culturescapes: there is now no place left in the world which is not subject to the activities of human beings. Even Antarctica is a place for human adventurings! Again a different use of Peirce's trichotomy of terms can make clear the diverse ways such human activities have transformed any landscape.

The most obvious culturescape transformations can be regarded as artificially created landscapes, as can be seen in the case of the parklands constructed in the eighteenth century by such landscape gardeners or better architects as William Kent (1684–1748), notorious for planting dead trees in Kensington Gardens so as 'to give a greater air of truth to the scene',[31] Lancelot Brown (1716–1783), who on seeing a great estate's layout would comment 'I see great capability for improvement here' and Humphry Repton (1752–1818) who transformed formal gardens into the 'picturesque'.[32] Such designs arose through the ⋆intentional activities of their architects, illustrating Peirce's category ⋆Thirdness once more, to produce what can be called artificially produced ⋆cultured landscapes.

In relation to the ⋆unintentional, we can consider landscapes which have been transformed by the accidental or chance consequences of human actions. Here *Firstness is invoked because we are considering the transformation of the felt quality of landscape through such contingencies. Consider the effect on the 1970s English landscape brought about by Dutch elm disease carried by imported timber into Avonmouth or the Thames Estuary ports, the colonisation of waste lands in California by garden crops such as fennel, parsnip and celery, natives in relation to Mediterranean Europe and the Ml which was found to have along its cuttings and embankments three hundred and fifty new species of plant besides the thirty or so species which had been deliberately sown or planted there less than twelve years previously.[33] These cases provide examples of what might be called *encultured landscapes, and if you dislike that particular nomenclature, it has been used only to invoke the idea of an introduced or invading species into an environment to transform the qualitative aspects of the landscape.

In relation to the *causal – the manifestation of *Secondness – we can consider a landscape physically as geographers do, i) as a site whose soil makes agriculture possible; ii) a rural environment deserving of its special landscape designation or iii) as an area suited for urban development. By the year 2016, over four million houses will be needed according to the Department of the Environment, up to forty per cent in the countryside, if the English continue to desire to live in the manner to which we are presently accustomed. Now the nature of the sheer physicality of a particular environment makes these developments possible. All three can be described as *cultural landscapes; fields subject to agricultural development can be regarded as cultivated landscapes, specially designated sites as culturable landscapes – capable of providing some aesthetic perspective in people's lives – and culturescapes proper, although Adorno would prefer to restrict that word's use to urban developments he finds aesthetically acceptable. So he calls urban development, which encompasses old architecture, beautiful because of its setting or because of the locally used building materials characterised in the local church and surrounding market place. What gives such culturescapes their validity and staying power 'is their specific relation to history.'[34]

Having tried so far to ascertain what landscape is, you may still wish to raise a further question: 'What endows landscape with value?' Use can be made once more of the distinctions already drawn between the *intentional, the *causal and the *unintentional in classifying claims in relation to arguments concerning the value of landscape or indeed of a particular landscape.[35] The painterly origins of the term 'landscape' suggest one way in which such value arguments can be regarded. The countryside can be regarded in an *intentional manner, as if it were created to sustain such views. Nature then comes to be regarded as a work of art. In this way Constable's remark 'I never saw an ugly thing in my life'[36] can be interpreted. But works require creators, so how was the creation found in nature to be explained? If, as Schelling claims, the 'modern world begins when man wrests himself loose from nature' then Christianity can be associated with that development to the extent that nature is no longer viewed without any sense of mystery, since what is infinite is treated symbolically in interpreting the finite. Whereas, too, for primitives and the Ancient Greeks, the finite 'counts for something in and for itself', for the Christian it can't count as anything in and for itself alone since it can only serve as an allegory for the infinite. Infinitude, representing 'eternal beings' of nature – gods having a higher, more immutable and enduring character than humans – rendered for primitives and the Ancient Greeks finite existence as fragile, whereas for the Christian, nature becomes mysterious to the extent that it serves 'only as an allegory of the invisible and spiritual world'.[37]

This standpoint comes out well in Weil's claim that the beauty of the world provides evidence for 'divine mercy here below': 'Stars and blossoming fruit trees: utter permanence and extreme fragility give an equal sense of eternity'. More explicitly she writes that a work of art 'has an author and yet, when it is perfect, it has something which is essentially anonymous about it. It imitates the anonymity of divine art. In the same way the beauty of the world proves there to be a God who is personal and impersonal at the same time and is neither the one nor the other separately.'[38] Such a defence for the protection of landscape, serving interpretatively as a microcosm of God's creativity, can be significant in another direction too; carried along by this 'belief in the divinity of nature'[39] was a 'faith in nature', a faith inspiring the activity of science itself. Whitehead expresses that faith in the following terms: 'Faith in reason is the trust that the ultimate natures of things lie together in a harmony which excludes mere arbitrariness. It is the faith that at the base of things we shall not find mere arbitrary mystery. The faith in the order of nature which has made possible the growth of science is a particular example of a deeper faith. This faith cannot be justified by any inductive generalisation. It springs from direct inspection of the nature of things as disclosed in our own immediate present experience.'[40] Since this kind of faith was manifested in the kind of scientific inquiry examined in the previous chapter, we need not examine this kind of claim further since other arguments press for our attention.

A further class of arguments reverses the kind of argument we have just been considering. The proponents of such arguments claim not that nature is to be regarded as a work of art but rather that works of art are akin to processes in nature. This kind of claim came to the fore as the prevailing view of nature made the place of human beings within it less certain in a post-Renaissance world.[41] What needs emphasising here, however, is the character of the ⋆causal processes making such activities possible. Here we are dealing with something more objective than a play of ideas released through some developed cognitive perspective[42] since it refers to something active – referring again to Peirce's category ⋆Secondness – in a genetic sense, something either inherited by humans, or acquired in the course of their historical development. In this context we can begin with the claim that there is increasing scientific evidence that 'there may be considerable correspondence across Western and some non-Western culture in terms of positive aesthetic responsiveness to natural landscapes' just as it can be shown that hospital patients recover from surgery more rapidly if they are able to enjoy such views.[43]

Such studies provide some confirmation for an assertion, made within the philosophic tradition of Rousseau's naturalism. This issues in Kant's claim that the 'superiority which natural beauty has over that of art, even where it is excelled by the latter in point of form, in yet being alone able to awaken an immediate interest, accords with the refined and well grounded habits of thought of all men who have cultivated their moral feeling'. In addition such evidence might also suggest that not only do views of a natural landscape entertain the mind with appropriate ideas or provide 'a feast for his soul in a train of thought he can never completely evolve'[44] but also may undermine a fashionable claim of Roland Barthes: 'regeneration through clean air, moral ideas at the sight of mountain tops, summit climbing as civic virtue etc.' are merely manifestations of an individualistic 'Helvetico-Protestant morality' seen as a 'hybrid compound of the culture of nature and of Puritanism'.[45]

Another element in Rousseau's naturalism, illustrating something causal in the nature of human consciousness, was anticipated by Spinoza when he advocated that men broaden their self-interest in relation to others through 'defending the possession of the land they inhabit'.[46] Such a stance might well be inspired by the way the land causes its inhabitants to have certain aesthetic feelings about it. This is because, as Aldo Leopold remarked, the land can be regarded as 'an instrument of self-expression' whereby society 'may some day paint a new and better picture of itself'. Indeed he invoked the cultivation of 'love of country' even if 'a little less spangled with stars' to enhance a sense of respect for the American landscape.[47]

Admittedly this view can once more be regarded as ideological, validating 'a cultural inheritance in the land' which can be 'linked to a right-wing legitimist politics'[48] rooted in the claims of *nationalism. The case for a possible defence for a form of nationalism has been examined elsewhere.[49] Sufficient here is it to say that the writings of Simone Weil do evoke a 'poignantly tender feeling for some beautiful, precious, fragile and perishable object' such as one's own environment disconnected from notions of 'national grandeur'. She ties this to the aesthetic dimension by speaking of a 'compassion for fragility' which 'is always associated with love for real beauty, because we are keenly conscious of the fact that the existence of the really beautiful things ought to be assured forever, and are not.[50]

Such remarks lead to the consideration of the value of

landscape not instrumentally, as a means to an end, but intrinsically in its own right. Weil's remarks indicate that such value is ascribed not because of the ideas which might arise out of experiencing a landscape nor because of the effects it may have upon its viewers but rather in relation to the felt qualities which might be enjoyed quite accidentally or *unintentionally* in viewing it. Sufficient has perhaps already been said about the iconicity of landscape viewing to clarify this way of viewing landscape in which consciousness can be transformed into a *reverie where 'soft as breezy breath of wind,/Impulses rustle through the mind'.[51] It may well be that in such a reverie, as Rader says in relation to Wordsworth's poetry, there is an 'Objectification of Spirit' where the imagination imparts 'its own dreamlike vividness and splendour to the things of sense', whilst, at the same time, there is a 'subjectification of nature' characterised by 'the sky sinking down into the heart, and the dead still water lying upon the mind even with a weight of pleasure'.[52] In this kind of experiencing then it is not possible to distinguish separately what the viewer brings to what s/he sees in separation from considerations about the perception of the landscape. Nor is it possible to distinguish separately what features belong to that landscape in separation from the viewer perceiving it. And the value in this kind of experiencing – where what is viewed is allowed to stand outside its normal context in an existence which serves some purpose in human intentional activity?[53] That value can be found in recognising the fragile contingency of a landscape's present *actuality, its likely redundancy in terms of instrumental value, where the possibility for the human being to acquire the truth about his or her own condition lies. In such a spirit, it can be said of appreciating a landscape what Benjamin suggested about taking a photograph:[54] in viewing a landscape an attempt can be made to redeem an environment from its more usual instrumental use. That can be done by granting it a kind of dignity through lending to our fleeting impressions of it an attention which would serve to bring that landscape within our own felt life of human experience rather than treat it merely objectively.

In this chapter, an attempt has been made to define landscape using a tripartite analysis provided by Peirce's categories. That attempt emphasised a landscape's defining features: the notion of a *human perspective* upon *a physical environment* capable, in being appreciated panoramically, of yielding an *iconic experience*. If the first feature represents something intentionally interpretative, the second something physical then the third represents the felt, qualitative dimensions in experience. After exploring the dangers of allowing any one of these elements to dominate over the others, it was discovered that each of them could provide rationales for why landscape-viewing is a worthwhile activity. The interpretative yielded the idea that nature could be regarded as God's creation, leading to the idea that our contact with nature can render hypotheses worthy of scientific investigation, as was seen in chapter VI. The causal implied that landscape viewing can be beneficial to health, affords pleasing ideas to the viewer besides lending to an inhabitant a sense of pride in his or her environment. Finally, the purely aesthetic attitude, in pointing to a sense of intrinsic value, can broaden human consciousness, enable insight into the fragile character of what constitutes our landscape viewings, besides making creative activities possible. It does seem then that in considering what makes landscape-viewing worthwhile, we have gone beyond the kind of anthropocentrism which dominated our opening chapters.

As far as the site at Tiptree is concerned, we have seen, in a previous chapter, the scientific argument advanced. In addition, it does provide opportunities for invigorating scents and scenes of earth and vegetation and can yield, through its panoramic views, a sense of place in a remoter part of Kent. Finally, its undulating fields decorated with odd, isolated trees do give a sense of felt-pleasure to the eye and the mind's inner eye. My first experiments in writing were induced by a walk on the ridge in this area. Whether such remarks are sufficient to establish that this site does possess sufficient qualities to be worthy of a local landscape designation is a question which can be pursued in the next chapter!

NOTES

[1]Glacier Park, Montana, 7 August 1991.

[2]Tiptree Farm Funton Appeal by South Eastern Minerals Ltd., Proposed Extension to Clay Quarry and Restoration by Infilling: *Landscape Issues*, Proof of Evidence by Geoffrey Bales DOE Ref. APP/W2200/A/95/25957, 8 November 1996, Kent County Council, Professional Services pp.5–9.

[3]The May 1993 'Third Review Deposit Plan and Proposed Modifications (August 1996): Environment', Appendix 1 of the Tiptree Farm Appeal, Appendices to Accompany Proof of Evidence of Mr N Hare, DOE NSF: APP/W2200/A/95/259578.

[4]Swale Borough Local Plan, Deposit Draft Written Statement (covering The Period 1991–2006), R J Harman, Director of Development Services, Swale Borough Council, 1996.

[5]*North Kent Marshes Study*, Applied Environmental Research Centre Ltd., Consultant's Final Report, Presented and Published by Kent County Council, May 1992 sects. 4.24 and 27(i) and 6.9–11.

[6]*The Kent Thames Gateway Landscape*: A Landscape Assessment and Indicative Landscape Strategy' prepared for Kent County Council and the Countryside Commission by Cobham Resource Consultants, Avalon House, Marcham Rd., Abingdon, Oxon OX14 lUG p.28, R4; p.28, R6 and p.46, R1a.

[7]The May 1993 'Third Review Deposit Plan and Proposed Modifications (August 1996): Environment', Appendix 1 of the Tiptree Farm Appeal, Appendices to Accompany Proof of Evidence of Mr N Hare, DOE NSF: APP/W2200/A/95/259578.

[8]'Tiptree Farm Funton: Appeal by South Eastern Minerals Ltd., Proposed Extension to Clay Quarry and Restoration by Infilling: *Landscape Issues*, Proof of Evidence by Geoffrey Bales DOE Ref. APP/W2200/A/95/25957, 8 November 1996, Kent County Council, Professional Services pp.5–9, 18–19 sects. 10.6–10.9 and pp.11–15 sect.7.

[9]Extension of Existing Clay Quarry, Restoration to Agriculture by Infilling, Tiptree Farm, Nigh Oak Hill, Funton, Sittingbourne, Appendix 6 to Item C1 – Consultee's views and further representations on the revised application 811/94/1137 – p.C1.48, the report of the county planning officer to the Planing Sub-Committee, dated 12 September 1995.

[10]Eric Hirsch, 'Landscape: Between Place and Space', Introduction to *The Anthropology of Landscape* ed. by N Hirsch and N O'Hanlon. Clarendon Press 1995, p.2.

[11]Hannah Weitemeier, *Yves Klein 1928–1962: International Klein Blue*, Cologne: Germany, B Taschen 1995, p.19.

[12]C S Peirce, 'Elementary Account of the Logic of Relations' 1986 in *Writings of Charles S Peirce: A Chronological Edition (1854–1856)*, vol. 5 ed. by Max H Fisch et al., Bloomington, Indiana Univ. Press 1993, p.380.

[13]Hannah Weitemeier, 'Yves Klein, op. cit. p.18.

[14]Alan R H Baker, 'On Ideology and Landscape', Introduction to *Ideology and Landscape in Historical Perspective* ed. by A R H Baker and C Biger, Cambridge UP, 1992, p.2.

[15]A Boersch, 'Landscapes: Exemplar of Beauty', *Brit. Journal of Aesthetics*, vol. 11, no.1, Winter 1971, pp.81–95, p.**88**.

[16]S Daniels and D Cosgrove, 'Iconography and Landscape', Introduction to *The Iconography of Landscape: Essays on the Symbolic Representation, Design and Use of Past Environments,* D Cosgrove and S Daniels, eds. Cambridge UP, 1988, p.1.

[17]P T Newby, 'Towards an Understanding of Landscape Quality', *Brit. Journal of Aesthetics,* vol. 18, no.4, Autumn 1978, pp.345–355, pp.352–3.

[18]Aldo Leopold, *A Sand County Almanac & Sketches Here & There*, (1949), Oxford UP 1987, p.47, *The River of the Mother of God & Other Essays*, Wisconsin UP, 1991, pp.135–136.

[19]N Green, 'Looking at the Landscape: Class Formation and the Visual', Eric Hirsch, op. cit. pp.31–42, p.33.

[20]Stephen Budiansky, *Nature's Keepers*, Weidenfeld and Nicolson, London 1995, p.71.

[21]Eric Hirsch, op. cit., p.5.

[22]C S Peirce, 'Elementary Account of the Logic of Relations' op. cit., pp.380–381,m 'Nevertheless, it will be found that in any instance of the use of a sign, there is one of the three characters which completely overshadows the others; and that this division is practically most important.'

[23]Raymond Williams, *The Country and the City*, London, Chatto and Windus, 1973

[24]Eric Hirsch, op. cit., pp.13, 22.

[25]Bryan Appleyard, 'A Flaw in the British Nature', *The Sunday Times* 12 January 1997.

[26]Holmes Rolston III, 'Does Aesthetic Appreciation of Landscapes Need to be Science-Based?' *Brit. Journal of Aesthetics*, vol. 35, no.4, October 1995, pp.374–386, p.378.

[27]N Green, 'Looking at the Landscape' op. cit., p.38.

[28]Holmes Rolston III, 'Does Aesthetic Appreciation of Landscapes Need to be Science-Based?', op. cit., p.381.

[29]Stephen Budiansky, *Nature's Keepers*, op. cit., p.98.

[30]I am putting on one side here the question as to whether there can be Natural Wilderness as such in these places. See articles by R Guha, J B Callicott and Holmes Rolston III in Section V 'Wilderness Preservation' in *Reflecting on Nature* ed. by L Gruen et al, Oxford UP 1994 pp.239–78 for the ongoing debate on this as well as Stephen Budiansky, *Nature's Keepers*, chapt. 6.

[31]Stephen Budiansky, 'Nature's Keepers' op. cit., p.30.

[32]J D Hunt, *The Figure in the Landscape: Poetry, Painting and Gardening in the Eighteenth Century*.

[33]Andrew Goudie, *The Human Impact on the Natural Environment*, 4th Edition, pp.71–76.

[34]T W Adorno, *Aesthetic Theory*, tr. C. Lenhardt p.95; tr. R. Hullot-Kentor p.64 as "is the expression of history"
[35]These distinctions are drawn out in another context in N E Boulting's, 'On Endymion's Fate: Responses to the Fear of Death' in *Rending and Renewing the Social Order*, Yeager Hudson, 1996 pp.367–387.
[36]K Clarke, *Civilization*, BBC/John Murray, London 1971, p.200.
[37]F W J Schelling, *The Philosophy of Art*, pp.59, 61–62, 66.
[38]S, Weil, *Gravity and Grace*, (1952), pp.100, 97, 136.
[39]K Clarke, *Civilization* op. cit., p.269.
[40]A N Whitehead, *Science and the Modern World*, Cambridge UP, 1926, chapt. 1, last para.
[41]Eric Hirsch, 'Landscape: Between Place and Space', op. cit. p.6.
[42]R S Peters, 'The Justification of Education' in *The Philosophy of Education*, ed. by R S Peters, New York, Oxford UP 1975 p.240.
[43]R S Ulrich, 'Views Through a Window May Influence Recovery from Surgery', *Science*, vol. 224 (1984), pp.420–1; 'Stress Recovery During Exposure to Natural and Urban Environments' *Journ. of Env. Psych.* vol. 11 (1991) pp.201–30 and R S Ulrich 'Biophelia, Biophobia and Natural Landscapes' in 'The Biophelia Hypothesis', p.97.
[44]Immanuel Kant, *The Critique of Judgment*, pp.158–159.
[45]J S Duncan (et al.), 'Ideology and Bliss: Roland Barthes and the Secret Histories of Landscape' in *Writing Worlds*, p.20.
[46]Spinoza, *A Political Treatise*, chapt. II sect. 15.
[47]Aldo Leopold, *The River of the Mother of God and Other Essays*, pp.180, 191 and N E Boulting, 'The Emergence of the Land Ethic: Aldo Leopold's Idea of Ultimate Reality and Meaning', *Ultimate Reality and Meaning*, September vol. 19, no. 3 1996 pp.168–188.
[48]N Green, 'Looking at the Landscape', op. cit. p.36.
[49]N E Boulting, 'Does Any Form of Nationalism Offer a Perspective on Ultimate Reality and Meaning?' in *Ultimate Reality and Meaning* September, vol. 18, no. 3 1995 pp.192–211.
[50]Simone Weil, 'The Need for Roots'(1949), p.164.
[51]Matthew Green, 'The Spleen', ed. by W H Williams (1036) p.26, lines 672–3, quoted in A Boersch, *Landscapes: Exemplar of Beauty*, p.92
[52]Melvin Rader, 'Wordsworth: A Philosophical Approach', p.177.
[53]N E Boulting, 'Edward Bullough's Aesthetics and Aestheticism: Features of Reality to be Experienced' *Ultimate Reality and Meaning* (1990), vol. 13, no. 3, Sept. pp.201–21
[54]W Benjamin, 'A Small History of Photography' in *One Way Street*, pp.240–57.

Chapter VIII

A LOCAL LANDSCAPE AREA DESIGNATION?

Speechlessness between former intimates leaves each in the right: Non-communication between nature and human beings puts both in the wrong. Yet what are we to make of nature's persuasive gesturings? Surely the skylark higher calling offers the prospect that our level of experience can always be transcended through an unintentional intention to reach beyond it. Here, a complete openness to experience, just because it recalls the cosiness of previous intimacy now gone, shakes the body with grief, as the caressing wind of the south strokes the cheek to welcome a return to where elongated bull-rushes whisper to each other as they run alongside these dykes. This elective affinity between consciousness and nature's gestures fails to yield the flat complacency which characterises the relationship between societal subjects who have to get on but, rather, pulls on the heart in unpredictable ways, as shafts of sunlight set between grey clouds select some farm shed for special attention or salmon-speckled clouds refract sunlight into a strange white glow. Yet as soon as the warmth of a May sunshine makes way for heavier black clouds from the North, such visions disappear upon recognition, to chill the psychic echo once found undistorted in a nature incapable of sustaining an actual or complete distortion.[1]

Large half-crown snowflakes fell on a slushy road as I cycled to Swale House from Sittingbourne Railway Station on Thursday, 9 January 1997. After being ushered into the public gallery I awaited the start of the inquiry into objections against Swale Borough Council's attempt to provide Local Landscape status to the hills between Iwade and Newington to include the Tiptree Site in its Policy E17: along with other areas it would 'be afforded long-term protection' because such a 'striking range of hills in an otherwise flat marshland landscape was highlighted in the 1992 *North Kent*

Marshes Study. They present impressive long-range views both inland and across the marshes'.[2]

Before that debate could begin, however, another objection to this policy was heard first. A substitute for the representative of the council for the Protection of Rural England (CPRE), who lived in Sittingbourne and so was familiar with the area, tried to extend the domain of three areas proposed as local landscape designations. One of these three – the other two involving the Lynstead Valley and the Syndale Park areas – referred to the Tiptree Site since the CPRE was suggesting that not only should the requisite hills be included but a larger area south of the M2 to extend as far west as the district boundary. In regard to all three areas this CPRE representative argued in a somewhat hesitant manner, without the use of maps for each party in this inquiry to see what was being proposed. His case was that such landscape areas should be preserved for future generations, for the grandchildren of our own children's grand children. After some debate, the Swale Borough Council's representative argued that the council had considered some of these proposals before but had decided against them. One of the reasons given would be relevant to the next objector, South East Minerals (SEM): 'We do not wish to use measures to create Local Landscape Designation Areas simply to prevent economic development in the Borough!'

The rest of the day was taken up by the QC's case for the appellant, SEM. The counsel for Swale Borough Council (SBC) called only one witness during his defence in the afternoon: the Director of Development Services for SBC. The QC for SEM called one witness in the morning before lunch, an independent planning consultant who had once worked for KCC in their countryside section responsible for mineral and waste disposal development control and planning. His second witness, called after lunch, was a landscape architect consultant in relation to landscape design and environmental planning with particular emphasis placed upon Environmental Assessments and surveys.

It would clearly take a number of chapters to narrate the individual blows struck on each side by the many arguments, which, as will be seen, often appeared philosophic in character. So, in the interests of brevity the inquiry's proceedings –

witnessed by only two people in the public gallery – that exciting debate will be condensed into six arguments which its reporter thinks are significant in the way the Assistant Government inspector will arrive at his decision. SEM's QC forwarded three weak and three stronger objections to SBC's case and these ran in and out of that Thursday's proceedings to provide the crux of the arguments! Let us start with the first of the weak objections.

Reference has already been made to Policy E17. SEM's first witness claimed that the policy was in conflict with another, E11. That policy statement begins:

> The countryside of the Borough, which is all the land falling outside the defined built-up area boundaries, will be protected for its own sake. Development in this area will be normally restricted unless:

and then eight conditions are listed. The first is the relevant one:

> it is demonstrated to be reasonably necessary to agriculture, forestry, the winning or import of minerals, or other land use which essentially demands a rural location.[3]

The latter statement recognises that there may be some economic development in rural areas yet this seems to oppose a policy, such as E17, concerned with long-term protection of a landscape. SBC's counsel argued that the designation under E17 does not stop the extraction of minerals in an emergency. Challenged as to how the clash between these two policies can be reconciled he replied that E17 would deal with what would amount to a balancing act between mineral extraction and the landscape interest. He then made an important point: E17 was not an overriding policy constraint in emergency cases.

The second weak objection was developed by SEM's QC. He argued that policies in the Kent Structure Plan gave the necessary protection for the countryside and even Swale Borough Council's local plan policy E11 incorporates a landscape consideration. He cited particularly Policy ENV1, which he considered protected the countryside satisfactorily, and ENV2 fortifying that protection. These two take the following form:

> The countryside will be protected for its own sake. Development which will adversely affect the countryside will not be permitted unless it can be demonstrated, to the satisfaction of the local planning authority, that there is an overriding need for the development which outweighs the requirement to protect the countryside. (Policy ENVI)
>
> Kent's landscape and wildlife (flora and fauna) habitats will be conserved and enhanced. Development will not normally be permitted if it would lead to the loss of valuable landscape features or wildlife habitats which are of scenic, historic or wildlife importance, or are of an unspoilt quality free from urban intrusion, unless it can be demonstrated that there is a need for development which outweighs these countryside considerations (Policy ENV2).[4]

At least three arguments were used by SBC's witness in the afternoon to destroy this objection. Firstly, he cashed the argument used earlier against the CPRE's representative. Local Landscape Area (LLA) designations were to be used with discretion to avoid debasing 'the currency' and their more extensive use had been resisted as was seen earlier in the day. Secondly, reference was made to a document of much significance in the inquiry. The Department of the Environment and Welsh Office produced it in January 1992 offering Planning Policy Guidance (PPG) entitled *The Countryside and the Rural Economy*. This document – PPG7, 1992 from now on – was cancelled after this 9 January inquiry by a new PPG7 published on 21 February 1997 by the Department of the Environment: *The Countryside – Environmental Quality and Economic and Social Development* – PPG7, 1997 from now on. The SBC witness referred to one of the claims in the PPG7: 'It is for local authorities to determine the more specific policies that reflect the different types of countryside found in their areas' (PPG7, 1992, 1.2; cf. PPG7, 1997, 1.5).[5] That was just what Swale Borough Council had done! Thirdly, on the basis of the QC's arguments there would be no need for (LLA) designations. But the fact was that such plans as the Kent Structure Plan provided a set of general policies requiring interpretation at the local level if more specific objectives were to be achieved. Policy ENV2 provided a good example since, in attempting to sustain the conservation and enhancement of wildlife habits and landscapes in Kent it leaves to it the local council to identify where the LLA designations are to be.

The last of the weaker objections was that there would be a serious long-term constraint on clay extraction from the site if the particular LLA designation was sustained, particularly as, under Policy E17, it was not clear what was meant by the phrase 'afforded long-term protection'. The SBC counsel's witness scotched that particular line of attack by arguing that LLA designation cannot be made on the basis of whether some landscape contained particular minerals, but on its landscape features. SEM's argument was fatuous since, from a logical point of view if it was valid, because the North Downs contained vast chalk reserves, it couldn't be designated an area of outstanding beauty! Moreover it was up to the Minerals Local Plan inspector to decide on SEM's challenge to the Kent Minerals Local Plan in relation to clay production. It was not a matter for this inquiry.

The QC's second witness provided the first of the stronger objections. He developed a line of attack presented in this book's earlier chapter 'The Landscape Question': there was no explicit mention in the *North Kent Marshes Study* of May 1992 of the hills around Tiptree as being covered by its phrase 'a striking range of hills'. *The Kent Thames Gateway Landscape* 1994 document fails to give the site the kind of designation given to Callum Hill: the former is labelled 'Undulating open arable farmland', the latter 'prominent wooded or farmed hills'. That is hardly surprising because the former lacks any kind of distinctive character worthy of a special designation and simply allowing for views over adjacent areas couldn't count as such a discernible characteristic in its own right. Furthermore, argued the QC himself, given that the report says of the 'Undulating, open arable farmland' that it has 'a relatively weak landscape structure',[6] then one should be very cautious in giving it an LLA designation especially in the light of a sentence which can be found in a publication issued jointly by English Nature, English Heritage and the Countryside Commission entitled *Conservation Issues in Local Plans* (CCP 485 from now on): 'The Secretary of State for the Environment, in a statement made in 1993, urged planning authorities to be cautious about introducing local designations because they may place unnecessary barriers in the way of economic activity' (CCP 485, 1996, 4.3).[7]

Again the SBC witness argued against this, the first of the QC's three stronger objections. Firstly the QC must not be tempted to quote selectively from documents. In fact the next sentence following the QC's quotation reads: 'The statement also recognised, however, that such designations may be appropriate where there is reason to believe that the normal planning policies cannot be applied to provide the necessary protection' (CCP 485, 1996, 4.3). His second argument was that from the top of Tiptree Hill it could not be denied that outstanding, good panoramic views could be enjoyed and this special characteristic provided a starting point for designating it an LLA, just as had been done in the case of the Thanet Local Plan in November 1996 where this kind of policy had been approved by the inspector. That was hardly surprising in view of what can be found in the countryside commission's document: 'Once landscape character has been described, it can provide a point of reference for many aspects of planning policy, as well as for detailed development control decisions' (CCP 485, 1996, 3.9). His third argument was much bolder. So as to fix that landscape character he claimed that the designation given by *The Kent Thames Gateway Landscape* document simply was an error. Tiptree Hill was to be regarded as a continuation of the prominent ridgeline to the north of Newington, described in the report as a 'unique landscape feature' creating 'an attractive backdrop for views from surrounding roads and nearby properties as well as being a contrast to this flatter more open farmland in this area'.[8]

The second stronger objection, pressed harder by the QC and his second witness, was that even if recognition is given to the higher elevated land form at the Tiptree Site that would not count as constituting a right of designation since it lacks any other further qualities. It simply fails to satisfy the following six criteria which are set out in the countryside's commission guidance document – CCP 423 from now on – for evaluating landscapes for designation: the *rarity* requirement – its importance for its representation of a landscape resource; *scenic quality* – pleasing forms and aesthetic factors; *unspoilt character*; *sense of place* – possession of a distinctive character giving a visual unity; *conservation interests*, satisfying historical, scientific or ecological concerns and the *consensual requirement* – professional and public agreement on its importance reflected in writings and paintings (CCP 423, 1993, p.25).[9] More specifically Tiptree Hill is marred by the clay extraction, it is spoilt by the presence of the brick factory and it lacks any conservation interest!

The SBC counsel's witness responded on all three fronts. First he used the possible admission that the higher elevated land form could be a characteristic by claiming that it was widely visible, it enclosed and defined the lower ground below it and served as a front to the undeveloped coast in the estuary. Secondly the mineral extraction area was not clearly visible and the brick-making factory was small and tucked away on the northern side. After much argument between the parties as to the meaning of the criteria used in the document CCP 423, the SBC counsel's witness insisted that for reasons already intimated Tiptree Hill was a landscape resource of local importance in shaping the skyline, it did possess scenic quality, it was largely unspoilt, it did possess a sense of visual and topographical unity and there was a general consensus between members of the public and professional people in regard to its importance. Indeed, particularly in relation to scenic quality, 'the essential and overriding factor in designations' (CCP 423, 1993, p.25), these claims are strengthened when the document from which these criteria are derived – *Landscape Assessment: Principles and Practice 1991* – is consulted.[10] Thirdly, the government office of the South East (GOSE) appeared to have been convinced of the special justification for the policy.

The SEM's QC made the most of his third stronger objection. He questioned his first witness as to what a public body should do in order to establish an LLA designation. He was told that there were government guidelines in PPG7 for so doing: there should be 'a systematic assessment of landscape character' and the 'countryside commission has published advice on landscape assessment techniques' (PPG7, 1992, 1.14). The QC emphasised the steps that would be involved in doing this; according to the commission's *Landscape Assessment Guidance* document CCP 423: Planning the Assessment; Desk Study; Field Survey and the Analysis and Presentation of Results. His second witness backed this contention referring once again to CCP 423: 'Any landscape evaluation exercise needs to commence with a landscape classification, dividing the landscape up into areas of common character. The value of each character area should then be carefully judged against the designation criteria. This is likely to require a combination of professional judgement on scenic quality, informed opinion as to special conservation values and known public preferences. The landscape evaluation should review fully and systematically the evidence for (and against) the designation on all of these counts' (CCP 423, 1993, pp.8–9, 25). Even if it was weak in not presenting its evaluation against designated criteria, he thought that *The Kent Thames Gateway Landscape* document provided a model for how landscape assessment should be done, which is more than could be said for the *North Kent Marshes Study*, since it obviously employed the judgments of professional people rather than the mere opinions of bystanders!

When it was pointed out to the SEM's QC on another occasion by a public gallery member that in logic there is a simple principle 'ought implies can; if one can't do something then there is no sense in saying that one ought to do it' and that since SBC were engaged in their strategies for drawing up their local plans in the early 1990s, they couldn't be expected to satisfy standards set by documents published in August 1993 and 1996, the QC soon snapped back. Advice on such techniques had been in existence since the countryside commission had published their landscape assessment document in 1987 and CCP 423 was simply a development of that approach. The SBC's witness, however, proved much more effective with a three part counter-attack. Firstly PPG7 also indicated an alternative approach to using techniques involved in a systematic assessment of landscape character, namely, as it is put in that document, 'non-statutory countryside and rural strategies in co-operation with a wide range of rural interests' (PPG7, 1992, 1.14). Secondly, there was no open chest of money to employ the kind of techniques that he, the QC, and his witnesses recommended. Council staff with local knowledge of local landscape characteristics in this vicinity had undertaken an assessment in this case and this process had had to face public consultation on two occasions so demonstrating a responsible approach to the exercise in arriving at a proper list of LLAs. Thirdly, GOSE had subjected the consultative draft stage of Policy E17 in particular to a precise commentary seeking further justification for this policy. GOSE was eventually satisfied with SBC's response.

The QC made it quite clear that such an approach was quite unsatisfactory. The countryside commission, aligning itself with the various government PPGs (CCP 485, 1996, 1.3) as well as PPG7 had made it quite clear how the subjective factor in such evaluations was to be overcome: 'In many cases local plans will inherit local landscape designations from structure plans, usually as either Areas of Great Landscape Value or Special Landscape Areas. Alternatively they may be identified as a result of a district-wide landscape assessment, usually on the basis of landscape types or areas which have strong, intact, and distinctive character, and special qualities and values that are considered particularly important' (CCP 485, 1996, sect. 4.18). Not only, as his second witness had argued, was the documentation for this policy E17 insufficient for making a systematic assessment to secure a proper landscape designation but there was no way of telling how it was done, probably by a bundle of people lacking landscape assessment qualifications thereby leading to an LLA designation without satisfying the proper means for doing so.

The SBC's witness employed a three-pronged defence upon this line of attack. Firstly, Policy E17 had emerged from the *North Kent Marshes Study* in 1992 not as a result of the application SEM were at present forwarding. Secondly, as the SBC counsel had made clear earlier in relation to PPG7 where 'policies and development control decisions affecting areas of outstanding beauty' favoured 'conservation of the natural beauty of the landscape' (PPG7, 1992, 3.9), the appreciation of beauty was always going to involve a subjective factor. Thirdly, admitting that staff with local knowledge had undertaken surveys during 1993 which led to the environmental policy for the local plan strategy being published subsequently in 1994, though they may not have followed the countryside commission criteria strictly, that did not make their activities invalid nor its result.

Then came the dramatic highlight of this inquiry. Suddenly the QC asked this witness for a copy of the details of the surveys: 'Where is it?' he demanded. 'Where is the documentary evidence for what was done in 1993?' The witness said he couldn't find it. He had looked in various files but it was missing. Now the QC triumphed: there was no evidence to show what the criteria were – used to establish the LLA designation – no details of the qualifications of those who did it and no way of establishing that it was not altogether an entirely subjective operation.

Two comments in the closing summarising speeches of the SBC's counsel and the QC for the appellant as well as a footnote end this report on the inquiry. In drawing attention to the validity of what SBC had done in making the LLA designation, SBC's counsel pointed out that there was no statutory requirement making the adoption of the countryside commission's recommendations on landscape assessment compulsory. Although their approach to integrated planning guidance had been around for some time, it had not been adopted universally so there was nothing wrong in what SBC did. Secondly he pleaded with the inspector to support the policy; SEM were the only objectors. Identifying a proper notation for Tiptree would make sense of the designation already given to Callum Hill.

The appellant's QC opened his final speech by referring to President Lincoln's decision in the face of his cabinet's reluctance to make war on the south: truth doesn't always lie with the majority. He didn't plead with the inspector. Rather he pointed out to him that the government was anxious that the kind of activity SBC engaged in did not happen, since it prevented such a policy as E17 from being tested. It was now up to the inspector to say whether the right procedure had been followed.

My footnote follows on the publication of the revised PPG7. Paragraph 1.14 in the old PPG7, which referred directly to the possibility of 'a systematic assessment of landscape character', has been completely removed. Reference is still made, however, in paragraph 2.9 to the importance of what are called 'non-statutory rural strategies, in partnership with other bodies and local communities' – an approach SBC claimed to have used! In addition paragraph 2.15, PPG7 now recommends to local planning authorities what it calls 'the character approach' when 'they have occasion to review their local countryside designation'. To get an idea of what that means the reader is referred to a future 1997 Countryside Commission document which 'will publish countryside character descriptions and English Nature will publish natural area profiles'. An anticipatory information leaflet from the Countryside Commission could be taken as providing the following as a clue: 'Landscape character exists everywhere whereas high quality does not'![11] More interestingly the new PPG7, in order to shed further light on this, also refers to a Bibliography in Annex J. Though the Countryside Commission's document *Landscape Assessment* (1987) (CCD 18) is listed, its updated advisory booklet *Landscape Assessment Guidance* (1993) (CCP 423) is omitted and its document *Conservation Issues in Local Plans* (1996) (CCP 485), which, as we have seen was used extensively in conjunction with old PPG7 throughout this inquiry, is not included either! Who'd be a government inspector?

A criticism could be made of the way these events have been presented to you. Insufficient attention has been given to the nature and origins of the three stronger arguments in favour of not granting an LLA designation to Tiptree Hill. It is an important criticism for two reasons. Firstly, that lack of attention has resulted in a failure to make clear the perspective from within which the QC's arguments arose. Secondly, for that very reason, a defence of a full-bloodied account of the anthropocentric point of view has not been elicited so far in this book. An opportunity now arises to correct that fault. To make the most use of such an opportunity I will not now employ the locutions 'you may object' or 'it could be argued' but rather allow the landscape architect consultant to speak for himself in a tone he might have adopted if he had read the contents of this book so far!

'As a landscape architect, who offers specialist advice with regard to landscape design and environmental planning, I am a practical person. I assess what you have written in relation to reflections on practice past and present rather than from an understanding of your arguments at a theoretical level. In considering your case, I will list the points where I can agree with your analysis and then go on to indicate your problem. In solving it I can not only highlight the important features of my proof of evidence at the Local Landscape Designation Inquiry but reveal to you why your approach is at best inconsistent and more importantly confused.

'Two key steps you employ must be understood in reading your presentation. The first is your adoption of four basic attitudes in the way nature can be considered, three of which you state are contained in Peirce's writings. Your second step does not provide you with a basis for opting for one of these attitudes as being more basic than the others but it does provide you with the means to keep a reader interested. That interest is carried because s/he thinks that you will be able to highlight an approach which will reveal what attitude is the more basic. In practice your attempt does little more than identify your problem. In addition, you appear to have been driven by the idea that the solution to the rationality problem – the grounding of the rational enterprise itself – will make an important contribution to the debate concerning the opposition between anthropocentrism and ecocentrism!

'Your arguments do make a clear distinction between these two ideas. I agree too with your delineation of the four basic attitudes and the impossibility – which you display – of justifying any of them. But my conclusion is different from yours: your idea of undermining the anthropocentric point of view which I take to be the point of your last two chapters – your ornithological and landscape chapters – appears to me to be at best misguided but more significantly incoherent. Let me explain.

'From an operational point of view, it is important to note that there is no specific mention or consideration of the practical nature of the work of Kent, Brown and Repton whom you refer to in your landscape chapter. It has to be understood that first and foremost William Kent was a mural and an historical painter. It is then no anomaly to find this Yorkshireman placing artistic objects – shepherd's huts, grottoes or Roman temples – as emblems, meant to suggest ideas from another age, in his constructed gardens, as in the case of Holkham Hall in Norfolk. Rather, the purpose of painting for him was to suggest a strategic vision for constructing an ideal landscape in a new style of gardening. So, the eighteenth-century essayist Joseph Addison referred to the possibility of making "a pretty landskip of his possessions" or of creating "a garden with a picture of the greatest variety". As you pointed out, Peirce claimed that such aesthetic creations envisaged nature as a picture and that was true for William Kent quite literally. You could say that the distinction between ideal and topographical landscape disappeared; conceptualised "ideal" landscapes became actual landscapes as Kent constructed emblematic gardens',[12] emblematic because they encompassed objects suggestive of imaginative possibilities from the past.

'A study in the detail of Lancelot Brown's creations reveals something quite different. This man from Northumberland – designer of the house and grounds at Dodington, in Derbyshire – was a consultant who made his constructions more like "natural" scenery than paintings, making more use of nature's forms rather than incorporating copied human creations. His gardens are less emblematic and more expressive of "the charms of nature gracefully combin'd".[13] Now the garden is not merely added to the house; rather the house, viewed from the outside, becomes the central point to be encompassed within a much wider visual scene!

'Finally, two key differences must be understood in assessing the practical activities of Humphry Repton, the Suffolk garden designer of Sheringham Hall in Norfolk. The first was his rejection of Kent's notion of the picturesque and Brown's naturalism in favour of what can be called the gardenesque so that 'terraces, flower beds and compartmented gardens' emphasised the social uses his landscapes could offer their owners. The second was his idea of making the house the central point of his design but now in the sense that views could be enjoyed by the spectator within it before moving out into the gardens themselves by means of convenient terraces, conservatories or gravelled walks.[14]

'I think therefore that I have demonstrated the inadequacy of your treatment of the work of these three men. Now although I am in agreement with what you state about the term "landskip" and its origins, I think it more sensible to highlight the development of the meaning of that term that you have done,[15] connecting it to my observations on the practice of those three pioneering landscape consultants. It should be noted that at the start of the seventeenth century the term "scape" was associated with "ship" as in township or the term "shaft", the former associated with shape – township, the shape or district of a town – the latter with something carved out of something else.[16] Landscape, then, is to be associated with the shape, extension or the delineation of a piece of land rather than the sea and it was a term used by painters to identify scenes on the land or sea as opposed to portraits. The practice of landscape gardening, following on the concern with the painterly, repeats this earlier approach. Kent's practice was to shape something scenic artistically or *aesthetically*.

'As has been noted, a key feature of Brown's contribution was to encompass the house within this "scape" – despite the wildness of its incorporated "natural" scenery – an absorption of everything into a single whole emphasised by the viewer having to move around the house by means of its "circling drives" allowing the mind "to read its own meanings and moods" into the "less specific scenery".[17] Now the idea of landscape appears to have been annexed to the concerns of *what* an eye can view in one sighting or many such sightings within single perspectives.

'Finally, the term has a more widely ranging usage to encompass not just painting but nature itself. By the eighteenth century, as Rousseau's works reveal to us, a clear distinction is made between nature and culture, the countryside and the town. But that idea of nature is constructed from within a town or the city by the eyes of its citizens looking forward to the Sunday excursion into the countryside.[18] So in Repton's work the 'natural scenery' is *read* by viewing it from a window, from within a domestic scene, into the landscape outside!

'I would therefore conclude that my definitional archaeology does sustain the definition of landscape as basically "a cultural image, a pictorial way of representing or symbolising surroundings" as Daniels and Cosgrove put it.[19] I leave it to you to go on about my three part delineation of the work of these three architects being a valuable basis from which to discuss Peirce's categories once more. Clearly the painterly can be associated with Firstness, the emphasis on what is viewed with Secondness and reading ideas into a perceived view with Thirdness. But what is really important is that all three presuppose a view taken on a piece of land by an outsider: "The outsider judges by appearance, by some formal canon of beauty".[20]

'Now what I will demonstrate to you is an argument underpinning the importance of that last sentence. My argument will be constructed in five steps. But before I begin I must refer once more to the work of two very practical men, two fifteenth-century Italians to be exact. The first was an outstanding architect, Filippo Brunelleschi, responsible for the dome at Sta Maria del Fiore, Florence's cathedral, a major feat of engineering. As a mathematician, however, he was interested in securing certain theoretical principles to underpin the laws of perspective. The second practical man is Leon Battista Alberti, another architect who not only wrote the first treatise on architecture but, because of his awareness of Brunelleschi's work, was able to define within painting the art of perspective in mathematical terms. If my argument is to be properly made, an argument based on the writings of Denis Cosgrove,[21] we should not skirt around the achievements of these two men. They demonstrated that a three dimensional world could be represented on a two dimensional surface, not as some strategic trick, but as a way of getting at the truth about their surroundings. Their work, then, is central to an understanding of cartography; map-making itself!

'Step 1: Just as the cartographer, in using the cartographic grid to provide a window frame for its user to map out or construct surveys on any object, can thereby *fix* space, so the landscape painter can *freeze* time, set things on the basis of how they appear. Both visual activities make practical actions possible: "Where cartography made possible an urban and regional planning in which the earth was bounded and reformed according to the quadratic net of the world, landscape painting provided a means of visualising the world which the landscape architect used to reshape the topography according to his designs".[22]

'Step 2: In accordance with both landscape painting and cartography a fixed relationship is set up between the subject and objects found in nature. That relationship puts the viewer outside, *at a distance from* the viewing or the surveying. Indeed the latter activity made printed maps – paper landscapes[23] – possible.

'Step 3: Both landscape painting and cartography do two things. They not only emphasise the idea of 'being at a distance' but *vision* too, as one of our senses, is privileged over the other four to sustain the claim that we live in a "specular civilisation" as Foucault has it.[24] During the rise of the Italian Renaissance, landscape becomes "...a particular way of seeing, the linear techniques of perspective developed in landscape painting at this time to create a "realistic" image parallel the development of practices such as cartography, astronomy, land surveying and mapping involving formal geometrical rules".[25]

'Step 4: Both landscape painting and cartography are also to be associated with *the rise of science* within modernism, following on the philosophies of Hobbes and Descartes, for which three things are central: physical seeing, inquiry – where nature is required to "answer" questions put to it by man – and the human control of the environment which that scientific enquiry ensures. All three types of activity privilege the specular image in the way ideas of space, place or the body of something can be represented to us today through the satellite image, the geographical information system, the aerial photograph, the microscopic plate, the distribution map or the photograph of Mars's surface.[26]

'Step 5: The created painting, the land mapped and the results of

> scientific enquiry put to use in *technological applications* or sophisticated products, become things to be bought and sold in the market place. Western capitalism, then, has developed as a result of the graphical as well as the geometrical representation of space – what has been called "the commodification of space"[27] – in which human activities have their place in a controlled and regularised manner. So Olwig claims: "Landscape was framed and reified as a cultural object, to be bought and sold as cultural capital on the burgeoning new art market, much as land itself was being divided up according to the geometric coordinates of the map, to be sold and traded on the property market".[28] Bender agrees: "Compilations, categorisations and a comprehensive mapping of Britain became increasingly important during the seventeenth and early eighteenth centuries. Refinements in surveying and mapping were part of a changing technology of power, integral to the development of mercantile capital and to the opening up of the New World. Control of knowledge, control of resources, control of nature".[29]

'In accordance with an understanding of this five-step argument, four corollaries can be drawn. Firstly, dependence on the specular image – embodied in the new technology to make discussion of virtual reality issues viable – and the current concern for the quantifiable, the assessable and the saleable continues apace. Indeed as talk of *postmodernism has intensified within the chattering classes so the full sweep of modernism continues unabated, so that no area of human activity remains untouched, whether that be in education, the welfare services or even the legal profession itself. Such a social state of affairs encourages, in the intellectual, psychic schizophrenia: it allows the postmodernist to object to the advance of modernism whilst he, like me, can continue to enjoy the benefits of it.

'Secondly, we can see the real origins of the four basic attitudes needed to make sense of the world in the nineteenth century. Those in the literate classes could adopt the idea of nature as something passive and detached "to be viewed and studied by the observer from the outside, as if by peering through a window"; nature as "a separate realm, offering no omens or signs, without human meaning and significance".[30] As something quite passive, nature can be viewed, as Peirce saw, either aesthetically as a picture or scientifically, as something whose ways are to be penetrated by active human enquiry! What he did not see so clearly was how such knowledge could be used instrumentally or sold through applications in the market place in carrying "on the business of the world" (Peirce: vol. 1, para. 43). His sense of practicality was relevant to the illiterate peasants, working increasingly as farmhands labouring on the land. As to your stress on the hedonistic, aesthetics brings pleasure just as can scientific enquiry and the use of nature for recreational purposes. For Peirce the scientific attitude is the one to opt for since it is concerned with truth for truth's sake. And why be concerned with that? After all one might still be a better man for not doing so (Peirce: vol. 1, para. 45)! We are not told.

'Thirdly, what I have just revealed through my use of argument is a world in which I find myself quite at home. From our more contemporary standpoint, nature as land can be regarded scientifically – as a Ramsar site, for example – aesthetically by giving it an NLA, SLA or LLA designation, instrumentally for agricultural purposes or again commodifically – to use your rather ugly word – where it can be logged or quarried for its resources. In essence, even the scientific attitude comes down to a human concern, so we are placed firmly in an anthropocentric dimension where – as in the Tiptree case – land can be regarded in utilitarian terms or aesthetically; pleasing to man's sensibilities. But it does not follow, once it has been so used for resource purposes, that its aesthetic quality cannot be restored. My work shows that it can be, as I will be happy to demonstrate at my next inquiry!

'Fourthly, in the case of Tiptree, I side with the resource case – the need for clay and void space – against the aesthetic one for three reasons. An earlier, independent enquiry regarded its status differently from Callum Hill, the latter thought worthy of LLA status. It fails to satisfy the six criteria set out in CCP 423 to ensure an LLA designation. Finally, the surveying methodology used to establish its claim for an LLA designation was not satisfactory. What more can a landscape architect say?'

NOTES

[1]On Not 'Settling Down', Perensey Ledges, Sussex, 30 May 1990.
[2]*Swale Borough Local Plan*, Deposit Draft Written Statement (Covering The Period 1991–2006), R J Harman, Director of Development Services, Swale Borough Council, 1996, p.48.
[3]*Swale Borough Local Plan*, op. cit., p.46.
[4]'Kent Structure Plan', Third Review: Deposit Plan and Explanatory Memorandum, May '93, Kent County Council pp.42–43.
[5]PPG7, 1992, 1.2 stands for *The Countryside and the Rural Economy*, Department of Environment and Welsh Office: Planning Policy Guidance, 7 January 1992, sect. 1 para. 2; PPG, 1997, 1.5 stands for *The Countryside – Environmental Quality and Economic and Social Development*, Department of the Environment: Planning Policy Guidance 7 (revised), 21 February 1992, sect. 1 para. 5.
[6]*The Kent Thames Gateway Landscape: A Landscape Assessment and Indicative Landscape Strategy* prepared for Kent County Council and the Countryside Commission by Cobham Resource Consultants, Avalon House, Marcham Rd., Abingdon, Oxon OX14 lUG, September 1994 pp.57, 58.
[7]CCP 485, 1996, 4.3 stands for *Conservation Issues in Local Plans*: London: Copyright English Heritage, Countryside Commission, English Nature, 1996, sect. 4, para. 3.
[8]*The Kent Thames Gateway Landscape* op. cit., p.45.
[9]CCP 423, 1993, p.25 stands for *Landscape Assessment Guidance*: Advice from the Countryside Commission prepared by Cobham Resource Consultants, Northampton: Countryside Commission; CCP 423, 1993, p.25 cf. Figure 21
[10]*Landscape Assessment: Principles and Practice*, A Report by Carys Swanwick of Land Use Consultants To the Countryside Commission For Scotland, Kirkcaldy: Inglis Allen 1991
[11]The four-sided leaflet published by English Nature and the Countryside Commission is called 'The Character of England: Landscape, Wildlife and Natural Features' and advertises a publication of that name for December 1996
[12]John D Hunt, *The Figure in the Landscape*, pp.39, 40–41.
[13]As an unknown poet said at the time; John D Hunt, *The Figure in the Landscape*, op. cit., p.189
[14]John D. Hunt, *The Figure in the Landscape*, op. cit.,pp.199, 219, 223.
[15]In what follows use is made, for my own purposes of Brian Stock's 'Reading, Community and a Sense of Place' in *Place: Culture: Representation*, chapt. 16 p.317
[16]K R Olwig 'Sexual Cosmology', chapt. 10 of *Landscape: Politics and Perspectives*, p.310.
[17]John D Hunt, *The Figure in the Landscape*, op. cit., pp.223, 242.
[18]This idea is vigorously defended by Nicholas Green in 'Looking at the Landscape: Class Formation and the Visual' in *The Anthropology of Landscape:*

Perspectives on Place and Space ed. by E Hirsch and H. O'Hanlon, Clarendon Press 1995 pp.31–42; cf. Jean-Jacques Rousseau 'The Confessions' Harmondsworth, Penguin Books 1963.

[19]D Cosgrove and S Daniels, 'Iconography and Landscape', Introd. to *The Iconography of Landscape: Essays on the Symbolic Representation, Design and Use of Past Environments*, p.1.

[20]Introduction to *Creating the Countryside* ed. by E M DuPuis and P Vandergeest, p.18.

[21]Denis Cosgrove, *Social Formation and Symbolic Landscape*.

[22]K R Olwig, 'Sexual Cosmology' in *Landscape: Politics and Perspectives* ed., p.328.

[23]Derek Gregory 'The Crisis of Modernity' in *New Models in Geography*, vol. 2, p.376.

[24]J Thomas, 'The Politics of Vision' in Bender, K R Olwig, 'Sexual Cosmology' op. cit., p.22.

[25]Christopher Tilley, *A Phenomenology of Landscape*, p.24.

[26]J Thomas, 'The Politics of Vision' in Barbara Bender, p.25.

[27]Derek Gregory, 'The Crisis of Modernity', op. cit., pp.376–377.

[28]K R Olwig, 'Sexual Cosmology' op. cit., p.331.

[29]B Bender 'Stonehenge: Contested Landscapes' in Barbara Bender, K R Olwig, 'Sexual Cosmology' op. cit., p.261.

[30]K Thomas, *Man and the Natural World* (1983) Harmondsworth: Penguin; quoted in Bender, Bender 'Stonehenge: Contested Landscapes' op. cit. p.259.

Chapter IX

THE INQUIRY REOPENS: CAN TWO UTILITARIAN ARGUMENTS ENTWINE?

Can consciousness ever be dependent solely on objects alone – an ideal condition which Rousseau unlike Sartre regarded as being in no way detrimental to freedom – so that their existence and the recollection of them are in a complete one-to-one relation? Such aesthetic contemplation, where thinking is frozen and awareness of time excluded, would allow the mind to expand fully to encompass what is presented almost as a gift beyond itself.

On the other hand, for an instrumental consciousness, the more concerns there are to occupy it – whether of a direct or an indirect kind – the greater the potential for confusion. But such confusion is not the result of not thinking clearly nor of simply entertaining self-contradictory ideas. Rather, different propositions subject to laws of thought which intention invokes, develop consequences unanticipated by such intentional activity, to yield just that tension which makes thinking have any point at all.

For such a consciousness one of two outcomes seems possible. Either *the original propositions, accompanied by these seemingly associated ideas, spin in a whirlpool of deeper intensification* or *this apparently mindless roundelay of disconnected thoughts suddenly does conform to some pattern so that their frenetic spin unwinds to yield a calmer, broader flow of ideas wholly integrated within themselves. Yet beneath this newly found harmony, produced by such conscious activity, there lies – constantly pushing upwards when least expected – elements of an earlier, repressed alternative mode of awareness encapsulating elements of feeling awaiting conceptual resolution which have defied so far the categories of any kind of rational thought at all.*

Now 'normal' mental activity – not to be confused with the busyness imposed everyday demands require – seems to operate between these two extremes in scurrying to and fro across the broad lake of consciousness neither attaining a restful, calm state – however much that may be desired – nor achieving a meaningful purposefulness which uninterrupted, free-flowing mental activity will allow.

Meanwhile, the original objects – inspiring either form of consciousness – play their part in nature's activity in an apparent discontinuity from human concerns. The sliding movement of an eighteen inch long grass snake through juicy green stalks, the gentle flapping of a large heron's wings as it rises gradually into the branches of an overhanging willow tree and the bright blue flash of a kingfisher's flight remind consciousness that it can never be at home completely identified with itself alone![1]

Tuesday, 18 November 1997 is the next important day in this narrative. It was the day when the inquiry into the appeal against KCC (Kent County Council)'s decision to refuse planning permission for the development of the Tiptree Site by SEM (South Eastern Minerals) reopened in Iwade Village Hall. The inquiry was to last nearly one hundred and twenty hours to include the nine hours of the previous year. Discussion of the matters generated would ensure several volumes; I need thereby to be very brief and extremely selective in what I draw to your attention. This account, then, can hardly avoid the charge of distortion.

The atmosphere of the inquiry on this first day was so very different from that of the previous year. SEM's QC seemed so much more confident. KCC's representative treated him with greater respect and talk of agreement with the appellant on behalf of the Environment Agency filled the air.[2] The better heated room had been arranged so that whilst the inspector remained on the stage, to his right were – closest to him – the KCC representative, then Swale Borough Council (SBC)'s solicitor and finally the representative for the Tiptree Action Group (TAG), whilst to his left the QC could face his adversaries, with microphone assistance for all! In addition, with the exception of written representations from the RSPB – following on correspondence with its president[3] – and the CPRE – undefended by Kent's planning officer and thereby given less weight during this inquiry's following proceedings[4] – the QC was able to point out that no objections to the development had been forwarded. He cited English Nature, the footpaths officer, the county archaeologist and the Ministry of Agriculture in this context. He made no comment about the RSPB's reservations nor the CPRE's rejection of the proposal. But his strongest card related to an event yet to be narrated to you.

Reconsider, from chapter I, that July evening in 1995 when a meeting took place at Wyvern Hall, Sittingbourne. Residents, you will recall, from all the villages were fierce in their rejection of the proposals to develop the Tiptree Site for clay extraction and waste disposal purposes. Yet a report on 12 September 1995 from the county planning officer to the planning sub-committee, recommended 'that permission be granted subject to conditions to be determined' by himself.[5] His recommendation was overruled by KCC, presumably because of the strength of the protest at that Sittingbourne meeting. In his report, however, he considered the need for clay and though he admitted that clay was required in Kent he didn't really conclude this site was necessary to satisfy that need! None the less he did emphasise the need for void space, though he acknowledged that Swale Borough Council was considering a local landscape designation for the area. You can imagine his embarrassment, later in the inquiry, when quotations from his own report were read back to him by the QC, as he tried to defend KCC's rejection of SEM's application. But his original endorsement may well have been a factor in the QC's quiet confidence now.

The QC's confidence was bolstered too by a ministerial keynote speech made on Thursday, 25 September 1997 by Michael Meacher. The QC kept referring to it in the cross-examination of his own witnesses. That speech, given to the Environmental Services Association's Annual Conference, concerned the place of landfill in the hierarchy of ways in which waste could be disposed. It is now a matter of common parlance, following upon EU priorities,[6] to assume that waste should, firstly, be minimised at source; secondly, reused or recycled where possible; thirdly, used for energy production; fourthly, reduced by incineration or by other engineering processes – such as were under consideration in 1997 during the inquiry by PowerGen at Kemsley – before landfill is even considered. Mr Meacher supported European directives to ensure waste disposal satisfied 'high environmental standards'. The SEM's QC was interested in what he considered to be the minister's back-pedalling on the incineration issue. Consider too, he said, what was implied about the waste disposal industry in the UK being 'actually at the forefront of well-managed landfill', that 'landfill can be a sustainable option for biodegradable waste' and Meacher's interest in 'allowing for landfilling where methane recovery meets high efficiency standards'[7] The QC was pleased to hear too that on Thursday, 27 November the *Government Office for the South East* (GOSE) had frozen the Kent Waste Local Plan,[8] a plan which down-played the need for landfill and which was used by KCC in its rejection of the appellant's case. Incidentally, the plan too which had been approved by an inspector at an earlier inquiry where objections to it by SEM had been overruled! No rationale was provided, at least before Christmas, for GOSE's decision.

In the following three weeks the QC did not have everything his own way. He was always dogged by TAG's representative's question 'Why were there no further boreholes to check clay levels in order to satisfy the requirements of the inspector's preliminary verdict on the 27 November 1996 after the previous year's adjournment: 'borehole testing would be needed to prove the local geology'?[9] The reply came that at the planning stage, four boreholes – generated before the inquiry began a year previously – were perfectly adequate to provide relevant information with respect to the hydrological and geological features of the site. But it was really SBC's solicitor who forced this issue by arguing that more boreholes would be required before the final levels of clay extraction could be agreed. As things stood, there were not sufficient boreholes to satisfy planning permission. SEM's reply that no further borehole testing was required, 'if you are dealing with such a well researched material as London Clay', seemed unduly complacent. The substitution of a Risk Assessment exercise instead, by a Consultant in Environmental Management, seemed, to a lay person at least, an evasion in view of what was found in relation to the four boreholes themselves. These indicated water rising continuously in the boreholes during testing, rumours of a spring originating at the top of the hill amid issues about water running from the fields as surface water onto the road beneath the hill. The Environment Agency had expressed no concern with these matters!

Unlike the KCC's representative, who rushed his questions thereby allowing witnesses to obfuscate important issues, SBC's solicitor was more ruthless and knew how to make use of silences at the end of a witness's answer. So SEM's Noise Evaluation witness confessed to not updating his evidence, despite the fact that a bypass had now removed most of the traffic, whose noise would have affected his earlier results, from the nearby Sheppey Way. The solicitor made at least two further witnesses admit the site could still be used for waste disposal purposes even if, initially, no clay was exported. That led them either in the direction of a 'land raising proposal' or worse, that the inquiry would then be dealing with 'a different animal'! From a second witness he gained the confession that all demands for clay could be met from other sites. Such dramatic moments still resonate in consciousness. When the solicitor rose to speak all present knew that a witness would be taken through a series of questions to some end point which s/he would not necessarily have anticipated. The moment of danger was always structured in this way. Far from looking at the witness directly, his eyes would wander to glance through the hall's windows at movements outside, as if he was hardly interested in the question which would freeze his witness's mind. Of course he did not always obtain what he wanted as an answer. So he had a most memorable phrase delivered with his nose in the air to register his discontent: 'I hear what you say!'

Four issues struck home. The first followed on the SBC's solicitor's question as to whether Biffa Waste Services or SEM controlled land opposite Combray Cottages to ensure the widening of Stickfast Lane. Following a negative reply, the inspector himself asked about land at the junction between Stickfast Lane and Sheppey Way allowing for lorries to turn right onto the A249. It wasn't clear which of SEM's plans referred to this lorry routing nor whether SEM had full control of the necessary land – by means of an agreement with the owner of Stickfast Lane Farm or with the owners of the house 'Balmoral' –

to ensure such routing. The other three occurred during the day allocated for evidence from the residents and TAG's witnesses on Wednesday, 3 December. The first, forwarded by TAG's chairperson, raised the issue as to SEM's status, a new company formed to supply materials to the construction industry! Its directors were those for Kerridgecourt too, but they also had, with their relatives, interests in other companies 'which closed owing money', in fact over half a million to previous creditors. Was this why Biffa, as a waste disposal company, had agreed to overall responsibility for working the site and to take up options on any benefits but not liabilities arising from such companies, leaving SEM simply to hire people and equipment for clay extraction only? Could SEM deliver its responsibilities? If costs were awarded against this 'shell company' at this inquiry, how would they be paid? No member of the SEM company ever appeared to answer such questions at this inquiry.

The second issue arose as a result of Biffa's proposal to transfer discharge of leachate from the proposed landfill site at Funton in a pipe to the junction of the main sewer at The Street and Vicarage Lane in Lower Halstow. As a result, the editor of Upchurch's parish magazine narrated at the inquiry the history of a series of serious drainage problems affecting the sewage outflow from Upchurch into this sewage system in its line en route to the Halstow pumping station for eventual transfer to the Motney Hill disposal site. Extra demand would entail upgrading the whole system to meet the new requirements from Tiptree.

The third issue concerned clay need. A representative on behalf of Wards Housebuilders pointed out that considerable quantities of clay were available as a result of building developments in the Swale area over the next twelve years in digging foundations, whilst providing drains, roads and ponds for housing developments. Moreover, his company at present, in building homes in Iwade, was having to pay to have that extracted clay dumped because no one wanted it. It finished up in a landfill site for the most part! The discussion of this issue, back to back with that of waste disposal, dogged much of the time spent at this inquiry's proceedings. Whilst his three adversaries could plague him with the fact – established by the inspector's report on the

Kent Minerals Local Plan Inquiry which took place during the summer of 1997 – that there was no clay need, the QC could mock their attempts to sideline the shortage of void space for waste in Kent. The QC may also have been encouraged by the following words in the inspector's preliminary verdict: 'It is not clear that – as KCC argues – this is now changed from being a clay-led to a predominantly waste-led application. It might always have been waste-led but nobody checked'.[10]

It was left to SBC's solicitor to clarify things in interviewing SEM's chief witness – overruling the QC's attempt to interrupt – on Thursday afternoon, 4 December. He made this witness face the issue as to whether the justification for a mineral extraction has to stand on its own grounds *not* on the basis of a need for void space. Three previous planning decisions – at Morton, Wirral; Haines Hill, Berkshire; Little Waltham, Essex – proved relevant here.[11] And as for shortages in the availability of void space for waste, in crisis situations Shelford Quarry, Canterbury – a landfill site recently granted an extension to Brent Waste Management – could be used. Against this the SEM witness countered by pointing out that short-term uses would undermine long-term availability. The SBC solicitor snapped back that, besides waste disposal facilities being available at Stangate, near Sevenoaks and Offham near West Malling, there were landfill sites at Shakespeare Farm on the Isle of Grain and at Brambledown, Isle of Sheppey and a Digestor at Ashford. Incineration sites were possible too, at Kingsnorth and Halling besides a Waste to Energy plant at Long Reach, Belvedere near Bexley. He then delivered a News Release on a Waste Processing Plant promised for Allington Quarry planned by Kent Enviropower. One went home that night with a sense that the inquiry – with the destruction of the Clay Need and the Waste Disposal arguments – was all over bar the shouting. But that judgement did not allow for further moves by the QC!

Such optimism was flattened on the last day before the Christmas adjournment. SEM's chief witness said the Belvedere project had been withdrawn by PowerGen. The year 2002 would be the year when the contract to pass waste from Kent to Chelmsford in Essex would expire; could it be extended to 2009?

An incinerator proposal for Richborough could only be sustained if a constant supply of the right kind of waste could be ensured over a twenty-year period. Kingsnorth, too, would require dedicated long-term contractors in a volatile market situation. As for the Allington incinerator, that was only being proposed to replace the Halling project, where 200,000 tons of household waste would have supplied electricity for 35,000 homes. The latter proposal had been withdrawn following the Secretary of State's calling in of this planning application and the one for Kingsnorth Power Station because of its proposal to disperse, on land, residues from this waste to energy plant at Hoo.[12]

In the subsequent cross-examination of his chief witness, the QC claimed that two of the cases – cited to illustrate that waste led applications could not drive clay extraction – were actually in green belt areas and thereby inappropriate examples.[13] At Belstead Farm it was judged that the waste need was insufficient. Through his cross-examination the QC was able to strike in five ways to sustain SEM's case: firstly, with London and the South East Regional Planning Conference (SERPLAN) 1992 report it is said that 'Within the region, county self-sufficiency for all wastes, other than those requiring specialised facilities of a sub-regional nature, should be sought'. In addition, *Policy Guidance 5* states that 'In preparing their waste local plans, counties should aim to provide for the disposal of their own waste arisings and to make an appropriate contribution to regional needs'. These statements are underlined in a later report RPC 2700 of December 1994. Secondly, this self-sufficiency principle provides underpinning for the proximity principle 'whereby waste should be disposed of (or otherwise managed) close to the point at which it is generated', rather than, for example, have vehicles carrying waste by road from the Medway towns to Essex! Thirdly, SERPLAN requires that London exports some of its waste to the *rest of the South East* (ROSE).[14] And in a recent London Planning Advisory Committee document, a moratorium on new incinerators had been placed upon the London Boroughs until the year 2002.[15] Fourthly, as the chief witness emphasised, there will always be a need for landfill since waste from incinerators has to be treated somewhere! Finally, the permission to extend the landfill at Shelford Quarry,

Canterbury is conditioned by the remark that 'no clay should be permitted to be exported from the site, unless a specific and acceptable use for it can be established.[16] This was hardly a pleasing 'Christmas' card for the TAG members, even if the following comforting words could have been written inside it by the SBC's solicitor: 'Since not all of Kent's waste can be handled within Kent, either it is "exported" outside the county to meet a short-term requirement or the environmental consequences of developing Tiptree will be incurred, providing yet another landfill site against EEC directives!'

On Tuesday, 27 January the inquiry resumed. The remainder of the representations from local residents were heard and an argument about bird statistics continued which had started before Christmas between TAG's expert witness – whose empirical findings inform chapter VI – and SEM's expert witness on Ornithology and Nature Conservation who had produced a major text on methodology in the former field. Right at the start, in response to a question, he had said in relation to concerns about pollution: 'I can't comment. You'll appreciate I have to be careful!' He compared and defended Ecoscope's scientific approach with that of TAG's expert witness pointing out that although the latter's methods had more constraints than Ecoscope's, the figures for bird recordings were similar. They differed chiefly on the issue of the significance to be attached to bird movements after sunset! KCC's ecology expert, however, scored in at least two ways. She claimed that golden plover and lapwing numbers had been underestimated because a proper assessment would have required five years. The QC's reply was: 'We live I'm afraid in a flawed world!' All that could be done was to do the best thing one could over two years and take on TAG expert's findings as a comparison! Her second hit was in relation to a photograph taken of seagulls over Shakespeare Farm. Despite SEM's expert witness's attempt to parry this attack by referring to the possibility of 'heavy weather' occurring on the day the picture was taken, she pointed out that the Environment Agency had asked that the issue of bird control at that site be re-examined. No new measures had been initiated at this site as a result, no evidence had been produced about the possibility of using effective measures from elsewhere and SEM's expert witness hadn't provided much in the way of gull migration studies generally!

In the New Year non-SEM witnesses were interviewed. The KCC solicitor tried to regain ground on the waste disposal issue. Good use was made by KCC's geological witness of an argument first cashed by the chairman of the Lower Halstow Parish Council and leading TAG witness: in response to the idea that a landfill site was needed to take pollutants from brown sites in the South East of England to assist housing projects in such areas, the question was put: 'Are we going, then, to create another brown site to do this?' But the waste disposal issue just wouldn't go away. The QC struck at KCC's chief expert witness, not only by reminding him of his decision in favour of developing the site on 12 September 1995, but by suggesting that even if Allington and Kingsnorth incinerators came on line to absorb 3/4 million tons of household waste, say, it was doubtful whether there would be a *sufficient supply of high-calorific waste* to encourage other incinerators to be built! Secondly, an incinerator needs waste of *a high calorific value to burn*. The more the high calorific material was taken by other incinerators, further down would go the calorific value of waste for other plants so there would be no place for a third incinerator say in the Richborough area. KCC's expert witness snapped back in two ways: these arguments had been put at the waste disposal inquiry and hadn't impressed that inspector whilst there was a great deal of high calorific waste in Kent. The QC had two replies: SERPLAN had made it plain that *the high calorific type of waste must be reduced* through 'greater levels of waste minimisation; re-use and recycling'[17] and even if all these KCC proposals for the future were taken together, there would still be a large cumulative deficit in void space capacity, an issue re-emphasised by the inspector at the recent Rosecourt Farm inquiry! Moreover, at that inquiry, the inspector had recognised that their own application had been waste led and had concluded: '...if the need for waste disposal overrides other objections to the scheme, the absence of a need for clay should not reverse that overall conclusion'![18]

The slight edge such arguments gave to the QC was sharpened, oddly, by KCC's landscape witness. Not only had he not revised his evidence – referred to in chapter VII – since 1996, but failed to make use of the appropriate literature in the field or the documents relevant to it but seemed, too, to be unaware of arguments forwarded at the Local Landscape Designation Inquiry! He confessed too that the site was relatively 'unremarkable'. He thereby agreed with SEM's landscape witness's landscape point of view, whose proof of evidence will be examined subsequently. The QC was delighted to receive this confession, though he couldn't believe he was hearing it! Indeed this KCC witness even admitted that SEM's proposed attempts to overcome the operational effects by landscaping the site in the future were 'not inappropriate', whereas other KCC and TAG witnesses had questioned them.

Some ground was regained. Firstly, the SBC solicitor cast a number of doubts on the exact legal status of the agreements drawn up between the different parties constituting the appellant's interest. Secondly, the most powerful witness of the inquiry – appearing on behalf of SBC – was very impressive once more and the QC kept his cross-examination of him short, since he had learnt from their skirmish in the Land Designation Inquiry, detailed in chapter VIII. This witness forwarded the prospect of an incinerator site to dispose of waste at Ridham/Kemsley pointing out that it was able to take in waste normally doomed to landfill disposal. Details of this scheme had been reported in the *East Kent Gazette* at the time. The incinerator produced, besides the flue gas, only ash which has a high value 'with a number of markets already established'.[19] This witness also cast serious doubt on 'the level of noise argument' put by the other side, in view of the fact that his evidence – supplied by SBC studies done during December 1997 – established Tiptree as a site in a quiet rural area! He also managed to undermine the QC's claims about the *self-sufficiency* and *proximity* principles in respect of waste. With regard to the former, the focus is in relation to the South East as a whole: 'The Secretary of State endorses SERPLAN's objective that the South East should aim to make adequate provision within the Region for the disposal of its own waste.' It is also recognised by SERPLAN that 'there will inevitably be some movement across Shire boundaries' and that regional counties at the fringe may receive 'imported waste from outside the South East' since it was possible 'for counties to jointly provide and use facilities'. With regard to the proximity principle, it, like the self-sufficiency principle, was an ideal too, since 'it will be necessary to make provision for some cross boundary movement of waste, particularly when counties have a landfill deficit'.[20] In addition, SBC's expert witness argued that if a further landfill site was created, this would undo the incentive to create further incineration techniques and the latter were to be preferred to the former in terms of the hierarchy of priorities for dealing with waste. Thirdly, and finally, continual correspondence and rebuttals in relation to the *Gabriel* editor's evidence ensured that the leachate and the associated sewage issues continued to float.

The site was to be viewed by the inspector on 17 and 18 February where he would be accompanied by the interested parties. Final speeches from each of the legal representatives would be made on Thursday, 19 February 1998.

This chapter began, as did chapter VIII, with an *aestival – an expression whose meaning can unfold on further examination – and both aestivals concern the problematic status of the idea of relating to nature, an ecocentric concern. In each case what followed were detailed arguments presented at both inquiries on behalf of anthropocentric interests. Indeed, in the last chapter, a rationale was provided for just such an intent. Again, at the second inquiry much use was made once more of the Countryside Commission's 1993 document *Landscape Assessment Guidance* (CCP 423) where those six criteria for evaluating landscape for any kind of designation were once more referred to: Landscape as a resource, scenic quality, unspoilt character, sense of place, conservation interests and consensus.[21] As a result of reading chapter VIII, you will know that the first criterion does not refer to use value in providing materials, agriculture or void space capacity! Rather it refers to its rarity or its being representative of local scenery. Consensus refers to professional and/or public agreement reflected in writings or paintings about it as a landscape. It was the discussion about the second criterion at this inquiry which proved interesting. The meaning of the phrase scenic quality is given as: *It should be of high scenic quality, with pleasing patterns and combinations of landscape features, and important aesthetic or intangible factors.*

The SEM's landscape witness, after defining landscape as 'a resource', admitted to TAG's representative that he could not think of a reason for not saying Tiptree Hill satisfied the criterion scenic quality. Under cross-examination by his own QC, he said he could not think of a reason for granting it that designation. Members of the public present in the hall thought he was, if not deceitful, then at least evasive. That opinion could be mistaken and to demonstrate why that might be so, a possible rationale to make sense of his position was included in chapter VIII. He tried to answer questions within the authority of his own discipline.

In defending my proof of evidence, I set out the arguments for and against awarding an LLA designation – aired at the earlier inquiry – which can be found in the first part of chapter VIII, and then continued by suggesting why SEM's landscape witness was forced to engage in a language game[22] – to use Wittgenstein's term – which was *incommensurable with – not related to – that used by TAG's objectors to SEM's case. His discourse was entirely quantitative in nature; he tried to deal with landscape entirely in descriptive terms, as he himself put it! His interest and manner of inquiry into these matters, as expressed in his cross-examination, was driven by an *ideology – made clear in chapter VIII's Rationale – that is manifested in the very documents he used to support his case: *The Kent Thames Gateway Landscape* document of 1994, *The Landscape Assessment Guidance* of 1993 (CCP 423 from now on) and *Conservation Issues in Local Plans* 1996 (CCP 485 from now on).[23] These documents treat landscape purely physically, entirely in geographic terms. SEM's witness was concerned with topological categories in relation to making descriptive statements about a particular landscape's character. The problem with this approach, philosophically, is how could he arrive at any kind of normative judgement, any kind of evaluation in relation to what he tried to describe? No matter how widely his approach may be used nationally, the decision about the quality of a particular landscape simply appears ungrounded, arbitrary: no number of *descriptive statements can yield a *normative one, which is just what an inspector or a government minister requires to guide decision-making!

The many issues arising from all this can be suspended for the moment since, before discussing them, one question must be settled first. Consider my claim that the rationale credited previously to SEM's landscape witness makes sense of the discipline's authority which he employed. Can that claim be defended? Let us then consider the veracity of that rationale attributed to him. To do this means returning to the documents he used or could have used in the two inquiries to sustain his case. Note, firstly, the significance of his dependence upon the idea of the specular image. This comes out in the way landscape is defined in CCP 423: 'The term *landscape* refers primarily to the visual appearance of the land, including its shape, form and colours' (CCP 423 p.4). That definition can also be found in the earlier Countryside Commission's *Assessment and Conservation of landscape character. The Warwickshire Landscapes Project* (CCP 332 from now on): 'landscape simply means the appearance of the land' (CCP 332 p.3). This definition is reiterated in the Countryside Commission's Technical Report *Environment Assessment*[24] (CCP 326 from now on), though it is admitted to be 'its narrow definition', but the mind boggles at the prospect that 'natural beauty is a larger concept than landscape' (CCP 326 p.12), How can the appearance of the land be separated from an appreciation of its natural beauty? Perhaps in order to avoid such confusions, this document does offer its reader a broader interpretation regarding landscape 'as a resource that encompasses far more than mere appearance of the land' recognising 'the wide range of physical, biological and social factors that are instrumental in shaping the landscape and our reactions to it' (CCP 326 p.11). The earlier *Warwickshire Landscape Project* provides the rationale: Britain's landscape has been formed by the interaction of the natural world and human activities over many centuries' (CCP 332 p.3). Indeed this last move provides the justification for adopting the definition of landscape offered in chapter VII, elucidated by an exploration of Peirce's categories: if landscape is not to be regarded merely physically – that is as environment – but rather defined as a human perspective upon a physical environment capable, in being appreciated panoramically, of yielding an aesthetic experience, then we would have three elements to consider in landscape discourse rather than just one. Such a definition stresses cultural concerns and the idea of seeing a perspective on something as a kind of picture, as well as seeing that thing physically, capable of scientific investigation for landscape assessment purposes!

That broader view is found in CCP 423: 'the landscape is not a purely visual phenomenon, because its character relies closely on its physiography and its history' (CCP 423 p.4). A clue is given as to the source from which this approach is derived in the Countryside Commission's landscape assessment *The Cotswold Landscape* (CCP 294 from now on)[25] in its attempt to define 'natural beauty': 'We have been guided by the definition of natural beauty given in the Wild Life and Countryside Act 1981[26] and by the Countryside Commission's own interpretation of natural beauty.[27] We conclude, that, although the emphasis should be on visual quality, geology, topography, flora and fauna, historical and cultural aspects are also relevant' (CCP 294 p.3). Yet the tussle between the earlier more restricted view of landscape as something visual versus the latter, broader interpretation continues when we are told there is a growing emphasis 'on the assessment of landscape character as opposed to landscape quality' though 'there is still a place for attention to landscape quality' (CCP 423 p.5).

The stages involved in landscape assessment clarify how this tussle is to be resolved: the first stage involves 'description of the objective elements of the landscape'; the second, the 'classification of the landscape into broadly homogeneous units or areas'; the third 'consideration of the subjective values associated with the landscape' (CCP 326 p.21). Indeed throughout the Countryside Commission's documents we find that, firstly, a landscape's physical characteristics are addressed, then its historical developments – buildings, land use, archaeology, past history – and only as a third issue its – aesthetic features! So we move from 'objective' factors to merely 'subjective' ones. Again the earlier *Warwickshire Landscape Project* gives the justification for this approach: '*Objective* studies are concerned primarily with aspects of the landscape itself – the 'object' of the viewer – and are often used to measure and quantify the various components in a landscape. *Subjective studies are more concerned with people's response to the landscape – the subject doing the viewing – and were particularly common in the eighteenth century when a great variety of written descriptions of landscape appeared' (CCP 332 p.5). The controlling function of the specular image emerges when we are given a list of tools which make the objective studies possible: 'maps, plans or aerial photographs indicating the landscape character of the proposed project site and its surroundings. Important views at ground level may also be used for illustration. These might, for example, take the form of photographic panoramas or annotated sketches'. The former were employed by SEM's landscape witness. In addition what are called 'zones of visibility' 'can be determined and plotted either manually or using computers' (CCP 326 p.22).

A closer examination of the Advisory *Booklet Landscape Assessment Guidance* (CCP 423) shows why a purely objective study approach breaks down. Firstly, how are 'important views at ground level' to be selected? How are sites for taking 'photographic panoramas' or making 'annotated sketches' to be chosen? Here is the answer: 'Good potential viewpoints many include ridgelines and river crossing points because they are sited respectively at the edge and at the centre of the 'visual envelope' (CCP 423 p.9). But in that case value decisions about what counts as good as a potential viewpoint have to be agreed before descriptive assessments can even begin to be made!

More seriously, the distinction the model requires between a *landscape character* and a *landscape characteristic* can't be sustained throughout the document. The former is defined as 'a distinct pattern or combination of elements that occurs consistently in a particular landscape'. An element is described *factually too as 'a component part of the landscape (e.g. roads, hedges, woods)'. But a characteristic is rendered *normatively: 'An element that contributes to local distinctiveness (e.g. narrow winding lane; ancient hedgerows on banks; vernacular building style)'. At the Tiptree Inquiry, for example, there was no agreement on the status of the issue of lanes any more than there was concerning the value of the hedgerows; still less about what the view of the vernacular building – Lower Halstow church – would encompass when viewed from the estuary with Tiptree Hill behind it! Indeed even in terms of landscape character, there was disagreement expressed at the Landscape Designation Inquiry and repeated at this one. SEM claimed Tiptree Hill was 'undulating, open, arable farmland' lacking anything distinctive, whilst SBC's expert witness regarded the site as the continuation of a prominent ridgeline; a unique landscape feature!

In respect of the document *Landscape Assessment Guidance*, however, in discussing landcover analysis, the term 'landscape character' is clearly meant to cover such labels as cropland, grassland and developed land, whereas the claim is made rather on behalf of landscape characteristics: 'After the initial coding parcels with similar characteristics are grouped together and coloured up' (CCP 423 p.13). Reference to the *Warwickshire Landscape Project* illustrates how this can be done (CCP 332 p.15). Indeed what is said next could hardly apply to the notion of grouping areas with similar characteristics rather than character: 'A degree of skill and practice in air photograph interpretation and in use of the coding system is required, but after a few days' training, the work can be carried out by a technician' (CCP 423 p.13). Later, in discussing the benefits of computer-aided landscape classification, we are told that this technique can also 'help to highlight the principal characteristics that distinguish one landscape type from another' (CCP 423 p.17) where presumably landscape characters again are meant! Later, the distinction between the two – character and characteristics – is drawn in line with the earlier definitions, but it is made clear that the former in relation to 'visual character, geology, landform, land use, and buildings and settlements' is required first and the latter only subsequently (CCP 423 p.33).

Emphasis is placed upon budget restrictions (CCP 423 pp.29, 31) and doing a quick landscape assessment. Importance is given to geological/ordnance survey maps yielding an overlay analysis to yield such character areas. Hence the importance of training people in such a way as to 'focus on ways of summarising and presenting information on the spatial patterns concentrations of visible features that often offer such vital clues to the present-day character of the landscape' (CCP 423 p.44). No training seems to be recommended, however, in relation to exploring the reasons why a landscape 'is valued by local residents and visitors and the ways in which landscape is perceived to be changing' though this special clause seems to apply more particularly to Areas of Outstanding Natural Beauty (CCP 423 p.39).

That last remark might be somewhat mean. After all valuations concern the activity of landscape evaluation rather than landscape assessment *per se*, and in respect to the latter 'a classification will be the main output, and there will be no need to evaluate the quality of the landscape' (CCP 423 p.23). In relation to the former, as we have already seen, the six criteria arise with special emphasis placed upon scenic quality. Yet the distinction between the descriptive and the normative or evaluative, thereby separating the concerns of landscape assessment from those of landscape evaluation, is not completely sustained throughout the Countryside Commission's documents. So we read in the Technical Report *Environmental Assessment* that landscape values 'are inextricably bound up with our perception of the landscape and are of particular importance in landscape assessment' (CCP 326 p.28).[28]

The first section of this chapter focused on the struggle to wind two utilitarian arguments together: the clay need and landfill arguments. By the end of the inquiry two other arguments – anthropocentric in character – were still vying with each other: the case for assessing landscape as a scientific activity and the case for designating it a value through some kind of landscape evaluation. In neither case were these struggles resolved, in the first case because no clay need could be demonstrated; in the second case because of the *bifurcation* between facts and values.

In dealing with SEM's landscape architect consultant's attempt to face the latter issue, the rationale offered for his position in chapter VIII was seen to make sense of the documents he used or could have used to defend his position at the inquiry. It has been shown how the Countryside Commission's documents are anthropocentrically, even utilitarian centred, in defining landscape as a resource. Even the aesthetic is so treated: landscape as human habitat 'holds a special meaning for many people as the source of numerous experiences and memories. Many of these are visual, but at times the landscape may also evoke other sensual, cultural and even spiritual resources' (CCP 423 p.4). In places in the Countryside Commission's documents we saw the historical dimension used to underpin the discourse of landscape anthropocentrically. The idea of landscape as a cultural image was forwarded in focusing upon the narrower definition of landscape as visual appearance or in the broader definition's assumption that it is formed through human activity upon the land. But in both cases 'an outsider' view was presupposed with the notion of the specular image well to the fore, reinforced by the tools employed to make the activity of landscape assessment viable. Inquiry at a distance makes the idea of objective descriptions possible whilst lending credence to the idea that aesthetic considerations are merely subjective. The former, the objective, is biassed towards the quantifiable, the assessable as opposed to the evaluative so giving rise to the bifurcation between facts and values. Finally, both the objective and subjective place the scientific and the aesthetic as ways of regarding nature above the instrumental, the sense of being related to or working on and with the land.

At least three philosophic problems arise from showing how the Rationale for SEM's landscape architect consultant's stance can be sustained within the Countryside Commission's documents: firstly, can it be shown that his discourse is incommensurable with that of the TAG residents; secondly, why can't a number of descriptive statements sustain a value judgement and thirdly, how did the six criteria for evaluating landscape arise and how can they be justified?

NOTES

[1]Conscious Flood-Waters, River Avon, Bradford-on-Avon, September 1992.

[2]Letter to Biffa Waste Services Ltd, from the regional solicitor for the Environment Agency, 24 November 1997, Ref: JLP/EP/5/1/119.

[3]My correspondence, based on the contents of this book's chapt. VI, was dated 2 and 10 October. The President's reply of 29 October 1997 included his conservation officer's formal written representation to the inquiry, dated October 1997.

[4]Proposed Clay Quarry, Restoration by Waste Disposal, and New Access Road, Tiptree Farm, Funton, Proof of evidence to be given at the public inquiry by the council for the Protection of Rural England's Planning Officer for the Kent Branch, Coldharbour Farm, Wye, Ashford, Kent TN25 5DE, November 1997. At the Dargate Landfill inquiry on Wednesday, 21 October 1998, its inspector pointed out publicly that written submissions, unlike oral statements, are regarded with less significance because they have not been 'tested' through cross examination.

[5]Report of Kent county planning officer to the Planning Sub-Committee, 12 September 1995, subject 'To Extend an Existing Quarry, Restoration to Agriculture by Infilling, Provision of Site Access Roads and Improvements to Local Roads, Tiptree Farm, Funton, Lower Halstow – South Eastern Minerals'.

[6]The London and South East Regional Planning Conference SERPLAN, *Developing the Waste Planning Guidelines, Advice on Planning for Waste Reduction, Treatment and Disposal* in the South East 1994–2005, December 1994, RPC 2700 p.7 para. 3.2.

[7]In response to enquiries, a spokesperson for the Government Office for the South East (GOSE) informed me as follows: 'Mr Meacher has made it clear on a number of occasions that his preference was to maximise recycling to divert waste from landfill; and that while he recognised the contribution that energy from waste plants could make to more sustainable waste management, he did not believe that they were invariably more sustainable than well-managed landfill.

Mr Meacher had been concerned by the European Commission's draft directive on landfill, which included unrealistic targets for reducing the landfill of biodegradable municipal waste; his view was that targets would effectively require local authorities to build incinerators, regardless of whether that was the Best Practicable Environmental Option for their area. Following discussion in the European Council, a more realistic set of targets has now been adopted in the draft directive, giving a greater scope for maximising recycling than the commissioner's draft. The directive is now due to be considered by the European Parliament'. Christine Kavanagh for the Kent Area Team of the Government Office for the South East, 30 April 1998.

[8]J Vaughan, Director for Kent of GOSE acting under the Town and Country Planning Acts of 1990 (sect. 44 (1)) and 1991 (Reg. 23) of the Development Plan.

[9]Ruling 1 of doc. CD1 of the inspector at the appeal by South Eastern Minerals Ltd; Site at Tiptree Farm, School Lane, Funton, Sittingbourne, Kent 18 November 1996, para. 13 sect. 1.

[10]Ruling 1 of Doc. CD1, op. cit., summary, 14d.

[11]At an inquiry in the summer of 1988 a proposal was rejected since to excavate 'clay reserves with the primary intention of engineering the landfill capacity is an inappropriate use of mineral reserves' on land at Carr Lane, Moreton Wirral (Wirral Borough Council (AM/JD/PD/A5/94F5 and PD/E4/206) versus North Wirral Brickworks 'Conclusions on Planning Merit' p.58 sect. 15. 4 (2)); the inspector for an appeal at Park Farm, Haines Hill, Berkshire decided on 19 November 1991 that it 'would be wrong if mineral working were to be justified on the basis of it being an enabling action so as to allow waste disposal operations' (T/APP/U0300/A90/169501/P4), Application no. 335626, p.31 sect. 12; on 4

January 1996 the inspector for an appeal to allow the extraction of sand and gravel and the subsequent infilling and restoration to approve golf course on land at Belstead Farm, Little Walthan, Essex concluded in the light of 'the thrust of Government policy' 'that the extraction of minerals must be justified on their own account, and not on the strength of the provision of void space to add to landfill capacity' ((APP/L1500/A/94/246849) p.26, sect. 9.33). See too the judgement of 17 January 1991 for an appeal to allow the extraction of peat, backfilling with inert waste material and restoration to agriculture at New Road, Bagshot Surrey where the inspector said 'I do not consider that the possibility of meeting the requirements for the disposal of waste amounts to a need to extract the mineral in the first place' ((Ref. T/APP/B3600/A/90/146629/P5) p.6 para. 25).

[12]See note number 7, above.

[13]This claim is in fact not true! In the Wirral case the inspector in fact ruled that in relation to Green Belt, 'I do not consider that objection would be sustained against a cogent need for landfill capacity in this locality' see 10 above (AM/JD PD/A5/945 and PD/E4/206) sect. 15 p.57 (281). In the Haines Hill case the inspector commented '...in the light of the policies it is clear that in principle there is no strong or overriding objection to mineral extraction in the Green Belt' see 10 above (App. no. 335626 sect. 9 p.30). The inspector remarked in the Bagshot case that mineral workings 'need not be inappropriate, or specifically incompatible with Green Belt objectives where they can meet certain criteria.' see 10 above (Ref. T/APP/B3600/A/90/146629/P5 p.1 para. 4 cf. p.9 para. 47).

[14]SERPLAN: *Waste Planning Guidance*, RPC 2266 *Waste: Its Reduction, Re-use and Disposal* (1992) p.4 para. 15 and p.3; *SERPLAN: Developing the Waste Planning Guidelines* – Advice on Planning for Waste Reduction, Treatment and Disposal in the South East 1994–2005 RPC 2700 Dec. (1994) p.5, 2.2 and p.6, 2.7 (v). For the proximity principle see RPC 2266 (1992) p.4 sect. 13 and RPC 2700 (1994) p.4, 5.3 (iii) and p.17 6.17. For ROSE see RPC 2266 (1992) p.4 para. 14 and RPC 2700 (1994) p.2, 1.9.

[15]London Planning Advisory Committee (LPAC) *Draft Supplementary Planning Advice on Waste*, July 1987, policy no. W10, p.17. This Planning Advice document was agreed during February 1998.

[16]The Planning Officer for KCC to the Planning Sub–Committee: Proposed Extension to Existing Landfill Operation Incorporating Materials Recovery Facility and Restoration to Public Amenity, Nature and Conservation, Woodland and Agriculture, Shelford Quarry, Canterbury, Brett Waste Management Ltd (MR 160 606)(CA/96/794) 1 Aug.1996 sect. 6.

[17]See note number 14, above, RPC 2700, p.4 para. 1.16.

[18]Sect. 9: inspector's Conclusions Appeal by BFI Ltd; Restoration of Mineral Workings by Landfill, extraction of Minerals and Demolition of Farmhouses and Outbuildings at Rosecourt Farm, Isle of Grain, Kent, 30 September 1997, sects. 26, 35.

[19]Benefits of a Combustion Plant forwarded to me by PowerGen in correspondence dated 27 March 1998 with the Team Head, Environ. Planning (Refs\:1998\jobs\grovehur\10327nrarlw) The proposed new combustion facility – as detailed in the documented Recycling: Improving Environmental Standards

seems to imply that it is now technologically feasible to handle all residues normally forwarded to landfill!

[20]The London and South East Regional Planning Conference, op. cit., (see note six above), RPC 2700 p.7 para. 3,4; p.13 sect. 5.9; p.10 para. 4.10 and p.17 para. 6.17.

[21]*Landscape Assessment Guidance* 1993 (CCP 423): Advice from the Countryside Commission prepared by Cobham Resource Consultants, Northampton, p.25 cf. Figure 21

[22]Ludwig Wittgenstein *Philosophical Investigations*, especially sects. 7, 47, 53, 77, 486 and pp.216, 225.

[23]*The Kent Thames Gateway Landscape: A Landscape Assessment and Indicative Landscape Strategy* prepared for Kent County Council and the Countryside Commission by Cobham Resource Consultants, Avalon House, Marcham Rd, Abingdon Oxon. OX14 1UG, September 1994; Landscape Assessment Guidance: Advice from the Countryside Commission prepared by Cobham Resource Consultants, Northampton: Countryside Commission (1993): CCP 423; Conservation Issues in Local Plans: London: Copyright English Heritage, Countryside Commission, English Nature (1996), CCP 485.

[24]*Assessment and Conservation of Landscape Character: The Warwickshire Landscapes Project* Countryside Commission (1991), CCP 332; Environmental Assessment Countryside Commission (1991) CCP 326.

[25]*The Cotswold Landscape*, A landscape assessment of an area of Outstanding Natural Beauty prepared for the Countryside Commission by Cobham Resource Consultants; Countryside Commission: John Dower House, Manchester (1990) CCP 294.

[26]Wildlife and Countryside Act 1981; sect. 53 (2), HMSO 1981.

[27]*Landscape Assessment: A countryside Commission Approach* (1987) CCD 18, Countryside Commission, Cheltenham.

[28]This denial of the bifurcation of facts and values is even sustained with the document *Landscape Assessment Guidance*: 'Meetings and consultations with people such as the county archaeologist, local historians and conservation officers therefore can be of special value when reviewing the physical and historical influences upon the landscape and when seeking to identify and understand particular patterns and concentrations of landscape features' (CCP 423 p.23). The same view is forwarded in the Warwickshire Landscape Project though more ambiguously on the place of such opinions with regard to landscape assessment or landscape evaluation: 'professional judgement is an essential component in the process of looking at and evaluating landscape'(CCP 332 p.5).

Chapter X

FINAL SPEECH DAY AT THE TIPTREE INQUIRY: UTILITARIAN TRIUMPH?

Do our intentions betray who we are historically? The more we seek to hide them from ourselves, the more they shine before us illuminating the requisite strategies and necessary objects we need to realise their fruition. Other possibilities – incited by ambiguity – are brushed aside, left to languish as mere flotsam to bob up and down meaninglessly on consciousness's sea-surface, whilst purposeful action stretches, like a black crow's neck, forwards towards what has so far succeeded in slipping beyond the domain of any creature's desire to control.

Can a sense of sheer openness to the smile of another human being or to nature's signs yet transform the incessant trod-plod of such means-to-ends predictability? These human or natural expressions, however, prove to be only fleeting, for just when the promise of some intimate communication hovers so as to release that special non-subjectivity either in the human subject – intent on actualising the content of this sudden unintentional contact – or in some natural object, such a possibility sails from view. What endures is that promise which, like the bright illumination of pink sunlight reflected on the sea at sunset, remains affixated merely in memory, no longer fully in view![1]

The inspector cracked a joke before the final speech day began on Thursday, 19 February. He spoke of his being 'impressed' by the sight of cyclists riding over the hill. He had enjoyed beautiful weather for his visit to the Tiptree Site and its surroundings a week earlier, so much so that even I had to suppress a desire to cycle around the hill for fear of being recognised. Rumour has it that just as the group of cyclists passed him and his cortège, a shout rose in the air: 'Say "No" to the proposed Tiptree development' or some such rallying call. As the QC put it later:

the inspector would have to draw his own conclusions!

TAG's representative started the day of the final speeches in fighting form. She showed how more than half of the clay to be extracted from Tiptree Hill would be used for reprofiling and engineering purposes arguing that, if the clay couldn't be sold, the hill would finish up a different shape. Along with the clay need and waste disposal issues, the design of the final top proposed for the hill had dogged SEM's case, as had the issue of the sight of the workings during the processes involved in constructing it. So SEM's landscape architect, for example, on Wednesday, 27 November, had avoided the suggestion that unexported clay would be seen. On Thursday, 4 December, such issues once more reared their head when comparisons were drawn between this application and that of Rosecourt Farm, Isle of Grain. Both applications involved clay extraction and could be regarded as waste-led, as the inspector concluded Rosecourt Farm was. Both were by the sea, both had similar conservation interests and neither offended against any particular agricultural constraints. But did either involve land-raising? It may have been so at Rosecourt Farm since there the inspector seemed to find it necessary to say that land raising was involved using the definition proposed at the Kent Waste Local Plan: 'disposal of waste to create higher ground levels'.[2] Whereas the SERPLAN definition makes reference to land raising in terms of what is necessary for doming,[3] the inspector judged that at Rosecourt Farm 'the quantity of fill proposed goes beyond what is necessary for that purpose alone'.[4] TAG's representative argued that this was also true for Tiptree Hill despite the fact that matters were complicated by not being able to tell from old national survey maps what the exact hill height was before clay had been extracted for the emergency flood conditions of 1953.

In addition, the capped methane gas burners would be visible during the process of working the site, especially at night. During the day, the use of machinery and vehicles would have a severe impact on the solitude and peace gained in this area, affecting especially the Saxon Way – a long distance footpath, passing through the appeal site – to say nothing of the noise, dust, litter and vehicle movements undoing the quiet, undeveloped rural character of this area. In that sense the photo-montages of SEM's landscape architect – concerned as they were with the site before and after development – were 'sanitised' in that they omitted reference to the many years during which an ugly impact would affect the area. So the main thrust of her case was carried by her claim that it would have to be demonstrated that the requirements for clay and landfill must be met by this site given '…a significant impact on amenities through noise, dust, heavy vehicles and litter' in this 'quiet, underdeveloped rural' area. These proposals would affect not only this site but also other affected villages over the duration of the development.[5] Even SEM's expert witness had accepted that there would be an impact upon the Callum Park Riding Centre – as its owner, supported by members of the British Horse Society, claimed – a centre only ten metres from the application site! Moreover, if the appellants – South Eastern Minerals (SEM) – did win, Callum Hill too could then be regarded as an 'appropriate site' for further clay extraction and landfill! How she hammered the point that SEM had not undertaken further borehole investigations despite the inspector's ruling over fifteen months ago to do so.

At this point the SEM's QC groaned, dropped his arms to his side as he slumped in his chair. He never really gained proper form for the rest of the day. None the less, he remained resolute as he had done throughout the inquiry. He had never attempted to argue with the inspector, never sulked or threw himself down in his chair petulantly claiming that others might like to interview his witnesses, nor was he ever rude to the residents, as members of the audience, but always treated them with respect.

The Swale Borough Council (SBC)'s solicitor then challenged the validity of the proposal to build a haul road rather than transform Stickfast Lane, which can't be widened without the consent of other interested parties. He drew out inconsistencies between the appeal on behalf of clay extraction and the different proposed transport arrangements. This was an important issue because it wasn't really clear what SEM were proposing. If the application did not make reference to the proposed new haul road, the various highway alterations and internal arrangements for plant, earthworks, offices, internal roads etc., there would be quite different from the nature of the application if the haul road was included. So, if the former were allowed, the haul road application would fall. If the latter were allowed then the former would fall since its structure did not concur with what was proposed in the case of the latter. None the less, whichever proposal might prove acceptable, the south-eastern section of Stickfast Lane would be altered. But what would happen to the rest of that lane if a new haul road was built? Presumably it would have to be closed for safety reasons. But where was the application to do this? And could it be fulfilled? After all, Stickfast Lane is supposed to be part of the new Sustrans (sustainable transport) cycle route – a National Cycle Network Project, the first national millennium project to be given lottery money. So another potential conflict of interest would arise here too.

He then explained how the Kent Minerals Local Plan (KMLP) inspector in 1997 had repudiated SEM's case for clay – 'engineering clay is not important nor essential in the same way as other minerals' – hammering the point that SEM's chief witness agreed that displaying a need for clay had to be demonstrated in itself not merely as a justification for creating more void space. Indeed, in this witness's cross-examination it was clear there was no need for clay at all but instead 'sites which may have a need for engineering materials of some kind at some indefinite future date'. He then demonstrated how SEM's proposal would undermine KCC's and SBC's development plans. But his real attack focused on the status of SEM itself. It owned no land, held no options and had withdrawn as site operators. 'Slowly but surely they have disappeared from view rather like the Cheshire cat in Alice in Wonderland except that SEM have not left even a smile behind'[6] he said, drawing laughter in Iwade Hall. Now it was all up to Kerridgecourt to adopt their options, with Biffa ready behind them, if they failed so to do. SEM had ceased to be appellants save in name only, thereby making this appeal quite unacceptable. Throughout this submission the QC looked most miserable, as the inspector was seen to be so focused on what the solicitor was saying particularly on his concern to protect the countryside for its own sake!

The KCC solicitor pressed the point that if SEM's appeal no

longer focused on clay extraction, because there was a sixteen to twenty-year-supply of clay, according to the earlier KMLP inquiry, then there was no warrant for digging the quarry. Throughout the inquiry panic measures had been adopted to sustain the clay need case. And the very farmer who had sought clay to deal with the untidiness of sea-defences on his own land was the very landowner who could have obtained clay from this site in the past to do so but had instead created that very untidiness, an activity KCC had had to curtail. The KCC solicitor continued to emphasise too the unenforceability of monitoring gull activity in the estuary, the unsatisfactory nature of lorry routing arrangements which involved up to three hundred and sixty heavy goods vehicles (HGVs) per day, one a minute. Not only would the road widening carry important landscape alterations, but also SEM refused to face the fact that the shortest route from the hill to say the proposed incinerator at Kingsnorth would be along the A2 or via other short cuts to and from the appeal site, since there is nothing to stop lorries using them. Indeed all the landowner or SEM or Kerridgecourt could do, would be to enter an obligation with another party – the contractor – and try to enforce that! But we would then have 'an obligation seeking the enforcement of a further obligation'. Whether there had been a breach in the obligation would then be a matter of interpretation between the operator and the contractor. Now might not that contractor further sub-contract? So where did/would ultimate responsibility lie? HGVs would be set loose on the local road network once it were understood that any such obligation in this regard would be unenforceable or at least unsatisfactory. This solicitor was at his best in showing how *Gabriel*'s editor's evidence on leachate disposal arrangements indicated that SEM had no adequate overall plan. Rather the relevant documents explaining their ideas on this matter had 'only been produced on a piecemeal basis'[7] in response to that editor's persistent questioning. Indeed, an adequate scheme for the disposal of leachate simply had not been presented.

The appellant's QC began his final speech in a confident and interesting fashion. He attacked any attempt to get at SEM's intentions by gossiping about what they might want to do with the site as opposed to what was set out in relevant SEM documents. In criminal law it was important to establish intent in relation to someone charged before a court. However, an applicant's intentions had no relevance to the reader of a planning application. He then struck hard in the direction of using KCC's expert witness's earlier claims in that 'infamous' 12 September 1995 recommendation to fuel his attack, noting that it was recognised that the application had a dual nature: clay need and waste disposal. To speculate otherwise was 'to introduce Alice in Wonderland reasoning to the process'.[8] Moreover, in that recommendation there had been no fundamental landscape objection! Since then things had improved: the extraction area on the hill had been slightly withdrawn from the Saxon Shore Way and the extent of the extraction reduced materially! Moreover, no domed landscape would be involved so that, whether the height of the hill be regarded as taken before or after the clay was extracted in 1953, the final shape of the land would not alter the cast of the present landscape. Now KCC were trying to claim that there would be such an alteration despite having in their hands proof, in that very 12 September recommendation, that planning consent had been recommended without such a reservation being made!

In addition he was able to use effectively the gaffs made by KCC's landscape witness, described in chapter nine, pointing out, too, that this witness agreed no alien landform would be created. That KCC witness's two reservations could easily be met: the scheme could be implemented as proposed because Biffa knew how to do so and secondly landfill practices and control techniques on such sites had generally improved. He made a point of emphasising how that witness along with SEM's landscape witness had adopted a quantification approach to the staged development of the proposal. Moreover, the former had not disagreed with the latter's claims about the viewing prospects of the operations and how long they would have to be endured. This KCC witness had admitted, too, that the flare stack compound had been inconspicuously located just as he admitted that the objectionable features of the proposal would be merely transient. In any case the area was not subject to a Local Landscape

Designation and the QC didn't seek to go over all that ground again, taking my statement as 'helpful' in this regard. Moreover new views could be opened in the restoration and wooded area, suggestions that could be modified to sustain, if desired, Tiptree Hill's present character. What made the Tiptree Site so appropriate for such a proposal was that effects upon landscape character would not be 'materially adverse' in this case.[9] As to nature conservation, English Nature (EN) did not object and despite the golden plover's and lapwing's use of the site, SEM's expert witness on these matters believed the site had little ornithological significance, so concurring with EN's view. Indeed, this witness, working with the Royal Society for Nature Conservation, was about to embark on a major study in the north-west of England on seagull behaviour with respect to landfill sites. As for the challenge from the photographs displayed in relation to Shakespeare Farm on 1 December 1997 by KCC's expert witness, clay would be available at Funton to cover waste material.

In relation to the road proposals, there were no disagreements, other than suggested minor alterations, no objections to their design nor to the improved routing. Much of what was said about lorry drivers taking short cuts not only ignored road curbing but failed to recognise that the official route was well connected, via Sheppey Way, to the motorway, free from obstructions, farm machinery movements, traffic calming measures, sudden bends and so on, characterising rural lanes. As to any lighter vehicles, there would be few of them. Obligations and duties would be placed on site owners and operators, and Biffa could make them work, especially as local residents often have an interest in policing them! Callum Hill horse riders had only the junction between the southern end of Tiptree Hill and either Stickfast Lane or the haul road entrance to overcome. But the site wouldn't be working on Saturday afternoons, Sundays or during the evenings and these were the times when most horse riding, cycling and walking was done. In any case Sustrans were aware of the Tiptree proposal when Stickfast Lane was put in the national route; the lane could be improved!

True, 1,285 metres of hedge would be lost but replaced by 2,985 metres! On its eastern end Stickfast Lane would be affected but the part hedged on both sides remained. This part may need to be closed, but that matter could be resolved later. All these objections, just like those about ground water pollution, leachate arrangements, noise, the dust danger, smell, litter and so on were weaker arguments so used to multiply issues with the vain hope that such a 'scattergun approach will bring down at least one pheasant, with all the waste of time inherent in that'.[10] As for all the dubious objections on so-called geological grounds raising fears about the accuracy of recorded borehole findings, basal heave, unproven geological faults, even earthquake dangers, all KCC were seeking to do was replace the impartial judgement of the Environment Agency with that of their own! The latter had withdrawn their objection of last year because they were satisfied with the Risk Assessment methodology. They had expressed no reservations at all with regard to the three methods which could be used to deal with leachate: treatment on site, lorry tankering or passage into a public sewer.

Suddenly, the QC's energy left him. He was asked if he wanted to take a break but he said he wanted to battle on and was soon forwarding the idea that the countryside can take change depending on its severity. It was 'intellectually paludal' to say, on the grounds of protecting the countryside for its own sake, that any change in it must be adverse. He continued by admitting, in relation to clay need, that it was something impossible to predict or assess. It was not unreasonable, however, for the inspector to bear three projects in mind even if they were ones internal to the interests of Biffa: the local farmer's sea defences and remedial work; the demand from Kingsnorth incinerator project and landfill scheme and finally at Patteson Court for facilitating its filling-in and restoration. He then said that 'whatever the outcome on the need for clay' the need for landfill was 'a material conditioned to be weighed'. The Rosecourt Farm decision was disanalogous because of its creation of an alien land form and it wouldn't have done much for void space availability. But everyone in the hall knew that more landfill was required in Kent and that Shelford couldn't meet the long-term need. Kingsnorth incinerator had not received planning permission nor had Ridham, Kemsley. The Halling proposal had collapsed and its sweet substitute Allington was merely a press release. Richborough floated as a mere conjecture. Any and each of these schemes would take at least five years to be operational. A support system no matter how temporary was required. He then concluded by saying that seeing as the application was as it was with respect to 'both mineral extraction and landfill, there is no legal requirement for the appellants to prove an overriding need for *both* clay and waste'. Even an absence of a clay need did not overrule 'the need for landfill' as 'a material consideration'. He emphasised that it was unlikely there would be more 'opportunities to secure a site where objection is so slight relatively and where the advantage of a clay deposit is to be found'.[11]

Needless to say, his speech was not well received. Hadn't he made the mistake of admitting that 'the need for landfill is a material condition' even if there was no clay need? He was forced to admit, too, as reported in the *East Kent Gazette*, that the operators were seeking to work the site for twenty-five not twenty years and he seemed caught off balance by the logic of the arguments set by the claims for legal costs from the three opposing parties. It was during this last hour, in defending SEM's claims for costs, not through the reading of his prepared speech, that he managed to use the word 'actually' 184 times as reported in the newspaper's 25/2/1998 *Gazette* gossip column. Even if at the start of the day SEM had enjoyed the edge in this inquiry, had this been sustained to the end of this inquiry's end?

The term 'utilitarian triumph' refers to what was the central issue on this final day. It wasn't any kind of aesthetic concern; it was the need for clay. In my own proof of evidence I distinguished the idea of need from demand. The notion of need has its home in relation to things that are required for human survival. You need air to breathe, water to drink and nourishing food to remain alive. So air, water and food are instrumental in keeping us alive. The notion of demand, however, is tied, as the appellant showed, to wants in the Market Place, where something is bought or sold at the right price. It relates to the idea of *exchange* value. One could have a need for something making a concern for a demand for it irrelevant. Yet there might be people willing to pay for something – thereby constituting a demand for it – without those persons necessarily having a need for that thing! Now had the appellant really demonstrated a need for clay; perhaps a demand for it could be provisionally or hypothetically grounded! The case for waste disposal seemed more difficult. There might be more than a mere demand for means of dealing with waste so the question arises: 'Is KCC's plan for waste disposal into the next century really satisfactory?'

To deal with that question I suggested, in my submission, a philosophical principle to decide between the claims for clay extraction and waste disposal on the one hand and landscape with conservation concerns on the other. It is called Heffernan's Principle: The survival interests of human beings ought to outweigh those of the rest of the biotic community and the survival interests of the rest of the biotic community ought to outweigh the nonsurvival interests of human beings.[12] Survival interests for humans or the biotic community refer to species rather than individuals, although that is open to philosophic debate! Furthermore, the notion of community has to be taken as referring to issues concerning the land. That is why landscape concerns can be tied to conservation ones!

The sad feature of this inquiry was that the decision, in relation to granting a Local Landscape Area (LLA) status, had not been given before this present inquiry took place. If granted, it would have appeared somewhat contradictory to have a waste tip placed upon an area granted an (LLA) designation, unless it had or could be clearly demonstrated – à la Heffernan's principle – that such a need had to be satisfied. But, as we saw in relation to the LLA designation inquiry, if such a status was granted, that in no way would prevent the removal of clay in the case of need, as occurred in the 1950s. I pointed out too that nothing in my proof of evidence related to the feasibility – geologically, legally, effluently or hydrologically – of proceeding with SEM's proposed development.

In relation to the three questions posed in the last chapter – the incommensurability of discourses; the descriptive/evaluative dichotomy and the six criteria for evaluating landscape – the final speech day did not carry us much farther. Apparently, against my argument questioning SEM's approach to Landscape Assessment, the QC indicated KCC's landscape witness's approval of the quantification approach, passing over the issue as to the disagreement on landscape evaluation in applying the six criteria. In my presentation, rather than in my proof of evidence, I had tried to overcome this disagreement by exploring where these criteria came from: the document *Landscape Assessment Guidance* (CCP 423)[13] makes reference to an earlier study (CCP 423 p.25) *Landscape Assessment – Principles and Practice*. This is a report by the Land Use Consultant (LUC from now on) published in 1991 for the Countryside Commission for Scotland. True, it too could be described as adopting a quantification approach, generating the distinction between objective facts and subjective value judgments. Indeed its approach seeks to 'reflect the "bird's-eye view" of landscape, as represented in maps and aerial photographs, and the view as seen on the ground'.[14]

The section of the document *Landscape Assessment – Principles and Practice* I tried to draw attention to at the inquiry was section 7 'Evaluation of Landscape for Designation Purposes' (LUC pp.23–26). Two features distinguish it from the other Countryside Commission Documents: it refers to American experience (LUC 7.2 p.23) and acknowledges explicitly the debate as to whether 'evaluation should be based on professional judgement or on public preference' (LUC 7.2 p.23). Scenic quality – the all important criterion of the six listed earlier and which was so hotly disputed during both inquiries – is given a detailed analysis (LUC 7.5 pp.24–5) in terms of i) a combination of landscape elements; ii) aesthetic quality and iii) intangible factors. I defended the first in the case of Tiptree Hill because of its 'particular mix of landscape elements' in relation to such special factors as 'sea', 'undeveloped coast', 'river estuary', 'steep slopes', 'grassland/pasture' and 'woodland'. Although the three important 'physical dimensions' are really significant for Areas of Outstanding Natural Beauty in Scotland, even at Tiptree there is a delightful contrast between 'landform/relief, naturalness/vegetation cover and presence of water'. It seemed to me too that aesthetic quality as a sub-criterion was also satisfied: the interaction of such 'visual characteristics' as 'form, line, colour, text, diversity, enclosure' etc. could be taken to satisfy the three requirements of 'vividness (memorability), intactness (wholeness of the scene) and unity (harmony of parts)'.

The third criterion – 'intangible qualities' – was defended by arguing it is 'a place special' for its ephemeral effects, 'light, weather or season, or a combination' of these with the intrinsic character of the landscape itself. Here it was possible to cash the claims for *prospect ('places which provide open views or prospects') and *refuge ('places of concealment from which to see without being seen'). Jay Appleton, I learnt subsequently,[15] defends these two further sub-criteria.[16] There is the opportunity too to experience a 'sense that the observer would see more if s/he were to move into the scene' (LUC p.25). Having made, then, a more detailed analysis of scenic quality, it became possible to place this defence within the context of regarding landscape phenomenologically – as defended in Christopher Tilley's *A Phenomenology of Landscape* – not objectively, from the standpoint of taking a bird's-eye view of it, but rather in the context of living within it. The latter emphasises for Tilley 'the physical and visual form of the earth' as surroundings and 'a setting in which locales occur', where 'meanings are created, reproduced and transformed'. Those who live in a landscape 'draw on qualities of landscape' which have significance for them in terms of their physical and biological experience on the land beneath the heavens. Such 'ontological import' emerges because people live on and through the land which is 'mediated, worked on and altered, replete with cultural meanings and symbolism'. On the other hand the bird's-eye view of landscape is more objective: 'The space created by market forces must, above all, be a useful and a rational place. Once stripped of sedimented human meanings, considered to be purely epiphenomenal and irrelevant, the landscape becomes a surface or volume like any other, open for exploitation and everywhere homogeneous in its potential exchange value for any particular project. It becomes desanctified, set apart from people, myth and history, something to be controlled and used.[17]

Are we not now ready to deal with the issue raised by this chapter's opening aestival? If communication between human beings and nature is problematic, isn't that true for any kind of genuine communication between people embedded in their own special language games? Isn't the language game of someone obsessed with the 'objectivist' conception of landscape incommensurable with that employed by someone who is living within its environment? But to do justice to the range of perspectives which generated such discourses at this inquiry, we need to return once more to Peirce's enunciation of the way nature can be viewed: aesthetically, instrumentally or scientifically. Tilley's dualism between the phenomenological and the objectivist leads him to sweep the first two dimensions together in order to contrast them with the scientific. He rejects the latter for its mental representational and cognitional stance towards landscape in favour of a discourse focusing on locales: 'places created and known through common experiences, symbols and meanings'.[18] But clearly someone could be close to the land, using its ground to grow crops, floating a boat out onto the high tide or ensuring the washing dries in the wind without having any aesthetic sensibilities about these things, taken for granted 'in his or her own backyard', at all! Similarly someone might be enlivened by what s/he gains through the senses in exploring the landscape at different times of the year through its changing sounds, colours, smells, textures and so on; be charmed by the shapes in the land, the movements of the long grass in the wind and the lines of furrows in the ploughed fields or, indeed, be inspired by the aesthetic meanings this locale incites in relation to what has been created in past writings about this place. One or all three of these forms of aesthetic awareness – whether giving rise to something qualitative (Firstness), referring to something formally existing (Secondness), or in relation to something symbolic (Thirdness) – such aesthetic awareness can be enjoyed without any reference to instrumental concerns either. At the inquiry there were clearly people who regarded Tiptree Hill within the instrumental perspective, others who had strong affections for it aesthetically and witnesses called – such as the SEM's authority on ornithology – who spoke about this area entirely from within a scientific perspective. Someone like the KCC's county planning officer to the Planning Sub-Committee can be described as a planner. In his professional work no particular perspective dominates his decision-making. Rather his activity is required to reflect, indeed perhaps balance, the representations put to him by people with different cognitive perspectives.

In setting up this four-part categorisation, use is being made of Peirce's categories. In chapter VI – The Bird Habitat argument – his method or 'road to inquiry' – what he called indagation – was examined. As the diagrams below illustrate we start with experience.

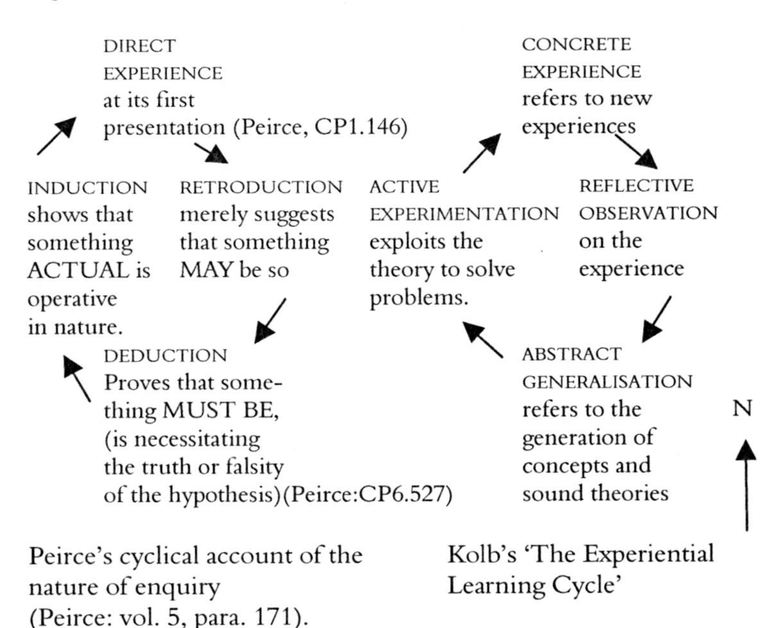

Peirce's cyclical account of the nature of enquiry (Peirce: vol. 5, para. 171).

Kolb's 'The Experiential Learning Cycle'

Peirce's first stage of inquiry was introduced – retroduction – where a hypothesis is suggested to an observer in solving a problem, say identifying a bird. The second stage – deduction – involved drawing out the consequences, cognitively, of the original hypothesis before the final stage completed this route of inquiry. That stage – induction – puts what has been deduced to the test, so that a new experience is originated, which could generate a fresh problem to be solved so that the inquirer begins once again this cycle of inquiry. This cycle is usually credited to David Kolb a psychologist, under the name 'The Experiential Learning Cycle'.[19] More recently, an advocate of Kolb's work, Gibbs, declared that the scientific method is fundamentally the same as the experiential learning cycle and recommends techniques in study commonly used in 'laboratory work'.[20]

Emphasis is placed on these stages now for a different purpose: to relate the suggestions of chapters I and II to this four-stage form of inquiry. For each cycle consider the four quadrants between each stage of the cycle the north-east, south-east, south-west and north-west quadrants. For Peirce, the person who reflects in a feeling way upon experience in the north-east quadrant is the artist. Kolb uses the label the diverger since his/her 'greatest strength' lies in an imaginative ability. For Gibbs the learning strategy of this person is reflection. The occupant of the south-east quadrant according to Peirce is the scientist. For Kolb this is the assimilator whose 'greatest strength lies' in an 'ability to create theoretical models' as disparate observations are integrated into a theoretical explanation. Gibbs refers to this person's learning style as that of theorising. The person occupying the south-west corner is the practical man for Peirce. For Kolb this is the converger, s/he who can practically apply ideas. Gibbs's label for this person's learning style – sheer practicality – is that of the pragmatist. Peirce does not have a name for the person occupying the north-west quadrant. On the basis of reasonings used in chapter I, it does not seem unreasonable to refer to him/her as the enured. For Kolb s/he is the accommodator, adaptable 'to specific immediate circumstances', likely to be someone who 'acts before he thinks – if he ever thinks''.[21] Gibbs describes this learning style in terms of *doing* or *responding*.

Why refer to these four types now? Because Porteous – an environmental aesthetician – has identified four basic but quite distinct even if 'interconnected approaches to the investigation of environmental aesthetics'.[22] Sustaining the same order set out already, he employs the following labels: humanist/purist;

experimentalist; environmental activist and environmental planner. The table below gives the following comparison of terms:

Lay-Person Label	Peircian 'Men'	Kolb's Types	Gibbs's Learning Styles	Porteous's Approaches
Experiencer	Artist	Diverger	Reflective	Humanist
Thinker	Scientist	Assimilator	Theorising	Experimen-alist
Technician	Practical Man	Converger	Pragmatistic	Activist
The Responder	Enured	Accommod-ator	Responding	Planner/ Manager

The first paradigm – labelled the humanist or purist – adopts a contemplative approach emphasising 'the surficially unique, whilst at the same time seeking universals intuitively'. The experimentalist paradigm derives from the extreme rigour of scientific theory-building: 'before we change the world we must first understand it'. Environmental activists use the 'basic concepts which experimentalists test' and in their endeavour to 'act now' thereby engage in political activity as 'protestors, environmentalists, lobbyists and legislators'. In Porteous's scheme the planner, one might expect, puts into practice the recommendations of the scientist. So, in the past, with 'their visionary schemes of engineering perfection' trees came down and rural life was destroyed with each planning development. So Porteous sees the environmental planner today *not* putting *any* particular cognitive perspective before any other but rather evaluating the inputs from the other three perspectives 'before recommending or undertaking action in the public domain'. Since s/he operates within 'a framework of legislation' outside her/himself, a by-product of past lobbying and protests by say environmental activists, s/he has to strike a balance between these different cognitive dimensions. In that sense planners are dependent upon the values expressed by humanists, users of the techniques of the experimentalist and undergo pressure by environmental activists.[23] But for all these people the landscape still remains a 'resource', a resource for spiritual uplift for humanists, scientific inquiry by scientists and an inspiration for political action by activists! The notion then of such a person seeking to strike a balance between the discourses of the humanist, the scientist and the political activist hardly shows that these discourses can be made commensurable with each other. But that isn't the only reason for attending to Porteous's conclusions. There is an additional reason provided by the question: 'What grounds the four basic approaches to 'assessing perceived landscape values'?'

Porteous renders a sense of the arbitrary in his answer to that question. In his examination of the requisite literature in the field, he indicates the nature of the methodology used to establish something like these four basic approaches: reviewing one hundred and sixty relevant articles in twenty or so Canadian, American and British periodicals, from which four 'major paradigms emerged'! His methodology was quite different from reviewing articles in the literature: 'extensive personal reading and intuition' plus an examination of 'the goals that modern scientists publicly espouse'.[24] Now we can appreciate the force of Peirce's intellectual work. What has been established in this book by now is that these four basic dimensions, constituting attitudes we can take towards the world, are experientially sustained in the way we relate to it (cf. chapter III: Confusion and chapter VII: The Landscape Question), logically grounded in the way we form inferences about our experiences of it (cf. chapter V: Clarification) and epistemologically justified in the way knowledge can be gained from it (cf. chapter VI: The Bird Habitats argument). Of course, as the landscape architect consultant might argue, any or each of these basic attitudes can be adopted and the decision to do so might be quite arbitrary, but his argument would not sustain Porteous's implication that each attitude in itself is not merely an arbitrary one, save in the case of the enured, an attitude which is purely subjective in its reaction to and in comparison with the other three! Of course, if each or any of these four is adopted to generate a particular discourse, we still have the problem how each or any of these discourses can be reconciled with any of the others.

This very inquiry, however, showed how that sense of incommensurability can be overcome. Each of the interested parties was able to deliver his or her own cognitive perspective and the inspector would then be left to reach a judgement based on the balance he was prepared to strike between them. Indeed the one thing all parties could agree upon was in relation to his work. His interventions were completely fair, his knowledge and control of the requisite documents – remember he had no secretary or team to support him – was largely in advance of those of the 'representatives' and his check upon possible 'preparation' of witnesses was extremely tight. So, at the end of my proof of evidence, the following point was made as to whether the appellant's appeal should be upheld: I awaited the full development of this inquiry to judge if there was really a need for a landfill site rather than a mere demand for one! That judgement hardly made me popular with some TAG members! In addition I wrote in our parish magazine *Gabriel* that because of the extraordinary manner in which the inspector ruled this inquiry – manifested in his control, understanding of the issues and sense of fairness – *all* the parties concerned should abide by his decision. Three questions are raised immediately by that claim. Firstly, what was his final decision? Secondly, could that final decision be counted as a fair one, and, if so, *procedurally but not substantively, *substantively but not procedurally, neither procedurally nor substantively or both procedurally and substantively? By procedurally is meant *how*, the way, the methods through which the decision was reached. By substantively is meant the *what*, the content of the decision itself. On the basis of what has been shown so far, it seems that the procedural requirement was met. But that still leaves the substantive question open. Thirdly, we have the philosophic question still unresolved within these last two chapters: is this the way the Fact/Value dichotomy is to be resolved? In other words, one party shows how they sought to derive a value judgement from factual claims, the other side how they derived an opposite value judgement from the same sources and an arbitrator decides, an arbitrator whose judgement is final? If so, someone adopting Thomas Hobbes's standpoint, as presented in chapter IV, 'Some

Semblance of Rationality' would be pleased!

NOTES

[1]Communicative Possibility, Strumble Head, Pembroke, 31 May 1995.

[2]Appeal by BFI Ltd; Restoration of Mineral Workings by Landfill, extraction of Minerals and Demolition of Farmhouses and Outbuildings at Rosecourt Farm, Isle of Grain, Kent 30 September 1997. The appeal was overridden not only because an alien landform would have been created but because there was no need for the clay extraction other than to make void–space available thereby harming this impressive and unusual part of the countryside p.59 sect. 9.78 and p.49 sect. 9.11.

[3]The London and South East Regional Planning Conference *SERPLAN, Waste: Its Reduction Re-use and Disposal* (1992) RPC 2266 p.14 para. 55.

[4]Appeal by BFI Ltd, op. cit., p.49 sect. 9.12.

[5]Closing Submission by Jeanne Taylor on behalf of Tiptree Action Group; and the Parish Councils of Iwade, Lower Halstow, Bobbing, Newington, Upchurch and Hartlip 19 February 1998 as the inspector, in his final report, recorded her conclusions. (See Appeals by South Eastern Minerals Ltd, Land at Tiptree Farm, High Oak Hill, Funton, Sittingbourne, Kent File nos. App/W2275/A/95/259578; APP/W2275/A/96/271958 and APP/V2255/A/96/2723060, 12 October 1998, sects. 7.9.3 and 7.9.2 on p.131.

[6]SBC Counsel, Final Submissions for Swale Borough Council on D.E.T.R. Appeal Refs: A) APP/W2200/A/95/249578 B) APP/W22 00/A/9627 1958 C) APP/V2255/A/96/272360 (19/2/1998) p.7 sect. 3.2; p.14 sect. 4.15; p.39 sect.6.l.

[7]Tiptree Farm Appeal: Submission of KCC 19/2/1998, p.7 sect. 23g, p.8 sect. 31.

[8]SEM's Final Submission: South Eastern Minerals Ltd/Biffa Waste Services Ltd Tiptree Farm, Funton, Near Sittingbourne, February 1998 p.2.

[9]SEM's Final Submission, op. cit., pp.6, 7.

[10]SEM's Final Submission, op. cit., p.20.

[11]SEM's Final Submission, op. cit., pp.26, 31, 34, 35.

[12]James D Heffernan, 'The Land Ethic: A Critical Appraisal' *Environ. Ethics*, vol. 4, no.3, Fall 1982 pp.245–7, 246.

[13]*Landscape Assessment Guidance* (CCP 423) Advice from the Countryside Commission prepared by Cobham Resource Consultants, Countryside Commission 1993.

[14]For the discussion concerning 'objective' facts and 'subjective' value judgments cf. LUC 'Summary of Key Points', 1–14, pp.ii–iii; sects. 2.2–4 p.2. For the two views of Landscape cf. 'Landscape is generally experienced at ground level and, ideally, the bird's–eye view should be complemented by the ground-level view which can only be provided through field survey using the sort of techniques described above.' i.e. those approved by the Countryside Commission and referred to in chapter IX! *Landscape Assessment – Principles and Practice* by Land Use Consultants (LUC) for the Countryside Commission for Scotland (1991).

---

[15]Gordon H Orians, 'An Ecological and Evolutionary Approach to Aesthetics' in Landscape, Meanings and Values, chapt. 2 p.9.

[16]Appleton, Jay (1975) *The Experience of Landscape* London, Wiley.

[17]Christopher Tilley, *A Phenomenology of Landscape*, Oxford: Berg Publishers, 1996 Introduction and chapt. 1, especially pp.25–6, 21.

[18]Christopher Tilley, op. cit., pp.18, 25.

[19]David A Kolb, 'On Management and the Learning Process' in *Organisational Psychology: A Book of Readings*, second edition, pp.27–42.

[20]G Gibbs (1988), *Learning by Doing: A Guide to Teaching and Learning Strategies*, Further Educ. Unit; cf. S W Weil and I McGill (1989) *Making Sense of Experiential Learning*, Open UP.

[21]David A Kolb, op. cit., pp.31, 30, 31, 33. In this context, the accomodator is referred to by Gibbs as not reflecting, not learning from his or her mistakes because s/he is simply not thinking! (See *Learning by Doing*.)

[22]J D Porteous, *Environmental Aesthetics*, p.13.

[23]J D Porteous, *Environmental Aesthetics*, op. cit., pp.14, 16, 18, 153, 191.

[24]J D Porteous, Environmental Aesthetics, op. cit., pp.12 and 13.

Chapter XI

TWO DECISIONS IN TWO INQUIRIES

Give a proper man a chainsaw and he loses consciousness. The sheer thrill of the uncompromising drill of this machine automatically seeks the bark of trees whose only offence was to set before human eyes a spectacle of life which might defy this offensive. Here, where two rivers meet, their natural co-habitation was preserved by a large lake over which these tall willow trees could acknowledge the existence of complimentary partners on the opposite bank by gently leaning towards them above the water. In this place, human consciousness itself could focus upon the lake's sheen to seek a new reflected reality by holding up to the existence of these delicate defenceless willows their own reversed image to masquerade for and add transcendence to their gentle, graceful actuality in a space whose silence was disturbed only by the quiet splash of a bird alighting upon this water's surface. Now, normality was to be restored: gone presumably would be that image and in its stead the shouted instructions of their owner to his men, the purposeful noise of mechanised activity and a huge burning fire with tall flames licking upwards towards the sun celebrating a job well done![1]

A week after the Tiptree Inquiry's speech day, Kent County Council (KCC)'s Planning Department received a letter. It was from GOSE – the Government Office for the South East – informing them that the Kent Waste Local Plan had been 'unfrozen' since the issues raised in it 'are not such that the Secretary of State's intervention would be justified'.[2] Indeed, at the same time, I received a letter from our local Labour MP informing me that a newspaper report was incorrect if it claimed KCC had been directed to reject its Kent Waste Local Plan. The rationale for its freezing was 'to allow further time for the Department to consider the issues raised by the proposal for waste to energy incineration in the plan',[3] The Tiptree inspector was then informed on 4 March by the county planning officer, responsible for that notorious 12 September 1995 recommendation, both that the Kent Waste Local Plan had been 'unfrozen' and that the Secretary of State had called in the applications by PowerGen and Biffa for Kingsnorth Power Station. He also informed the inspector that KCC had adopted the Kent Waste Local Plan on 2 March.[4]

A further letter from GOSE on the 22 April 1998 confirmed the above, as did details of developments on the incineration front. No application had been received for the Halling site. Because of the controversy the Secretary of State had called in planning applications both for the 'materials recycling facility, energy from waste plant, residual disposals building' and the proposal to dispose on land 'residues from the energy from [the] waste plant at Kingsnorth Power Station'.[5] The developers had since withdrawn both applications. This flurry of correspondence was completed by an interpretation of Michael Meacher's stance taken on the issue of well-managed landfill sites referred to in chapter IX.[6]

On 23 July 1998, the verdict of the inspector at the appeal by South East Minerals (SEM) against Swale Borough Council (SBC)'s attempt to provide a Local Landscape Area (LLA) designation to the hills between Iwade and Newington to include the Tiptree Site, an appeal detailed in chapter VIII, 'A Local Landscape Area Designation', was published. The decisions reached in relation to the hearings on Thursday, 9 January 1997 were that 'no substantial case' (para. 6)[7] had been made for the council for the Protection of Rural England (CPRE) for an LLA designation for the dry valley features at Lynstead Valley nor, that in relation to the Syndale Park area, should the council revise the proposed LLA boundary around the Park and Valley (para. 7). As for the large area south of the M2 to extend as far west as the district boundary to provide a countryside gap between the Medway towns and Sittingbourne, the inspector recommended that all or part of this area be regarded for consideration under the council's Policy El 2 (para. 5):

> The Borough Council, through its development control function, will seek to protect important countryside gaps between settlements in order to retain the individual character and setting of settlements.

Now let us turn to the inspector's decisions on the six arguments – the three weak and the three strong arguments – for refusing an LLA designation for the line of hills between Iwade and Newington. He registered no comment on the claim that there was any kind of inconsistency between SBC's policies but rejected the idea that sufficient protection was provided by the council's other policies (para. 1). As for the claim that the LLA status would prevent extraction of minerals, the inspector judged that to be an irrelevant consideration (para. 10). The inspector, however, supported the third of SEM's weaker arguments: the attack on the idea conveyed by the last three words given in Policy E17:

> The Local Landscape Area, as shown on the Proposals Map, will be afforded long–term protection.

This statement implied that the area was worthy of a designation for Outstanding Natural Beauty (AONB) or as worthy of a Special Landscape Area (SLA) status (para. 3).

The inspector disagreed with the first of the stronger of SEM's arguments – that Tiptree lacked any kind of distinctive character to be worthy of an LLA designation – ruling that it did provide 'a backdrop to the North Kent Marshes SLA' and thereby warranted an LLA designation' (para. 11). Against SEM's second stronger argument that the site offered no special particular historic, scientific or ecological interest, the inspector admitted that some of the LLAs 'do not have a uniform landscape characteristic, but it is evident that many of them encompass areas that are physically prominent, whereby development within them would be particularly conspicuous in the locality. Hence, the need for Policy E17 is supported in principle.' (para. 2). Finally, he granted the force of SEM's third stronger claim that in the case of this proposed LLA designation there was an 'absence of a published landscape appraisal and the failure to adhere to Countryside Commission guidance on landscape assessments'. But he added that it was apparent that 'the council conducted their own appraisal taking into account a number of existing studies including *The Thames Gateway Landscape* assessment prepared by consultants for Kent County Council and the Countryside Commission' (para. 2). He then remarked that he sided with SBC's decision since 'there is considerable local support for the LLA designation of the line of hills between Iwade, Newington and Lower Halstow' (para. 11). Policy E17 was supported, reworded as:

> The Local Landscape Areas, as shown on the Proposals Map will be afforded protection from development that would undermine their integrity and character.

The inspector's decision in relation to SEM's main appeal against the granting of planning permission for the Tiptree Site was published on 12 October 1998. The document extends to 183 pages.[8] It is divided into a number of sections setting out the background facts, and the cases made by SEM, KCC, SBC, the Environment Agency (FA) and Tiptree Action Group (TAG) plus evidence presented by other interested persons. Attention will be given, in what follows, to the inspector's Conclusions since the main arguments in the preparatory part of his report have been aired in previous chapters.

In his introduction, against SEM, he noted 360 letters and petitions appealing against the development at the application and 600 at the appeal stages (9.1.2),[9] that there were no applications for stopping-up Stickfast Lane and that policies to protect the internationally significant Lower Thames Marshes were important (9.5).[10] For SEM's case he noted the notorious 12 September recommendation (9.1.7), that SEM's proposal was and still remained a composite one (9.3.3) and that the EA did not object to the proposal. Again, he noted on the one hand that at the close of the inquiry the SBC Local Plan had yet to be decided upon by its inspector – especially in relation to that LLA designation – but he saw too that the Kent Waste Local Plan (KWLP) had been frozen (9.5). As to the appeals overall, he judged that the application for a haul road was a county council, not a borough council matter, and was part of the application for a clay/landfill scheme, even if the haul road application had lacked 'clarity of both content and intent' (9.2.5), at least initially! Now we can turn to his treatment of the individual arguments.

One of the most important features of this inspector's report is what he had to say about the need for landfill. It was important because what he said was constantly referred to in a later inquiry arising from a waste tip application at Dargate, an area between Faversham and Canterbury. Although he criticised SEM's figures since they 'took no account of any future landfill permissions' – how could they? – he drew attention to the Kent Structure Plan (KSP). In that plan it is emphasised that *additional provision for waste disposal must be provided in Kent*. Though a shortage of landfill availability might incite other methods for waste disposal it would be very risky for such a shortage to be deliberately contemplated to encourage other waste disposal methods. Indeed, there was no evidence of a causal story here and it had to be borne in mind that any pre-waste treatment – such as recycling of paper for example – would reduce the high calorific value of waste (9.7.5). Even if it is now recognised that landfill is at the bottom of the hierarchy of methods for disposal, there was still a significant short-term need 'for the disposal of waste by landfill' (9.7.8).

A *second area of agreement with the case of SEM* was with their arguments concerning the area's *Geology and Hydrology*. Despite his acknowledgement of water still rising in boreholes used to measure clay depth levels, he noted the EA's conclusion that a shallower excavation would enhance ground water protection: 'The judgement of the EA is to be relied on' (9.11.1–6). He dismissed *Gabriel*'s editor's arguments about the local sewage system: Southern Water had accepted 'the principle of discharge to the sewer' (9.11.8).

SEM's case was sustained, too, by the inspector *in a third area of controversy: the Ornithological and Nature Conservation arguments*. Though he attached little importance to SEM's suggested Environmental Trust he noted no objection from English Nature (EN) (9.12. 3 and 2).[11] The latter suggested that if any protected species were likely to be affected by the working activities, SEM would be obliged to do something about it. Moreover, with the exception of the provision of protection for owls, outside 'the sightings of some species of birds and excluding any use as a support site' the site was 'of limited to medium nature conservation value' (9.12.8). In the case of the golden plover and lapwing he noted there was no evidence that because this site would be out of action ornithologically for say twenty years or so, the ornithological values of the estuary would diminish (9.12.10–11). No use need be made of Heffernan's Principle,[12] which he referred to in his report (8.1.5). His only reservation in this fourth area was with respect to monitoring birds off site even if deterrent measures for scavengers and systematic recordings of the breeding habits of avocet and little tern populations could be agreed.

In relation to two further areas the inspector seemed to be undecided. In the case of the Highways and Transport argument, he pointed to the traffic statement agreed between KCC and SEM but not by SBC and TAG and *he judged in favour of the former* (9.9. 1–2). He approved of the road designs to sustain HGVs on the right route; agreed that the improved road proposals to Sheppey Way would satisfy traffic routing rules, and found against the Bobbing villagers: Sheppey Way was suitable for the proposed traffic and restriction on speeds through the village were well in hand (9.9.3 and 6). He was satisfied the lorry routing arrangements would sustain on the whole (9.9.11). Limited use was made of the Tiptree Hill–Stickfast Lane area by cyclists, horse riders and pedestrians (9.9.7). He admitted that such routing would not be so pleasant a road for such users and, indeed, as far as the Sustrans policy was concerned, it wasn't clear whether there was an alternative lane for the National Cycle Network (9.9.10). But minerals have to be extracted where they can be found and can create void space. Landfill sites in such layers of clay were obviously beneficial hydroecologically. Yet his overall endorsement was checked by his awareness of the distance between the site and Sheppey Way, as well as the necessary alterations required for parts of Tiptree Hill and Stickfast Lane. There was, too, the problem of objections to Stopping-Up orders for the latter and the doubts as to whether SEM controlled the necessary land to facilitate its transport proposals, but presumably there was the possibility of 'realigning the proposed highway alteration and for the purchase of other land' (9.9.14–16). The haul road would have sufficient capacity to take HGV traffic too (9.10.1). And, with respect to the recreation and tourism evidence, despite the harm done whilst the site was in operation, the loss of views over Tiptree Hill from the start of the work, and the harmful impact upon the Callum Park Riding Centre so that recreational activities would be affected, he *agreed with SEM*. The affect upon recreational activities would be lessened if the haul road access scheme was made available (9.14.3–9). In addition only a small amount of agricultural land would be lost. The Ministry of Agriculture, Fisheries and Food in July 1995 had had no objection to the development.

In opposition, however, to the three specific areas of tempered agreement with SEM, there were three clear areas of disagreement. These concerned the clay need, the Landscape and visual impact and the noise, dust, litter and odour pollution problems. With respect to clay need, SEM had put forward very similar arguments to those rejected by the inspector for the Kent Minerals Local Plan. None of those cases put for three 'internal' uses of the clay by SEM, together or singly, made a convincing case for digging clay at Tiptree. There was then 'no need for clay extraction at this site' (9.6.7–11).

On the Landscape and Visual Impact arguments, Policy E17 – as part of the Swale Borough Council's Local Plan – and other policies were not of a kind such that 'no weight' should be attached to them, as SEM sought to claim. He recognised the case for and against the LLA designation as reported in my notes – cast into the first part of chapter VIII, 'A Local Landscape Area Designation?' – and judged the appeal site to be part 'of an attractive and prominent ridgeline'. He saw the force of the argument for an important landscape feature in the rural gap between Sittingbourne and the Medway Towns, despite KCC's Landscape Expert's judgement that as a landscape it was 'unremarkable' (9.8.2–3). The inspector acknowledged the local landscape impact of the operations over twenty or so years and the removal of trees; developments which were not only 'seriously harmful' but in conflict with so many local plans (9.8.4–5). Litter, dust, and smell problems, along with attempts to check scavenging birds, would all draw attention to the hill, and its impact on the North Kent Special Landscape Area should not be overlooked (9.8.7). The inspector suggested that land raising was indeed proposed over a large area of the site but that it would not constitute an alien feature in the landscape. Yet, with the additional tree planting, the development would obstruct views over the estuary and the marshes from Tiptree Hill Road. And the beneficial efforts of restoring the site would only come into play after operations had ceased (9.8.9–10). Moreover, any widening or realignment of roads would undo attempts to protect parts of lanes which would be damaged in such activity whilst the haul road would be harmful to the countryside, 'which is to be protected for its own sake'. It would cause intrusions into the rural landscape and would be noticeable at junctions with existing lanes and footpaths (9.8.11–13).

Finally he agreed with much of what SBC's solicitor argued in relation to noise, dust, litter and odour pollution. He judged the council's figures for noise levels were to be accepted for such a quiet rural area! Screen fencing would be needed to protect Tiptree Bungalow from noise adding to further visual clutter. Though he ruled against the claim that the development threatened Iwade's expansion, he did think that incidents – even if isolated – regarding dust, litter or smell would draw attention to the hill.

Three specific areas of agreement are then set out – even if tempered – in favour of SEM's case. These are then balanced against three areas of strong concern against their proposals. How can a fair decision be reached? In his Conclusions he found that in proposing to develop the site there had been 'compliance with the relevant Environment Assessment Regulations' and whilst the appeal against the application for the haul road against SBC was invalid, the one forwarded against KCC stood (9.16.2–3). Now how did the inspector weigh his decision between the three areas of agreement with SEM and those counter arguments? What he did was emphasise particularly two elements associated with the arguments about the use of this site: its relation to developmental plans for the area and the landscape and visual impact. He emphasised again Policy E17 of the emerging SBLP, particularly the countryside for its own sake case and the difficulty in sustaining long-term protection for the North Kent Marshes (SLA), even if for 'limited periods', in addition to protecting 'scenic, heritage or scientific value' of the undeveloped coast and estuaries' (9.16.4–6). As to the claim that the landscape would be restored subsequently, such benefits would not outweigh the harm done during the operation, the fact that the hill requires no 'improvement' and that there are other ways of securing tree and hedge planting. And even if the haul road would be less damaging environmentally than altering Stickfast Lane, further intrusion into the countryside – needed to be protected for its own sake – would occur (9.16.7–8).

His remarks on the case made by SEM on highways and transport are now, in his conclusion, more tempered by his concern for the clash with the National Cycle Network proposals; his rejection of the idea that this site is well related to the primary road network, and possible difficulties in obtaining Stopping-Up orders for parts of Stickfast Lane (9.16.9–12). The arguments on geology and hydrology, on ornithology and nature conservation as well as dust litter and odour pollution, are simply summarised, as are the points about recreation and tourism along with the agricultural argument (9.16.13–17). Such benefits as employment opportunities during the site's working or the landscape restoration on the completion of the project were outweighed by the many disadvantages already indicated.

Clay could not be extracted just to create void space capacity; that constituted a waste of mineral resources and added to the objections against the applications (9.16.18–19). So would the need for landfill tip the balance as SEM's QC hoped? After all there was the suitability of the location in clay strata, the satisfaction of the proximity principle and the need for a landfill site in Kent to meet 'the waste management needs of the South East region'. But such considerations, the inspector decided, did not outweigh what he considered to be the substantial damage the development would have on a quiet rural area in Kent 'offering space and recreational facilities' (9.16.20–1). He dismissed then SEM's appeal but in case the Secretary of State differed, he then commented on the conditions and obligations drawn up by the interested parties. The Secretary of State, however, endorsed his decision!

There were, however, three interesting features related to that endorsement. It appeared that the Secretary of State for the Environment (SSE) received the inspector's report and his decision of 31 July 1998. Secondly, he notes that SBC drew the attention of the present inspector to the Local Plan Inquiry inspector's decision in regard to the LLA designation. Thirdly, the SSE remarks that this and 'other relevant policies in the Swale Borough Plan do not affect the conclusions of the appeal inspector'. Neither, in his view, did the withdrawal of the planning application for the Kingsnorth Waste incinerator![13]

Great satisfaction may be felt by you, the reader, that eleven oak trees would not now suffer the same kind of fate bestowed on those willow trees referred to in the aestival opening this chapter. But two philosophic problems remain and perhaps even a third one. Let me explain.

At the end of chapter X a number of issues remained outstanding. Would the inspector's decision, even if you agreed with it, be fair substantively? It has been argued it was fair procedurally, though you might argue that it was procedurally satisfactory only in relation to the way the respective cases were handled at the inquiry. There still remains the question whether procedurally in his report his decision followed or was consistent with the findings he drew together in that report, meant to guide his decision. To make the point clearer, consider again that notorious recommendation of 12 September 1995 of Kent's county planning officer.[14] It might be possible to read his report and find that in no way could his recommendation be regarded as following consistently upon the report's twenty-seven clauses!

We have the philosophical problem too of trying to solve the Fact/Value dichotomy. Thirdly there is the adequacy of the Hobbesian solution: where there is disagreement between parties over, say, some evaluative decision, each can set out its case and an arbitrator decide between them, a decision which is final. A further philosophic problem can be labelled the Imperative fallacy. More about that later but let us begin with the Hobbesian solution.

Despite an aura of similarity, this arbitrator analogy soon breaks down. Though the opposing parties agreed to take part in the procedures of the inquiry, neither chose this particular inspector. Again, a further arbitrator, the SSE himself, had the right to challenge the decision. The inspector has to make the rationale for his decision public and justify it. In Hobbes's case the arbitrator is agreed upon by the parties, but not the kind of grounds he may employ to establish his decision. So, in the case of the inspector, it can be asked whether his decision is a just one, whereas in the Hobbesian case, it must be just since he, the arbitrator, made it! By this rather short circuitous route, then, we have been led back to our first philosophical question: 'Was the inspector's decision fair?'

Here we do seem to have a problem. Outside the Highways and Transport and the Recreational/Tourism arguments, he forwarded SEM's case on the following grounds: waste disposal; their case on hydrology and geology and the Ornithological and Conservation arguments pointing out, too, that there was no agricultural counter-case! Of course you might want to argue that in his Overall Conclusions he does angle more his impartiality with regard to the Highways and Transport argument (9.16.9–12) but in his earlier rationale, it might be argued that SEM's case could be given the edge.[15] Again, the same applies to the Recreational/ Tourism argument (9.16.16).[16] Only in three specific areas does he clearly oppose SEM's case: the clay need rationale; on landscape and visual impact, and in regard to matters concerning dust, litter and odour pollution. The latter can hardly be a significant matter since such concerns are common to any landfill site. If completely valid, permission would never be granted for any such development! So we are left with two central grounds for his overruling SEM's appeal: no clay need was established and the landscape and visual impact argument. Two features of the presentation of the latter argument indicate the inspector's concern with this issue: the way his Overall Conclusions are cast in its light and the fact that six further paragraphs are devoted to it (9.16.3–8).

SBC's solicitor could challenge this last conclusion: the

Inspector had no choice but to find in the way that he did because of the legal precedents provided by the inspector's decision at Carr Lane, Moreton Wirral in 1988, Park Farm, Haines Hill, Berkshire in 1991, Belstead Farm, Little Waltham, Essex in 1996 and New Road, Bagshot, Surrey[17] in 1991, noted in the inspector's summary of Swale's case (5.4.1), though not carried through to his conclusion that extracting clay to create a void space for landfilling 'would be a wasteful and non-sustainable use of mineral resources' (9.16.19). So even if there are doubts about the way his Overall Conclusions are drawn procedurally from the evidence used, there can be no doubt that his substantive conclusion was sound. But does that last judgement indicate the way the Fact/Value dichotomy can be overcome and what about the Imperative fallacy?

Taking the dichotomy first, in chapters II and III a way was suggested for justifying value claims: the *activity of *inquiring into matters of fact presupposed certain value claims. If the pursuit of truth is of significant concern, such a pursuit cannot be undertaken alone; it requires the possible checking by others in the interests of error elimination. Indeed within any inquiry such as we have been discussing, written testimony is regarded with more weight if it is or could have been tested in a public defence.[18] Now in order for such checking to take place it is a prerequisite that inquirers make themselves clear to others, a moral claim for *intelligibility; that inquirers respect points of view different from that of their own, a moral claim for *fairness and *impartiality; that inquirers can only advance and share knowledge if they seek to speak truly, a moral claim for *truthfulness and *sincerity, and only through the latter can what is the case be established between those involved in inquiry, a moral claim on behalf of the value of *truth. As far as inquirers are concerned, then these four moral claims are legitimate in the sense that any or all of the inquirers have to argue that they remain 'in force' if any kind of consensus about the truth of the matter is to be achieved.[19] In addition such a Peircian interest in sustaining these four moral claims or preconditions makes sense of the concern and care which can emerge through such shared cognizing, that is to say 'a very special kind of intimacy'.[20] In this way we can make sense of

Peirce's claim on behalf of a logical socialism': 'He who would not sacrifice his own soul to save the whole world, is, as it seems to me, illogical in his inferences, collectively. Logic is rooted in the social principle' (Peirce: CP 2.654).

Earlier, criticisms were considered of this *logistical way of grounding value claims, a way which emphasises Peirce's category *Thirdness! It risks a *conceptual imperialism on the one hand or a *cultural relativism on the other'.[21] If the cognitive claims of the scientific or institutional community counted solely, then their ideas would be regarded as supreme. If different kinds of community claims were tolerated then how could cultural relativism be avoided since each community might be applying valid but different standards. We will return to this issue in dealing with the *Imperative fallacy. Again, the logistical way can only apply to inquirers, to those engaged in language use and to those whose cognitive interest lies in seeking a common resolution to disputes: 'In public discourse, conviction is an impediment to consensus'.[22]

If the logistical way proves unsatisfactory, Peirce provides another route for overcoming the Fact/Value dichotomy, considered in chapter III, through a concern with the aesthetic dimension. Even those not concerned with inquiry or cognitive matters at all could act morally in relation to what exists by considering the latter aesthetically. If each human being sought to exercise an undistorted will to achieve the value of future experience based upon what is experienced qualitatively in the present, then ethics would presuppose aesthetics. Peirce considered this aesthetical approach, employing his category, Firstness, when, in 1903, he remarked that the problem of ethics 'is to determine by analysis what it is that one ought deliberately to admire *per se* in itself regardless of what it might lead to and regardless of its bearing upon human conduct. I call that inquiry aesthetics...' (Peirce: vol. 5, para. 36). But what is the nature of this admirableness, this aesthetic ideal which he comes to identify with 'an ultimate aim' to guide ethical conduct (Peirce: vol. 5, para. 132–134)? Peirce speculates that such an aim 'should accord with a free development of the agent's own aesthetic quality' rather than that of other persons, the community or even the biotic community! At the same time 'it should not ultimately tend to be disturbed by the reactions upon the agent of that outward world which is supposed in the very idea of action' (Peirce: vol. 5, para.136).

Can these two features – the concern for the individual's free development and the effects of experience upon the subject – be reconciled? If these two cannot be reconciled, then adopting the first alternative would lead either to hedonism or decisionism. *Hedonism – the doctrine that pleasurable living constitutes the good life – does deserve consideration Peirce claims but is to be rejected (Peirce: vol. 1, para. 614). Hedonism, as a doctrine, emphasises the static, fixes something as a 'stationary result' to be regarded as admirable in itself. *Decisionism is not so easily rejected as a Peircian option. Consider, for example, Peirce's rationale for his choice of the 'experiential method' (Peirce: vol. 5, para. 406), which was considered at the end of chapter IV, over the three other methods for gaining the truth. On the basis of that passage, Peirce simply adopts his method, as a decisionist, since no grounding can be provided for his choice of method.[23] One adopts some stance as reflecting, say, what is desired – whatever that may be, in Peirce's case the quest for truth – and then subjects that choice to criticism. On this view value statements or claims pertaining to ethical standards are created by us in the light of factual claims which are accepted by us in the generation of such standards. But even if Peirce was tempted by this stance in 1877, presumably he rejected it in adopting his logical socialism subsequently!

Peirce's further remarks might indicate another way in which his aesthetical approach can be redeemed. He claims that the two conditions can only be reconciled if 'the esthetic quality toward which the agent's free development tends and that of the ultimate action of experience upon him are parts of one esthetic whole' (Peirce: vol. 5, para. 136). He points out that this is a *metaphysical issue and is explicated in 1893 in the following way:

> The aesthetic quality is the total unanalyzable impression of reasonableness that has exposed itself in a creation. It is a pure

> Feeling but a feeling that is the Impress of a Reasonableness that creates.[24]

His metaphysical stance is to be understood in the following way. For him, the laws of nature emerge as 'results of evolution' (Peirce: vol. 6, para. 13). Such laws can be conceived as arising from absolute chance, by fortuitous variation in the spirit of Firstness, *tychastically; by mechanical necessity, determined in the spirit of Secondness, *anancastically; or through divine love in accord with Thirdness, *agapastically.[25] Here we can see Peirce creating his metaphysics analogically to his epistemology. Just as 'creative love's sympathetic empathy' (agapism), in forming habits of the universe,[26] mediates between fortuitous chance variations (tychism) and selection through 'force of circumstance' (anancasm) (Peirce: vol. 6, para. 302–307) so, in the habitual search for truth, Thirdness mediates between spontaneous qualitative intuitions (Firstness) and the checking of them by 'brute facts' (Secondness). On this view, the claim that ethics has now become 'the Application of Metaphysics' (Peirce: vol. 5, para. 121) comes out in his assertion that 'the saving truth is that there is a Thirdness in experience, an element of reasonableness to which we can train our own reason to conform more and more (Peirce: vol. 5, para. 160). And the pursuit of that sense of reasonableness would lead to an extension of one's own concerns 'to seek aesthetic optimization for the community', as Charles Hartshorne puts it. Here a community is now understood to include life as a whole rather than 'limit value to humanity'.[27] Only this kind of metaphysical stance can make sense of Whitehead's claim that 'the ultimate "ground" to which all probable judgments must refer can be nothing else than the actual world as objectified in judging subjects'.[28] Making sense of those last eight words involves employing a metaphysic. Whether that be a Peircian or a Whiteheadian one would be the subject of another kind of inquiry.

Peirce could be read, however, as offering a further method for overcoming the Fact/Value dichotomy by insisting that moral conduct is connected to habitual actions *not* to theoretical reasoning. Accordingly, instinctive or common sense reasoning is more reliable than any kind of rationalistic morality such as we might have been considering so far, so that in matters of business, family 'or other departments of human life' especially in relation to 'certain *dicta* of my conscience', 'the dictates of Instinct' are to be preferred 'to those of reason herself' (Peirce: vol. 2, para. 177).[29] Such a stance follows from his view that it 'is the instincts, the sentiments, that make the substance of the soul. Cognition is only its surface, its locus of contact with what is external to it' (Peirce: vol. 1, para. 628).

Now in the context of considering 'matters of vital importance' where 'reasoning is at once an impertinence towards its subject matter and a treason against itself', since it can only be employed adequately in theoretical not practical matters, Peirce speaks of *casuistry. The latter is defined as, 'the determination of what under given circumstances ought to or may be done'. Peirce regards causitry in a sympathetic way even though it may be 'in ill-repute and somewhat deservedly so' (Peirce: vol. 1, para. 671; vol. 1, para. 577; vol. 1, para. 666; vol. 2, para. 153). Unlike the other methods for overcoming the Fact/Value dichotomy considered so far – the logistical, the aesthetical or the metaphysical – this method is directly inductive in character, thereby invoking a sense of Secondness. So Peirce claims, for example, that sexual mores are 'an instinctual or sentimental induction summarising the experience of all our race' (Peirce: vol. 1, para. 633) and leads him into endorsing a 'true conservatism': 'not trusting to reasoning about questions of vital importance but rather to hereditary instincts and traditional sentiments' (Peirce: vol. 1, para. 661). Casuistry, however, is not concerned with a general rule to govern moral outcomes but, rather, with exemplary cases of conduct which can serve as a *paradigm for other cases. Recently the methodology employed in this approach has been unpacked in terms of six steps: 'the reliance on paradigms and analogies, the appeal to maxims, the analysis of circumstances, degrees of probability, the use of cumulative arguments and the presentation of a final resolution'.[30] Clearly any moral maxims arising from this approach would develop consensually through a period of time and thereby represent a 'sensus communis' or a 'common sense' 'which is the resultant of the traditional experience of mankind' (Peirce: vol. 1, para. 654). Such common sense becomes, then, built into our sentiments or instinct – he treats them synonymously (Peirce: vol. 1, para. 634) – so that they can't be regarded as natural inclinations. Unlike the latter, instincts can grow and develop along lines 'parallel to those of reasoning' (Peirce: vol. 1, para. 648).

In advocating 'sentimentalism' he is not advocating casuistry directly. In the past, a direct case-based method had initiated corruption into a situational ethics by so emphasising special circumstances in relation to the distinctions and exceptions it employed, that separating right from wrong became an almost impossible task. Today, we have become more aware of how 'common sense' is not shared by individuals within a community, never mind between communities more broadly understood. Indeed, in relation to the former the sheer range of cognitive perspectives, enjoyed by people with different appreciations of 'common sense', was explored in chapter X. In addition this current cultural situation is not assisted by a tendency, anticipated by Peirce, to substitute moral judgments about human actions by calculative reasonings in relation to desired behavioural outcomes where theoretical knowledge, generated by experts, is imposed upon a depoliticized population through its direct application. But Peirce could be read as admiring the emphasis casuistry puts upon moral action rather than on ethical theorising, its concern with what must be done in some existing situation morally and its emphasis on traditional mores and historical precedents.

Two issues do emerge, however, from these somewhat cursory remarks on casuistry. It could be said that during the Tiptree Inquiry itself there was recourse to this approach. SBC' solicitor adopted a case-based approach to establish that a mineral – sand and gravel, clay or peat – could not be removed to create void space. Such cases were challenged by SEM's QC by saying some of them referred to decisions with regard to green belt sites and so on. Of course, even though these be legal examples, we gain a practical sense of how casuistry can be employed in decision-making. Notice, too, that in the cases situated in green belt areas – Carr Lane (Moreton Wirral), Haines Hill (Berkshire) and New Road (Bagshot) – it was claimed that the demand for the creation of void space clashed directly with green belt concerns, just as there was at Tiptree between the need for landfill and the interests of an LLA designation. Now the issue of the Imperative fallacy can be faced.

It was argued that since in Kent there is a void space shortage for waste disposal, it is *right* that adequate provision be made. So the Tiptree Site *ought* to be developed for such purposes. It was also argued, particularly by myself, that because it was right that the site gain – as it did – an LLA designation, it ought not to be developed as a landfill site. What has now appeared is something different from the Fact/Value dichotomy: the right/ought dichotomy, the latter labelled the *Imperative fallacy. Clearly, at this moment, for example, it could be argued that it is right for this writer a) to find out the news from the radio to keep himself informed, b) to write this chapter, c) to phone a lonely relative, d) to write an overdue letter, e) to prepare a meal downstairs and so on! Now on what grounds ought this writer to pursue one of these *right* actions rather than one of the others? To assume that something is *right* and therefore it *ought* to be done, commits the Imperative fallacy since no grounds have been given for justifying that particular action over other right courses of conduct. Again, it could be and was argued that it was right to prevent the development of the Tiptree Site because it would disturb the survival needs of its present biotic community. Against that view, it was argued that because a waste disposal site was needed, it was right to develop the site. How is an imperative ought to be generated? I suggested Heffernan's principle so as to provide a way of overcoming this Right/Wrong dichotomy. The inspector rejected that tactic since he was not persuaded by the ornithological case.

In this chapter we have concentrated on the question: 'Was the inspector's decision fair?' Given all the facts presented at the inquiry, was his decision one that ought to have been made, that is to say, did it overcome the Fact/Value dichotomy appropriately? Three strategies can be found in Peirce's writings to meet this problem philosophically. The logistical emphasises Thirdness since it is a cognitive approach based on the idea that a moral decision can only arise out of the investigations of a community committed to inquiry. The aesthetical emphasises Firstness since it presumes that moral conduct can only be justified by enhancing the value of future experience. If a straightforward version of either hedonism or decisionism is to be avoided, then these approaches may have to seek recourse in the metaphysical. Finally, an approach emphasising Secondness – casuistry – was considered, a case-based method which stresses certain paradigm cases to provide-rules-to cover what ought to be done. It was noted that by making use of this strategy – applying the rule derived from previous cases that minerals can't be extracted simply to provide void space – the inspector's decision has to be the correct one. Yet it was delivered in terms of the significance of the landscape and visual impact argument, a case which in the past, for previous green belt cases at least, was not regarded as of overriding importance!

Two comments might be made on this chapter. Firstly, the Fact/Value dichotomy has not really been overcome since the three strategies for solving it are each weak in themselves: the logistical, because it focuses on cognition alone; the aesthetical, because the metaphysical concern was merely described not justified; the approach of casuistry, because of difficulties arising in its application. Secondly, these methods have not been reconciled and do not appear to be so within Peirce's writings themselves. Indeed, he could be regarded as offering his reader different philosophical directions without reconciling any of them.[31] Yet Peirce, in 1868 does, in a crucial remark, indicate a way out of both these difficulties:

> Philosophy ought to imitate the successful sciences in its methods, so far as to proceed only from tangible premises which can be subjected to careful scrutiny, and to trust rather to the multitude and variety of its arguments than to the conclusiveness of any one. Its reasonings should not form a chain which is no stronger than its weakest link, but a cable whose fibres may be ever so slender, provided they are sufficiently numerous and ultimately connected.' (Peirce, CP 5.265)

How that cable can be constructed is a subject for a further chapter!

NOTES

[1] Insanity in sanity', River Avon, Bradford–on–Avon, Wilts., 1 June 1994.
[2]Letter to the Chief Executive, KCC from the Government Office for the South East 27 February 1998.
[3]Letter to Derek Wyatt, MP for Sittingbourne from the Parliamentary Under Secretary of State for the Environment 16 February 1998; ref. PT/RD/PSO/977/98.
[4]Letter to The Planning inspectorate from the Head of Planning Applications, 4 March 1998, Ref: PC/MH/SAM/SW/94/1137.
[5]Letter to N E Boulting from the government office for the South East, 22 April 1998.
[6]See chapt. IX, 'The Inquiry Reopens': Can Two Utilitarian Arguments Entwine?' see note 7.
[7]Para. 6 refers to para. 6 of 1–17 paras. of the inspector's Conclusions to be found in *Report of a Public inquiry into Objections to the Swale Local Borough Plan*, Date of inquiry 3 December 1996 to 17 February 1997. The Planning inspectorate, Tollgate House, Hamilton Street, Bristol B52 7DJ, 23 July 1998 pp.100–104
[8]Town and Country Planning Act 1990; sect. 79 Appeals by South Eastern Minerals Limited, Land at Tiptree Farm, High Oak Hill, Funton, Sittingbourne, Kent File nos. APP/W2275/A/95/259578; APP/W2275/A/96/271958 and APP/V2255/A/96/272360, 12 October 1998.
[9]Town and Country Planning Act 1990, op. cit., where 9 in 9.1.2 stands for the ninth part of the report, 1 for the sect. 'Introduction' and 2 stands for the para. number.
Town and Country Planning Act 1990, op. cit., where 9 in 9.5 stands for the ninth part of the report and 5 for the section 'Policy Background'.
[11]The inspector adds: 'but it, along with the Kent Trust for Nature Conservation, Kent Wildlife Trust and the RSPB, has expressed detailed concerns and it recommends a number of safeguards. Town and Country Planning Act 1990, op. cit., (9.12.2)
[12]'The survival interests of human beings ought to outweigh those of the rest of the biotic community and the survival interests of the rest of the biotic community ought to outweigh the nonsurvival interests of human beings.' J D Heffernan 'The Land Ethic: A Critical Appraisal' *Environ. Ethics* vol. 4 no. 3 Fall 1982, pp.245–7 p.246.
[13]Town and Country Planning Act 1990, op. cit., the accompanying letter of the Planning Directorate, Minerals and Waste Division, 12 October. 1998.
[14]Report to the county planning officers to the Planning Sub-Committee, 12 September 1995 from the county planning officer. Subject: 'To Extend An Existing Clay Quarry, Restoration to Agriculture by Infilling, Provision of Site Access Roads and Improvements to Local Roads, Tiptree Farm, Funton, Lower Halstow – South Eastern Minerals Ltd (MR 878.669)'.
[15]He adopts the agreed statement between SEM and KCC; HGVs could be restricted from turning on leaving Tiptree Site by proper road design; improved

roads would have adequate capacity for HGVs; he opposes the Bobbing residents case; though the arrangements would make for a 'much less attractive and pleasant route for pedestrians, cyclists and the equestrians from the Callum Park Riding Centre' he recognises there were few of the former and that for the latter there would be little in the way of any real 'reduction in its existing use or any loss of current employment' (9.14.7); lorries in the main would use their proper routes; the proximity principle is satisfied; SEM might purchase land for realigning the proposed highway alteration etc. His only reservations concern the NCN, the connection to Sheppey Way since difficulties with Stopping-Up orders for Stickfast Lane form the basis of a condition agreed by all parties (9.9.1–16).

[16]Consider his case on the Callum Park Riding Centre (note 15), his claim that recreational activities in the estuary are not likely to be affected significantly by noise and his claim that 'the effect of the scheme in various forms of recreational activity would be damaging. The harm would be less, although still significant, if the alternative haul road access scheme were provided' (9.14.9).

[17]See chapt. IX endnotes 11 and 12.

[18]See note 4 chapt. IX; note too the inspector's comment at the New Road inquiry, Bagshot, Surrey (cf. note 11, chapt. IX), in assessing the case for the extraction of peat I have three reports before me – in attempting to balance the conflicting information it is unfortunate that these were presented in the form of written reports only and without any opportunity to test them' (T/APP/B3600/A/90/146629/P5) p.6 para. 27.

[19]James B Sauer 'Discourse, Consensus and Value: Conversations About The Intelligible Relation Between The Intelligible Relation Between The Private and Public Spheres' in *Political Dialogue: Theories and Practices*, pp.143–66, p.150.

[20]E S Read, 'Knowers Talking About The Known' *Synthese* vol. 12, 1992 p.97.

[21]Cheryl Misak 'A Peircian Account of Moral Judgments in *Peirce and Value Theory*, pp.39–48, 46.

[22]James B Sauer, 'Discourse, Consenses and Value' op. cit., p.152.

[23]Noel E Boulting, 'Hobbes and the Problem of Rationality' in *Hobbes: War Among Nations*, pp.179–189, 181–182.

[24]Quoted by H Parrett in his 'Peircean Fragments on the Aesthetic Experience', Cheryl Misak, 'A Peircian Account of Moral Judgments', op. cit., pp.179–190 p.183.

[25]Noel E Boulting 'Charles Sanders Peirce (1839–1914): Peirce's Idea of Ultimate Reality and Meaning Related to Humanity's Ultimate Future as Seen through Scientific inquiry' in *American Philosophers' Ideas of Ultimate Reality and Meaning* ed. by A J Reck (et al.)Toronto UP 1994 chapt. 3, 2.3.

[26]To appreciate the full significance of this dimension in Peirce's thinking see Carl R Hausman 'Eros and Agape in Creative Evolution: A Peircian Insight' *Process Studies*, vol.4 no.2 Spring 1974 pp.11–25.

[27]Charles Hartshorne, 'Beyond Enlightened Self–Interest: The Illusions of Egoism' in *The Zero Fallacy and Other Essays in Neoclassical Philosophy*, ed. by M Valady Chicago, Open Court, 1997 pp.198, 199.

[28]A N Whitehead *Process and Reality* ed. by D R Griffin and D W Sherburne, New York, The Free Press 1979 p.203.

[29]cf. '...we have an instinctual theory of reasoning, which gets corrected in the course of our experience. So, it would be most unreasonable to demand that the study of logic should supply an artificial method of doing the thinking that his regular business requires every man daily to do...'(Peirce, CP 2.3).

[30]A R Jonsen (with S Toulmin), *The Abuse of' Casuistry*, Berkeley, California UP, 1988 p.251. Mary B Mahowald relates Peirce's ideas to Casuistry in her 'Collaboration and Casuistry', pp.61–71.

[31]'There are at least two reasons why the notion of strands of system is appropriate to Peirce's life's work. The first is his claim that philosophy proceeds not from a single premise or set of premises along a single thread of reasoning but inductively, gathering from experience what it can and braiding it into a cable of belief. –The second reason lies in the fragmentary nature of Peirce's philosophical writing' D Anderson, *Strands of the System: The Philosophy of Charles Peirce*, p.26.

Chapter XII

A QUESTION OF COSTS

Like ripples propelled in various directions over some smooth surfaced lake, fashionable conceptions simply come and go for those whose thinking is generated by TV screen appearances, to prevent consciousness focusing on its own deeper set of ruminations. But might such chic notions as the significance of the global economy or the importance of having trained personnel be any more than verbal accompaniment to the increasingly global-mode of production which legitimises discussion of these concepts, even if as idle chatter? If not significant in itself then in their constant repetition, like the persistent rush of leaves in trees standing solemnly by this water's edge, these notions demand attention because of the new forms of consciousness they incite which, though never explicitly referred to, gain their legitimacy through such reinforcement: self-interestedness and the beneficial effects of greed. Consequently, he who even seeks to define these new forms of mentality finds himself offset, left to carp, like some black crow persistently croaking over some rural English scene about possibilities which, for those with whom he communicates, simply fail to exist. And if these latter 'thinkers' do occasionally beat against the walls of their 'own' apparently consistent set of thoughts which imprison them, then, he who-would-be-conscious, simply endures reflections which circle around these walls, justified only by the immediacy of felt recollections from a distant past, now judged irrelevant in this constant contemporary chase for what are called 'new ideas'.[1]

A further document was published on 27 October 1998: the inspector's report and recommendations on the applications for costs. But the idea of costs extends beyond his findings. The memorial service of Lower Halstow's parish councillor took place in Upchurch's St Mary's Church, five days after the inspector's dismissal of the appeals was published. Hence he, as one of Tiptree Action Group (TAG)'s leading activists, would not have known the inspector's decision before he died of cancer five days previously, though he may have been aware of the decision drawn by the LLA designation inspector published in July. Then there is the consequence of the dismissal of SEM's appeals to be considered: the likelihood of a landfill site being developed elsewhere in Kent. Finally, there are some broader issues, perhaps of concern to a layperson 'who-would-be-conscious', in relation to costs, which need to be considered. This chapter, then, has three sorts of concerns to constitute its contents plus reflections on these. Let us begin with the inspector's report on costs.

The report on the application for costs[2] runs to fifteen pages. In addition there is a three-page covering letter and two pages of guidance notes. Again you will have to forgive your informant for truncating these proceedings in what follows. The important features in the accompanying letter, besides the inspector's five recommendations, lie in two claims. Firstly, that costs can only be met in the face of 'unreasonable behaviour' which results in unnecessary expense for any party; otherwise each is required to meet their own expenses. Secondly, since Biffa Waste Services Limited were not, as a distinct party, represented at the inquiry no specific costs could be awarded against that company! In relation to this last feature in SEM's response we find the following passage containing a view enunciated on speech day by the QC: *Biffa has no legal obligation to pay any costs that might be awarded but if an order were made, Biffa would not disclaim it, even though the order could only technically be made against SEM* (6.10).

After the Introduction (sect. 1), the report is divided into the following sections; submission by Kent County Council (KCC) (sect. 2); by Swale Borough Council (SBC) against SEM (sect. 3); by TAG against SEM (sect. 4) and by SEM against KCC (sect. 5). There then follow two responses, one by SEM (sect. 6) and the other by KCC (sect. 7). The inspector's Conclusions follow (sect. 8) and the report ends with his 'Overall Conclusions and Recommendations' (sect. 9). Attention will be focused upon his Conclusions since there is overlap in the content of the three submissions made against SEM.

First of all, an award of total costs is not made against SEM. Perhaps you thought they might have been. The inspector at

Kent's Minerals Local Plan (KMLP) inquiry rejected SEM's case on clay need, accepting the plan's approach as 'entirely reasonable' whilst pointing out that there was plenty of clay in Kent for engineering purposes. KCC then asked SEM to withdraw its appeal against the Tiptree application on 12 March 1997. SEM refused: the KMLP was adopted (sect. 8.2). The inspector, at the Tiptree Inquiry, gave three reasons legitimating SEM's behaviour: the KMLP is concerned with generalities not specifics (Thirdness not Secondness!) in focusing on the overall strategy for the county; it fails to adjudicate on 'site specific matters' e.g. the landscape issue at Tiptree, and thirdly it does not consider landfill needs (8.4).

The inspector again was in favour of SEM with respect to its borehole evidence. Though it was a pity the evidence could not have been presented before 13 November 1996, the provision of the new evidence did not constitute unreasonable behaviour (8.5). He again supported SEM with regard to environmental information. Even if inadequate, as a result of the publishing of the new borehole evidence in 1996, it was adequate by the time the inquiry restarted a year later in the light of a Supplementary Environmental Statement (ES) on Landfill Engineering and Hydrogeology which surfaced in April, followed by another in August 1997. He rejected, too, KCC's claim that any new matters, such as changing the application from a 'clay led' to a 'waste led' one, were introduced between the adjournment of the inquiry and its reopening a year later, save for matters consequential following the development of issues already flagged previously, such as the adoption of the KMLP (8.8).

Judgments were made against SEM in two ways. Firstly, the inspector decided that their Supplementary Environmental Statement (ES) submitted to KCC was late and incomplete, so that the inquiry could not be resumed on 17 June 1997 as planned in the previous November (8.9). Secondly, besides the restart of the inquiry being put off for a second time until 18 November 1997, there had to be, consequent on that first ES, a procedural meeting on 27 August to ensure clarification of the proposals and some other details raised by the inspector in his preliminary report the previous November. Moreover, SEM delivered a further ES at this procedural meeting (8.10). It did not seem unreasonable then, to charge costs for this meeting to SEM.

He was not concerned with whether or not SEM was a quasi-appellant. He did think, however, that both third parties – SBC and TAG – should be awarded costs in the light of the procedural issues on which he had judged (8.9–10), raised above (cf. 8.11–12). He scolded KCC for not accepting the Environmental Agency (EA)'s hydrogeological judgement, on the new excavation levels and the preliminary risk assessment report done for SEM, as final. Dragging in a further witness in an attempt to undermine the EA's judgement so as 'to substitute its own judgement for that of the agency' the inspector deemed unacceptable, particularly because he claimed SEM had to fund a further witness to challenge KCC's attempt, an attempt which aired matters 'not raised in detail by any other party' (8.13).

Now his 'Overall Conclusions' can be listed. Because of SEM's continual speculations over clay, which KCC had to keep refuting as a result of SEM's resistance to the KMLP inspector's judgments, a partial award of costs was made in favour of KCC against SEM (9.1.i). A partial award of costs was made in favour of SEM against KCC because of KCC's non-acceptance of the EA's judgement (9.1.v). Partial awards of costs were awarded against SEM and for KCC, SBC and TAG because of SEM's failure to deliver the ES on time and because the 27 August procedural meeting had to be initiated.

In relation to our first concern, it would be rather odd to speak of the death of one of TAG's leading activists as a cost of the inquiry. Rather, it might be said that the Tiptree fight kept him going longer than might have been otherwise expected. Reference is made to his memorial service on 16 October 1998 now because of the way his son described his father: he was a man interested in knowledge for its own sake gained through an enormous amount of book reading and that he exercised a sense of duty for its own sake both towards the concerns of his community and towards the surrounding natural environment in the Lower Halstow and Tiptree areas. This phrase *for its own sake* was used on two further occasions. It was used by SBC's solicitor in his 'Final Submission'[3] and it was also used in relation to Kent Structure Plan Policy,

ENV1:[4]

> The countryside will be protected for its own sake. Developments in the countryside should seek to maintain or enhance it. Development which will adversely affect the countryside will not be permitted unless there is an overriding need for it which outweighs the requirement to protect the countryside. (ENV1)

Again it appears in the first sentence of SB's Local Plan, Policy E11:

> The Countryside of the Borough, which is all the land falling outside the defined built-up area boundaries, will be protected for its own sake (E11).

It is a phrase which is used too in the inspector's Conclusions in his report of 12 October 1998 (9.8.3, 13, 9, 16.6.8). How is this phrase to be construed?

Our second concern can best be mediated in the following way: Imagine you are driving from London to Dover down the M2. When the turn off for Canterbury is reached instead of branching right along the A2 up the hill, proceed as if going to the Isle of Thanet i.e. to Whitstable and Herne Bay. About two miles along on your right-hand side to the east, a line of hills appears. On this landscape it is proposed to create Europe's biggest landfill at Lamberhurst Farm, Dargate. The inquiry into the appellant's appeal against KCC's rejection of the landfill application opened in March 1998, was adjourned and reopened during November 1998 in Boughton's Village Hall. It was then adjourned again before reopening in 1999. Superb views are to be enjoyed from the farm which seems to consist of a number of industrial units. But it is impossible to proceed further up a track to the top of the hill; rumour has it that great difficulties were undergone in gaining permission to examine archaeological remains which are supposed to be of some historical significance there.[5] No plan is made for digging minerals to ensure a hole for taking waste as at Tiptree, some seven miles to the west of Dargate, and the site does seem to have convenient access to the main London–Thanet road system. The waste will be tipped into a slight dip in the hills with the consequent danger that the kind of strong winds, which can be enjoyed on this northern coastline of Kent, will simply blow the lighter materials straight into Whitstable! In addition it is proposed that leachate, after being generated by the waste in the site and collected at the bottom of the tip, will be returned to the top and passed through the rubbish again in order to decompose the latter, thereby causing that leachate to become more toxic and a possible environmental hazard! A cycle ride around the villages in the area – Boughton Street, Staple Street, Herne Hill and High Street near Yorkletts – along with views from the villages on the north-western side of the main road to Thanet, soon make a cyclist, at least, aware of how much of the activity on this hillside can be viewed far and wide. At this inquiry there was no local borough representative. Was that because Dargate fell between the Swale and Canterbury areas? There was a solicitor acting for the local residents and the KCC solicitor was the same person who set out the case for KCC at the Tiptree Inquiry. Now we can turn to some broader issues in relation to the question of costs, our third concern!

Swale's Labour MP raised an important issue at the Tiptree Inquiry. He claimed that a large company, such as the appellant represented, had no upper limit on the money they could spend at inquiries such as those at Tiptree and now at Dargate. On the other hand, there was a limit which residents could afford through representation by such a protest group as TAG.[6] Rumour had it that behind Biffa stood Severn Trent Water Company. If so, then the money such a company could raise through charging its customers, even if they be different from those residents supporting TAG, could be regarded as being used to sustain the appellant's efforts. Such a position is hardly fair, especially as in the Dargate case residents 'must live with the knowledge that an amended application can keep the spectre looming' of a further application if an appeal against KCC fails.[7] Surely a law could be passed to ensure that *any company engaged in developing land in such a way as to lead residents, a borough council and a county council to protect itself against it, should pay something towards the cost of the resident's campaign.* The Tiptree campaign cost residents no less than £20,000. Such a law would ameliorate a moral injustice.

A second issue on the matter of costs separates the case of Tiptree from Dargate. Cleanaway – 'Don't let them get clean away with it!' is the Dargate protesters' slogan – is officially the appellant at Dargate. But who was the appellant at Tiptree? KCC claimed Biffa had funded the appeal in its entirety, supplied witnesses at this inquiry, would gain from the waste disposal application and enjoyed options over the land covered by the application. KCC tried to show legally, then, that Biffa was really 'the appellant' (2.1.2).[8] The inspector rejected such arguments: 'Biffa was not represented at the inquiry as a separate party in its own right' (8.11). Fortunately Biffa did undertake to cover SEM's liabilities. But can it be right in either a moral or legal sense for a company with outstanding debts, whose status is gradually withdrawn during an inquiry[9] and which provided no witnesses on its own behalf, to be allowed to continue to forward such an inquiry at such public expense?

A third issue arises from a distinction made in the inspector's report and recommendations, a distinction already drawn in this book between procedural and substantive matters. In his summary of SBC's submission for costs it is pointed out that costs are hardly ever made in relation to 'the substance of a case' and only partial costs relating to 'procedural matters' (3.2). The same distinction is used again in summarising TAG's submission though here claims were made for full award of costs relating to the case's substance (4.3) In the case of the latter, TAG's representative had argued that SEM's proposals ran against development plan policies and so had no reasonable hope of success. We have already indicated the inspector's view on this matter, but let us consider again the contribution Lower Halstow's parish councillor made in this inquiry.

Attention has already been drawn to this councillor's concern for issues related to value 'for its own sake'. In addition, both in written (Doc. 12) and spoken evidence he referred to the possibility of a spring on the hill and water continually running from the site over parts of the shore road below.[10] Given this evidence and the acknowledgement by the inspector of ground water rising in boreholes when measurements of clay levels were taken (9.11.3), it seems odd that in his costs report he should have chided KCC for airing matters which 'were not raised in detail by any other party' (8.13). He could defend his case by saying that the representative for TAG did not emphasise sufficiently her own witness's evidence in rendering TAG's final submission, for this issue to be given further consideration. At most, this issue might constitute a procedural oversight rather than a substantive moral blunder on the inspector's part!

Any layperson must be concerned, however, with two obvious implications flowing from these two inquiries. One is a substantive matter: what does any land designation constitute? The other is a procedural matter: if a landfill is required in a county such as Kent, is the system of a series of inquiries dealing with each site on a piecemeal, individual basis the fairest way of deciding where it is to be located? In regard to the latter, over the last three years, first of all there was the application by Brett Waste Management to extend an existing landfill site at Shelford Quarry, in Canterbury. Permission was granted in October 1996 under the proviso that no clay be exported simply to create void space, unless there is some legitimate need for clay.[11] The application by BFI Ltd for a composite proposal for clay extraction and landfill at Rosecourt Farm, Isle of Grain followed and was refused on the 30 September 1997 because it created an alien landform and there was no clay need, making it a 'waste-led' application.[12] Now we have had the Tiptree decision and Dargate appears next in line![13] After that there is the prospect of a landfill application near Oare Marshes Nature Reserve, north of Faversham.[14] The next will probably be for Richborough in East Kent.[15] If there is a landfill need in Kent, is it not more rational for a body such as KCC to consider the most suitable locations rather than have this individualistic, piecemeal approach where the final site may not be decided on landscape or ecological grounds but by the sheer weight of local public opinion. Perhaps with such worst-case prospects in mind, we can now appreciate the reason for KCC's planning officer's notorious decision of 12 September 1995;[16] he might well have sought to end this process at Tiptree itself, but his decision was overruled because of the sheer weight of local public opinion against it!

The former substantive issue is raised by the question: 'What significance is to be attached to the label Area of Outstanding Natural Beauty (AONB)?' The amazing common feature in regard to the decisions with respect to mineral extraction followed by landfill at Carr Lane, Moreton Wirral in 1988, New Road, Bagshot Surrey in 1991 and Park Farm, Haines Hill, Berkshire 1991[17] is that in all three cases the rejection was because mineral extraction, without a clear need for it, could not justify the creation of void space! In all three cases the fact that these areas were in green belt districts was not judged to be relevant! In addition, evidently mining is allowed in an AONB in the Brecon Beacons! This state of affairs is especially confused in the case of Tiptree because, on the one hand, the inspector at the LLA designation inquiry criticised SBC's wording of Policy E17 because it appeared to carry the weight of an AONB status yet, on the other, the inspector at the main inquiry in his 'Overall Conclusions' gives more space to the LLA issue than any other![18]

It is clear from what has been said that moral issues still haunt this inquiry's outcome and the issues which arise consequent to it; issues of both a procedural and a substantive nature. In addition we have the question: 'What is meant by something being valued or protected for its own sake?' In chapter XI it was suggested that three methods can be distinguished in Peirce's writings for arriving at a moral decision to overcome the Fact/Value dichotomy: the *aesthetical* – suggested by the category Firstness – since an action would be described as moral if it enhanced the present value of experience for the future; the route offered by *Casuistry* – emphasising Secondness – where agreed paradigms from the past indicate rules for what ought to be done and the *logistical* – issuing from a concern with Thirdness – where a moral decision can emerge from the cognitive inquiries of a community of investigators committed to ascertaining the truth of the matter. A model will now be suggested for bringing these three strands of argument together to form a cable of reasonings, as Peirce (Peirce: vol. 5, para. 265) put it, to enable a valid moral judgement to be created. After setting out this model we can then see where the notion of valuing something for its own sake fits into the picture!

The aesthetical dimension provides the initial fibre for this model, as Hartshorne, in a Peircian moment, suggests: 'Since the intrinsic value of experience is by definition aesthetic value, and since goodness is the disinterested will to enhance the value of future experiences, ethics presupposes aesthetics'.[19] But what is it to experience aesthetically? Peirce suggests that it is to exercise that rare faculty 'of seeing what stares one in the face, just as it presents itself, unreplaced by any interpretation, unsophisticated by any allowance for this or that supposed modifying circumstance. This is the faculty of the artist who sees for example the apparent colours of nature as they appear' (Peirce: vol. 5, para. 42).[20]

You may seek to object in a way Habermas has already proposed: for epistemic purposes, can we 'expect to be able to use the experiential potential gathered in non-objectivating dealings with nature'[21] this aesthetic perspective provides? A rationale can be provided for claiming it is not possible to do so. Ordinary things acquire the aesthetic qualities they appear to have through 'projection from subjective impressions – awakened by other properties of these objects, discoverable through sense and intelligence'.[22] Passmore makes the same point in quoting Lotze on human creations as to whether:

> ...the feelings we then admire are supposed to be those of the artist or whether we admire our own feelings anthropomorphically ascribed to inanimate objects – 'we transform the inertness of a building into so many hubs of a living body, a body experiencing inner strains which we transport back into ourselves' – the suggestion is in either case the same, that nothing inanimate can have aesthetic properties in itself, that nothing is 'beautiful' except the human spirit.[23]

Such a claim follows in the tradition of an aesthetics developed since Kant for whom a disinterested concern underpins an aesthetic awareness, that is to say, what is contemplative in relation to the character of a possible object either humanly created or natural. This view is indifferent with respect to whether that aesthetic character exists or not! Schiller, following Kantian principles in his *Aesthetic Letters*, implies the same: 'The reality of things is the work of things themselves; the semblance of things is the work of man; and a nature that delights in semblance is no longer taking pleasure in what it receives, but in what it does'.[24]

Peirce, too, admitted that when he was once 'a babe in philosophy my bottle was filled from the udders of Kant' (Peirce: vol. 2, para. 113) and that he was deeply influenced by Schiller's *Aesthetic Letters* (Peirce: vol. 2, para. 197).[25] Peirce adopts a Kantian approach, viewing nature as if it were a picture thereby relating aesthetic awareness to simply appreciation of fine art (Peirce: vol. 1, para. 281). And as we saw in chapter II, Peirce's treatment of aesthetic matters, at least until 1902, is very human mind dependent, making it anthropocentric in character.

In taking Peirce's more phaneroscopic approach seriously, however, sense can be made of his desire to treat 'Phenomena in their Firstness' significantly, for in experiencing them 'in themselves as phenomena' (Peirce: vol. 5, para. 122) it makes no sense to separate what might be brought to it by a subject on the one hand, and what belongs to the existent on the other. In fact how can an object be viewed in separation from the viewer perceiving it whether instrumentally, scientifically, unconcernedly or aesthetically, to use distinctions drawn in chapter I? But in aesthetic perception we are not, as we are in the case of scientific viewing, completely distanced from the object as in a laboratory, but rather participants in a greater whole, the gestalt of an aesthetic experience.[26] To claim, with Passmore et al., that within this perception there is an unreality or semblance of things, is already to presuppose either the scientific or even the instrumental, or indeed the unconcerned stance, as the dimension upon which to judge the aesthetic perspective as being somehow unreal!

In chapter XI we considered the status to be bestowed upon this aesthetic charm that can saturate our life 'which, however despised by some and neglected by most, yet exhales an aroma so ethereal that it hardly obtrudes itself upon attention, though we may at once become sensible of its loss',[27] as Bullough puts it. If either the celebration of an individual Hedonism or the endorsement of a private decisionism is to be avoided – temptations considered by Peirce – then we might see why he thought metaphysics was needed to account for this type of awareness. But that issue concerns the grounding of the aesthetic dimension rather than its status in a normative model required for reaching a moral decision.

So far we have been explicating the aesthetic dimension of experience in terms of a remark of Adorno's which would do justice to Peirce's phaneroscopic position: 'To yield to the object means to do justice to the object's qualitative moments'.[28] Here Adorno follows Benjamin's lead in focusing upon the contingent, the discarded, the non-instrumental elements in experience[29] where the possibility lies for humans to gain something as to the truth of the fleeting actuality of existence constituting their human condition every day. So Adorno admired Benjamin's tendency to be attracted by 'the petrified, frozen or obsolete elements of civilisation, to everything in it devoid of domestic vitality no less irresistibly than is the collector to fossils or to the plant in the herbarium'.[30] Within an experience constituted by a concentrated attention upon something one is unintentionally drawn towards, the human subject can realise, within a careful observation of that thing or entity, something transcending its occasion of particularity. But to avoid a private Hedonism or Decisionism methodologically rather than *existentially, that is to say, in order to see how this aesthetical dimension can contribute to the generation of a normative model for action rather than what it suggests as to the nature of ultimate reality, we need to consider what can be done with this insight or spark of awareness realised in the contingency of what exists.

The second fibre is constituted by a concern for Thirdness, sustaining the logistical approach. The individual insight yielded through aesthetic awareness can be made subject to the scrutiny of critical examination between fellow moral inquirers through sign interpretation. In engaging in such activity participants imply their commitment to the idea of 'an ideal communication community'[31] where each member seeks to sustain an ethic embodying those 'dispositions of heart which a man ought to have in searching for the truth. These are Peirce's three regulatives identified as St Paul's famous trio 'Charity, Faith and Hope' (Peirce: vol. 2, para. 655). If such dispositions do indeed inform the habits governing the way humans act and such habits do indeed issue from beliefs considered to be true, then such agents can be said to embody reason – Thirdness – practically in their everyday lives. In this case they can in principle, at least, become members of such an ideal communicative community. But striving to attain such an ideal community does presuppose an anthropocentric concern for the survival of the human race. Such striving presupposes 'choices affecting the environment and our relationship with it' since what we do now has consequences for that environment which we see today increasingly having counter-effects on our own species. Hence such sentiments have to be extended beyond our existing human society towards a humanity more broadly understood[32] if the human race is to survive.

Note that this second strand of the normative model under construction is procedural rather than substantive in character, whereas the latter describes the nature of the resultant insight from this model's previous aesthetical strand. That procedural character allows possible distortions, which could be associated with the way the insight gained from the aesthetical dimension is mediated in language, to be overcome or at least transformed through a mutual sharing of awarenesses. After all, much of the language we use to elucidate what is experienced is already preformed, constructed within a framework of unexamined cognitive claims. Again, much of the everyday experience our use of language sets before us is itself mediated through legitimated channels of communication, whose value-laden character[33] can be revealed in critical communicative discourse. Indeed once these forms of possible distortion are revealed for what they are, they may be offset, displaced, enabling the human subject to return to experience itself to grasp it differently, perhaps aesthetically in a novel way. Rather than these two aspects of the normative model being, as it were, one above the other, we can see how they may relate to each other. So whether procedural concerns or substantive issues have the higher priority in our normative model depends on which fibre of our cable is being considered. The latter, the substantive, concerns the insight that provides the content for what ought to be: the procedural, the spirit in which discourse about it proceeds.

Given that what was once a spark of awareness can become a

critical insight into what is to be done, the latter is ready to become embodied in action to form a third fibre in the chain. Whereas, with respect to critical insight, attention was placed upon an ideal of conduct – an ethical concern – the present focus is upon something 'thoroughly deliberate' – a moral matter – what can and must become 'a moving cause of action' (Peirce: vol. 1, para. 574). Whereas the second strand emphasised Thirdness, reasoning in generalities 'commonly free from the nauseating custom' of being useful, this third fibre stresses application – Secondness – within a specific situation, the very issue Peirce characterises as casuistry: the activity of determining 'what ought to be done in various difficult situations'. Such 'instrumentality of cognition' can reach what Peirce calls the 'soul's deeper parts', the sentiments and instincts, so that 'the very core of one's being' can be reached through cognition's 'slow percolation' and thereby 'come to influence our lives' (Peirce: vol. 1, para. 667, 666, 648).

This normative model meets three Peircian requirements. Firstly it employs the three requisite strands which inform this model: If an action (Secondness) is to be regarded as conduct and thereby as 'thoroughly deliberate (Thirdness), the ideal (of conduct) must be a habit of feeling (Firstness) which has grown up under the influence of a course of self-criticism and hetero-criticisms' (my interjections in brackets). Secondly, this cable of reasonings has to be understood holistically: 'According to this view, esthetics, practics and logic form one distinctly marked whole – …the question where precisely the lines of separation between them are to be drawn is quite secondary. It is clear, however, that esthetics relates to feeling, practics to action, logic to thought' (Peirce: vol. 1, para. 574). Thirdly, the human subject is not to regard the issue of metaphysics as something separate from him or herself as a kind of science of being or ontology, spelling out what there is or exists theoretically in separation from his or her own experience. Rather that human subject comes to contribute to the metaphysical nature of what it is to be actual. How? By recognising 'a higher business than your business', that is to say 'a generalized conception of duty which completes' – notice 'completes', not replaces – 'your personality by melting it into the neighbouring parts of the cosmos' (Peirce: vol. 1, para.

673). So, whereas for a person who has not considered these matters and whose motive for doing something is simply a 'purely passive liking for a way of doing whatever he may be moved to do' (Peirce: vol. 1, para. 594) so that reasoning is subordinate to sentiment, for someone who is to be moral, 'the very supreme commandment of sentiment' is that the human being 'should generalize', 'should become welded into the universal continuum, which is what true reasoning consists in': 'In fulfilling this command, man prepares himself for transmutation into a new form of life, the joyful Nirvana in which the discontinuities of his will shall have all but disappeared' (Peirce: vol. 1, para. 673).

What Peirce is advocating here is the reformation of the life of sentiment, and such an activity is to be informed by 'the theory of the deliberate formation of such habits of feeling' which he thinks ought to be identified with aesthetics. His notion of the aesthetic is to be cashed as 'the mood of simply contemplating the embodiment' of quality (Peirce: vol.1 para. 574; vol. 5, para. 132) for its own sake and not for some ulterior purpose. But now we come to that question: 'What does it mean to speak of valuing something for its own sake?' Unless this notion can be clarified, the tripartite nature of reasonings involved in constructing Peirce's 'cable' can hardly begin since, as we have already seen, any such model, enabling moral judgments to be drawn, must be rooted in an aesthetic perspective.

One way of facing this last question would be to consider Whitehead's remark, '"Value" is the word I use for the intrinsic reality of an event'. He claims that it is 'an element which permeates through and through the poetic view of nature.' Peirce identifies this dimension with Firstness: 'The poetic mood approaches the state in which the present appears as it is present', 'regardless of past and future', what 'is such as it is quite regardless of anything else' issuing the imperative 'Go out under the blue dome of heaven and look at what is present as it appears to the artist's eye' (Peirce: vol. 5, para. 44). For Whitehead, to realise such a 'poetic rendering of our experience' is to appreciate something's 'being an end in itself', 'of being something which is for its own sake.[34] To make the possibilities here clearer some analytic work of Eugene Hargrove[35] may help to set us straight, though the way his distinctions are used in what follows may not be entirely to his liking!

It should be a matter of commonplace by now that intrinsic value – something being prized for its own sake – can be distinguished from instrumental value; something rated for the sake of something else! Again, in the interests of *realism, it does not seem unreasonable to distinguish the anthropocentric from the non-anthropocentric. After all, as long ago as Aristotle, human beings learnt that living entities had ends of their own, independent of human valuations. We have then four possibilities which are relational in character. These four can now be indicated in the following diagrammatic representation below:

NON-ANTHROPOCENTRIC

The relationship of harm or benefit between plants and non-human animals in an environment	Living organisms have ends of their own which Aristotle cast teleologically
INSTRUMENTAL	NON-INSTRUMENTAL
The instrumental benefit or harm of things to humans; chemicals in water	Human beings judge that any entity can have non-instrumental value as a cultural existent

ANTHROPOCENTRIC

For the purposes of the present discussion two of these possibilities can be put on one side. In chapter III, the anthropocentric instrumental sense of value was examined and characterised as the paradigm sense of the instrumental exemplified in Heidegger's sense of the ready-to-hand. Furthermore, the non-anthropocentric instrumental sense of value can be put on one side too, since we are concerned there with how organisms can either thrive or decline as a result of the presence of other entities. That particular issue was raised in chapter VI where the degree to which an increase in gull population might affect the habitats for other birds was considered. We have, then, two senses of the non-instrumental left: the anthropocentric or cultural sense of value and the non-anthropocentric sense. Both are relational, the latter with respect to something's present actuality being in relation to what it may become. A return can now be made to Peirce's categories.

Thirdness relates to what conforms 'to a general rule' (Peirce: vol. 1, para. 126) governing what occurs in the future. Thirdness gives meaning to human beings since our concern is with the sayable, the cognizable, the representable. Clearly some natural existents are appreciated because they stand for something for human beings and are valued as such, culturally not instrumentally. This type of value can be labelled ***inherent** value. And why should that concern us? Mark Sagoff's remarks concerning the non-instrumental anthropocentric value of caves or rock formations provides the rationale: 'These values we cherish as citizens express not just what we want collectively but what we think we are: we use them to reveal to ourselves and to others what we stand for and how we perceive ourselves as a nation. These values are not merely chosen; rather, they constitute and identify we who choose'.[36]

Secondness refers to the factual; definite actualities existing independently of human beings. Animals and plants have goods or ends of their own. Such goals can be discovered factually, just as 'we can discover facts through scientific research that such and such kinds of organisms do or do not instrumentally use other parts of nature for their own ends'[37] in establishing a non-anthropocentric instrumental sense of value. An ***innate** sense of value has now been identified. Rocks can't possess it, even if they can be credited an inherent sense of value, because they lack ends in themselves and can't grow or develop in relation to their own strivings! And why be concerned with innate value? Because we may indeed undo ourselves as actualities in undoing the ends of existences which have the ends of living things.

Firstness, as non-relational, represents a possibility 'such as it is regardless of aught else' (Peirce: vol. 1, para. 25). It refers to the qualitative, the felt and is characterised by 'freshness, life, freedom'. These are all 'may-bes', each realised by a feeling having 'its own positive quality which consists in nothing else, and which is of itself all that is; however it may have been brought about' (Peirce: vol. 1, para. 302, para. 306). Because our focus is on the positive internal characteristics of something, we can refer to that thing, as having *internal or *intrinsic value. Now can mountains or rocks, say, be ascribed this sense of value? We can speak of 'mountainous' or 'rockiness'. This notion can be explicated in one of two ways: the first would be to speak, as the earlier Whitehead quotation might be read as implying, of the reality a given thing has for itself, independently of what it might hold for an observer.[38] This sense of **internal** value stands in relation to the natural events making that entity what it is as revealed say through scientific inquiry, a sense of value then to be at least objectively, perhaps ontologically, understood! But once human beings experience such an entity a different sense of value is manifested. Here, in a more 'holistic' yet unintentional way, something can be experienced transforming a human being by means of a communication received through a sheer attentiveness to the entity, as such, in a phenomenological sense of experiencing. Why be concerned with this intrinsic sense of value? Walter Benjamin provided the answer: without that awareness human beings would have 'a perception whose sense of the sameness of things' develops to such a point 'where even the singular, the unique, is divested of its uniqueness'. Benjamin showed what induced this phenomena: reproducibility of images! Humans would then live merely in a monoculture![39] The importance of this sense of valuing will be reconsidered in the final chapter.

After considering, in this chapter, the monetary costs to the different participants at the Tiptree Inquiry, the issue of its costs were broadened. One cost of the Tiptree inspector's decision was that the search for another site for waste tipping in Kent would continue. Related to this consequence were issues concerning whether or not a community in the selected area for a tip should pay all the cost of its own defence; what constituted the idea of a proper appellant at such inquiries and the significance of the distinction between procedural and substantive matters in relation to the award of costs was also considered. In addition, the idea that the inquiry's proceedings cost the life of one of TAG's representatives was dismissed but that his concern for the outcome of the inquiry in relation to his community and its environment for its own sake has provided a focus for the present chapter. And why? Because without a satisfactory account of what it means to speak of valuing something for its own sake, the grounding for an adequate normative model to account for the overcoming of the Fact/Value problem in moral philosophy, cannot be provided.

At the inquiry, the idea of something having value for its own sake was defended in a number of different ways. The late Halstow parish councillor defended its inherent value, what it meant to the local residents along with its possible historical associations. Orthinologists, were concerned especially with its innate value as an ecological site. The site's internal value – its objective existence more geologically understood – was referred to in the defence of Swale Borough Council's representative in the stress he placed upon Tiptree Hill being part of a prominent ridgeline. I defended the **intrinsic** value of the site, its aesthetic perspective which grasps in a much more 'holistic' way the play, the contrasts and interweavings between these three other value dimensions; the inherent, the innate and the internal. But to make proper sense of this aesthetic perspective and exactly how it can be grounded, more attention needs to be given to a further issue: the aesthetics of nature. That subject matter provides the content for this book's next and final chapter.

NOTES

[1]*Ultimate Redundancy*, River Avon Bradford–on–Avon, Wilts., 23 September 1996.

[2]Local Government Act 1972 – sect. 250(5) Town and Country Planning Act 1990, sects. 78, 320 appeals by South Eastern Minerals Limited: land at Tiptree Farm, High Oak Hill, Funton, Sittingbourne, Kent: Application for costs, 27 October 1990 Ref:APP/W2275/A/95/159578, APP/W2275/A/96/271958 and APP /V2255/A/96/272360

[3]SBC Counsel, *Final Submissions for Swale Borough Council on D.E.T.R.* appeal refs: A) APP/W2200/A/95/249578, B) APP/ W2200/A /96271958, C) APP/V2255/A/96/272360 (19/2/1998) p.27 5.2.3.

[4]cf. too ENV2 'Kent's landscape and wildlife (flora and fauna) habitats will be conserved and enhanced development will not be permitted if it would lead to the loss of features or habitats which are of landscape, historic geological or wildlife

importance, or are of an unspoilt quality, free from urban intrusion, unless there is a need for development which outweighs these countryside considerations.' 'The May 1993 Third Review Deposit Plan and Proposed Modifications (August 1996): Environment' Appendix 1 of The Tiptree Farm Appeal, Appendixes to accompany Proof of Evidence of Mr M Hare Doe ref: APP/W2 20 0/A/9 5/259 578.

[5]Evidence has come to light off medieval tile kilns which may have been used to fire pottery during the reign of Henry VIII. *Sheppey Gazette* Wednesday, 14 October 1998, p.7.

[6]Town and Country Planning Act 1990; Section 79 Appeals by South Eastern Minerals Limited Land at Tiptree Farm, High Oak Hill, Funton, Sittingbourne, Kent File nos. APP/W2275/A/95/ 259578. APP/W2275/A/96/271958 and APP/V2255/A/96/272360, 12 October 1998. The inspector's reply: 'In planning appeals, the parties are normally expected to meet their own expenses, and costs are awarded only on grounds of 'unreasonable behaviour', resulting in unnecessary expense' see 2) above para. 3 of the accompanying letter.

[7]Mike Peirce 'Dumping on Democracy', see also note 5, above p.7.

[8]'C8/93 does not give guidance on a case where a party acting as appellant is not the appellant. However, by virtue of section 25 of the Local Government Act 1972 (applied to local planning inquiries by sect. 302 (2) of the Planning Act) the Secretary of State has power to 'make orders as to the costs of the parties at the inquiry and as to the parties by whom such costs are to be paid...' The power is accordingly unfettered although tempered by Circular 8/93. notwithstanding p.12 of Annex 1 of C8/93, Biffa is for all intents and purposes 'the appellant' (2.2). Local Government Act 1972, op. cit., p.3.

[9]SBC Counsel, *Final Submissions*, op. cit., p.39 sect. 6.1.

[10]'There has been slippage of clay on the northern slopes of Tiptree Hill behind the Funton Brickworks (D12 p.2). There could be further slippage arising from building heavy bunds near the top of the hill and then landfilling. This could cause cracking and breakdown of the sides of the void, allowing leachate to escape down the hill into the aquifer and the estuary. There is a spring on the north side of the road over Tiptree Hill.' Town and Country Planning Act 1990, op. cit., p.126, 7.5.15.

[11]Proposed extension to existing landfill operation incorporating materials recovery facility and restorations to public amenity, nature conservation, woodland and agriculture, Shelford Quarry, Canterbury, Brett Waste Management Ltd (MR 160606) File ref. CA/96/794, item C3, date valid 1/8/96, KCC Planning Officer, para. 6, p.C3.4.

[12]Town and Country Planning Act 1990 – sect. 78(1) Appeal by BFI Ltd (formerly Drinkwater Sabey Ltd), restoration of mineral workings by landfill, extraction of minerals and demolition of farmhouse and outbuildings at Rosecourt Farm, isle of grain, Kent Ref. APP/W2200/A/95/260723, 30 September 1997, government office for the south east.

[13]Part of the reason for the delay in publishing this book is that its author hoped to render to its reader an answer to the question, 'And where was the landfill site finally deposited in Kent?' Rumour has it that the report for the inspector for the Dargate Inquiry has sat on the minister's desk for the last eighteen months, two

years since the inquiry closed in March 1999. Note that the Tiptree Inquiry report was published in six months. It might be said that the UK's waste disposal industry hangs on this decision! (Subsequent to the publication of the first edition of this book, the Secretary of State, on the 30th August 2001, ruled that there was no pressing need for the Lamberhurst Farm project at Darget.)

[14]The original proposal was rejected by KCC. It is planned to resubmit a new application accounting for some of the wildlife issues 'Landfill – Success, Threats, Opportunities and Responsibility' in *Wildlife News.* The newsletter for Kent Wildlife Trust Members Autumn/Winter 1997 p.2. The resubmission was put but withdrawn in the Spring of 1999.

[15]Here there are seven kilometres of dyke rich with different species of creatures, twenty hectares of grazing land besides lagoons useful to sixty pairs of reed warbler. These lagoons are used too by wintering birds as well as those of passage! See also note 13, above, the wildlife losses means Kent Trust ill oppose.

[16]Report to the county planning officers to the Planning Sub-Committee, 12 September 1995 from the county planning officer. Subject: 'To Extend An Existing Clay Quarry, Restoration to Agriculture by Infilling, Provision of Site Access Roads and Improvements to Local Roads, Tiptree Farm, Funton, Lower Halstow – South Eastern Minerals Ltd (MR 878.669)'.

[17]See chapt. IX, 'The Inquiry Reopens: Can Two Utilitarian Arguments Entwine?' notes 11, 12.

[18]Of course, personally, I may be pleased the inspector did so. But here I am trying to be impartial in seeking a consistent set of rules for making decisions of this kind.

[19]Charles Hartshorne, *Creative Synthesis and Philosophic Method*, p.308.

[20]Alfred N Whitehead, 'Direct experience is infallible. What you have experienced, you have experienced. But symbolism is very fallible, in the sense that I say induce actions, feelings, emotions and beliefs about things which are mere notions without that exemplification in the world which the symbolism leads us to presuppose'. *Symbolism: Its Meaning and Effect*, 1827 New York, Capricorn Books, 1959, chapt. 1, sect. 4, p.6. Much of the argument in what follows can be found in Boulting, N E Boulting 'Edward Bullough's Aesthetics and Aestheticism: Features of Reality to be Experienced' *Ultimate Reality and Meaning*, September 1990, vol. 13, no.3, pp.200–221 especially sect. 7 and 'Grounding the Notion of Ecological Responsibility: Peircian Perspectives' in *Religious Experience and Ecological Responsibility*, pp.119–142.

[21]Jurgen Habermas, 'A Reply to My Critics' in *Habermas – Critical Debates*, pp.224–225.

[22]K Price, 'The Truth about Psychical Distance,*Journal of Aesth. & Art Criticism*, vol. 35 (summer 1977), pp.411–423, p.420.

[23]J A Passmore, 'The Dreariness of Aesthetics', (1951) in *Contemporary Studies in Aesthetics*, pp.427–443 p.**441**.

[24]F Schiller, *On Aesthetic Education of Man in a Series of Letters*, p.193.

[25]For Schiller's influence on Peirce see Omar Calabrese's 'Some Reflections on Peirce's Aesthetics from a Structuralist Point of View' in *Peirce and Value Theory*,

Herman Perret (ed.), J Benjamins Publishing Company, Amsterdam, Philadelphia, 1994 pp.143–151.

[26]See N E Boulting 'Edward Bullough's Aesthetics and Aestheticism', note 19) above p.218.

[27]E Bullough, *Aesthetics – Lectures and Essays*, 1972 p.65.

[28]Theodor Adorno, *Negative Dialectics*, p.43.

[29]Walter Benjamin *One Way Street & Other Writings* and *Illuminations*.

[30]Theodor Adorno, *Prisms*, p.233.

[31]Karl-Otto Apel, *Towards a Transformation of Philosophy*, p.282.

[32]L Johnson, *A Morally Deep World*, p.231.

[33]Don Ihde, *Technics and Praxis*.

[34]Alfred N Whitehead, *Science and the Modern World*, p.93.

[35]Eugene C Hargrove, 'Weak Anthropocentric Intrinsic Value' in *The Monist* vol. 75, no.2, April 1992, pp.183–207.

[36]Mark Sagoff, 'Ethics and Economics in Environmental Law' in *Earthbound: New Introductory Essays in Environmental Ethics*, pp.147–77 p.175.

[37]Eugene C Hargrove, op. cit., p.187.

[38]Jay McDaniel, 'Physical Matter as Creative and Sentient' in *Environmental Ethics*, vol. 5, Winter 1983 pp.291–317 p.315.

[39]Walter Benjamin, 'A Small History of Photography' (1931) in *One Way Street and Other Writings*, London, New Left Books, 1979 p.250; cf. Boulting N E, 'Between Anthropocentrism and Ecocentrism' *Philosophy in the Contemporary World*, vol. 2, no.4 Winter 1995, pp.1–8 especially p.6.

Chapter XIII

CONCLUSION: A CONTINUING DEBATE

Are words merely the money counters men reason by? And if so surely this can apply on to language found acceptable in the ordinary communications of agreement and debate. Yet, consider those words used in intimate discourse and private speech: like steps which issue from a body seeking a path between tall slender trees stretching upwards to point to a black sky above, where no moon nor stars do shine, expressive phrases seek an assurance that will allow their latent content – promised in the felt desire to utter them – to be developed over ground whose existence cannot be guaranteed. So, each phrase, like each step, has to seek its foothold in that discursive space where the other's response may yet indicate a proper direction still to be found.[1]

Imagine the promise held in the following possible event. Such an event, as a coincidence, might occur in ordinary life but if it appeared in fiction we usually accuse its author of contrivance. The local university – more in the spirit of times past than in the interests of a cost-effective present – is hosting a conference entitled 'Environmental Discourse, Landscape Issues and Aesthetics', A landscape architect consultant is presenting a paper entitled 'Scenic Quality: Applying Theory in Practice.' A local borough council representative will lead a workshop session on 'Environmental Issues in Constructing a Local Plan for the New Century'. An officer from the county's planning department will speak on 'Creating the County's Countryside: Tensions between Government and Local Policy'. A visiting philosopher is of an idealistic persuasion: he believes matter to be nothing more than 'a specialisation of mind'.[2] He is not known to the other three speakers but will lecture on the subject of 'Aesthetic Considerations in appreciating Landscape'.

The first three persons are all known to each other because of their participation in the appeal against the county's decision to stop an attractive and prominent ridgeline'[3] in Kent's Swale area, being a site for the removal of clay and its use for landfill purposes. After their initial discussion about the conference arrangements, the visiting philosopher joins them just as the inspector's decision with regard to the appeal is being discussed.

LANDSCAPE ARCHITECT CONSULTANT (LAC): It is my opinion that the inspector at the Tiptree Inquiry was unduly influenced by the considerations of landscape and visual impact. He gives fourteen paragraphs to this in his Conclusions and goes on to write a further nine paragraphs in his Overall Conclusions. He does not appear to consider sufficiently the landfill need in Kent even if he does mention it. Considering other possible sites in Kent: Dargate, where the workings will be far more public; Oare Marshes where serious ecological damage could occur, and Richborough which may not be suitable. Tiptree would obviously have been the best choice!

BOROUGH COUNCIL REPRESENTATIVE (BCR): It is quite clear from his report that the inspector could do no other than decide against Tiptree's development. Central to that decision had to be the landscape case made in our representation both to that inquiry and the Local Landscape Area, the LLA Inquiry. I accept your point about the importance he attached to this issue but I was more surprised by what the Secretary of State said about a late representation we made, drawing attention to the inspector's decision at the earlier LLA inquiry. Evidently the Secretary of State took the view that the inspector's decisions at that earlier LLA inquiry and other policies in our Local Plan 'do not affect', he wrote, the Conclusions of the Tiptree inspector.[4] In bringing forward the earlier inspector's Conclusions we hoped these would affect the later inspector's recommendations. Of course the minister sees no conflict between the two sets of recommendations. The very existence of the arguments at the LLA inquiry had to be incorporated into the later one, so there couldn't be a conflict between them. I am doubtful too whether there was no communication between the two men!

COUNTY PLANNING OFFICER (CPO): In accordance with standard practice any inspector gives due weight to the arguments heard at the inquiry. The minister's remarks make it clear that there is no inconsistency between the two reports, that's all! Obviously the proposal would have meant demonstrable harm to that part of the county's landscape. Clearly, it was the landscape case that was of considerable and of overriding concern both to the people in the area and to the inspector, but it was odd how one interested person tried to make a case for sustaining the landscape as it was.

VISITING IDEALISTIC PHILOSOPHER (VIP): Please don't leave us completely in the dark as to the details of the argument…

CPO: It was stated by someone who said he was a philosopher and a local resident and yet, at the same time, said he was not speaking for the Tiptree Action Group or TAG as it was referred to at the inquiry. In somewhat emotional terms, he presented an argument which appeared to contradict itself. The first half of his speech indicated that he was concerned to defend the idea of the hill as having scenic quality and that the inspector should have regard to a landscape assessment document drawn up for the Countryside Commission for Scotland in 1991. It is suggested there that in respect of those intangible qualities which constitute a criterion for scenic quality, there is a clear preference for Jay Appleton's prospect–refuge theory: the prospect of a landscape refers to the provision of open views whilst the idea of refuge indicates concealment where views can be enjoyed without the viewer being seen![5] Yet in the second half of his speech, he afforded a much higher priority to the idea of a landscape being lived in rather than being viewed from a detached point of view.

VIP: Weren't his ideas challenged? After all Tiptree is hardly in Scotland and it would surely not have been very difficult to show that necessarily his ideas were self-contradictory.

CPO: A silence fell on the room. I don't think people knew how to question him!

VIP: Consider a little more closely what he might have wanted to say. There are many accounts of what counts as aesthetics in the market place of ideas and I hope to address some of these in my lecture. You refer to Appleton's *Experience of Landscape*. He seems to have in mind primarily the issue as to how an interest in aesthetic appreciation arises. As you intimate, two important variables are drawn to our attention in seeing landscapes, prospect – landscape viewing accessing lines of perception – and refuge, the opportunity landscape offers for concealment. Prospect assisted in the hunting of some creature for food whilst refuge guaranteed the hunting creature would not be seen before the final act of possession. These two – prospect and refuge – are fundamental to discussions of aesthetic matters since they concern the issue of survival biologically for earlier forms of man, and may well account for our emotional response in aesthetic matters. A form of perception which appears to be enjoyed for its own sake may be in fact an enjoyed satisfaction arising just because human beings have been released from the concerns of mere survival. There may well be some truth then in the claim that, as the strategic value of a perspective ceases to be crucial for survival, so its value emerges in aesthetic experience.[6] However, to substantiate the claim about living in the world, giving another sense of the aesthetic, would be a quite different kind of undertaking.

LAC: I assess the basis for that latter analysis to be based on his advocacy of Christopher Tilley's book *A Phenomenology of Landscape*. This book will allow anyone to claim that any kind of landscape is of value because its inhabitants experience and understand their world through living in it. There was and is no dispute between TAG members and me that they enjoy an attachment to their area which provides perhaps a stability of meanings underpinning their way of life. But many locales are, of course – through certain common experiences, and shared understandings – created and known to their local folk in this way.[7] To use that argument against development is only to further the Nimby case: 'Not In My Back Yard'! But say we ignored the Nimby case. What we have left would not necessarily constitute a proper argument for a different way of viewing landscape either. Central to Tilley's case is nostalgia. We can see that in two key steps he makes, pushed by his quotation from the German thinker Martin Heidegger. It is fair to say that the first step pictures human beings in a medieval environment[8] – peasants living in farmhouses within the Black Forest whose recognition of time, for example, depends on the sun's position in the heavens[9] – a world long since past. Tilley's other key step is to characterise these people as employing a craft-technology, not one employed today. We live in a world now subject to the great advantages brought to us by the screen in our own homes. The purpose of Appleton's case, on the other hand, is to suggest a strategic vision of what it is to be a human being on scientific grounds giving us a view of man which can be understood in terms of habitat theory. This focuses upon the importance of what I called 'the specular image' now embodied in the new technology but having its origins in Alberti's efforts to define the art of perspective in picture-making in mathematical terms, proving of enormous importance in the art of map-making itself![10]

VIP: A good deal more could be said about your evident dismissal of Heidegger's phenomenological approach to landscape matters, but with your too narrow approach determined by your obvious obsession with what you call the 'specular image' there is the constant danger of impoverishment if not sheer omission in relation to what you say about the history of landscape painting. Convenient ignorance is a method much condemned in the sciences and you completely ignore the work of

Svetlana Alpers in *The Art of Describing*. There it is demonstrated quite clearly that for Alberti a viewer is posited – looking through a frame provided by a window, say – thereby at a distance from what is viewed, to produce a substitute or pseudo-world expressing some story to interpret reality. For the artists in the Dutch Renaissance it was quite different. For the latter, painting was meant to be descriptive, recording a great deal of information and knowledge about the world. Whereas the former kind of art instructs us to 'read' a painting in terms of a fiction, the latter celebrates a proper culture of vision where a set of values is 'seen' not 'read' within an existing, living society. For the former, man – as artist – remains the measure of all things: for the latter the problem is to secure and define existing things in what is seen. Whereas Alberti defined a picture in terms of the rectangle or framed window imposed upon a world, for Dutch painting there is no such prior frame in the attempt to mirror a world as it is seen in natural vision, Whereas in the Italian case, the picture with its frame comes to exist as a thing in the world, in the Dutch case the picture actually replaces the eye so that it becomes necessarily a controversial issue as to where the viewer is standing. The viewer is taken into a world, not allowed to remain at a distance from it. It should be noted too that it is that preoccupation with what exists in the world that gives rise to the activity of map-making even if, in the Dutch case, it gave rise only to the art of 'decorative maps', at least at first![11]

LAC: The sorts of thing you have just said are not in themselves sufficient to do justice to what I say about the importance of the 'specular image'. Even if Alpers's work does make an important contribution to the study of art, her argument would not necessarily constitute a refutation of my claims. It just depends on your point of view. She is interested in art, I am fascinated by the history of the consequences of these two artistic movements which share many common features of both. Firstly, both schools are concerned with seeing, with vision. Secondly, she admits that the Italian school has, by and large, determined our concept of art in the Western world so that our understanding of Dutch art has been reworked in the image of Italian painting. Otherwise Dutch art has been associated more with photography and thereby been under-emphasised. Thirdly, she points out that these two movements are not to be regarded as distinctly separated in practice; Rubens, whom she refers to as 'a northerner steeped in the art of Italy', made use of both traditions. Indeed, she herself admits she does not wish to slight the idea of a continuous interrelation between the art of picture-making in different cultures.[12]

CPO: It seems to me that this conversation has become highly esoteric. It is evident that we are miles away from the Tiptree Inquiry.

BCR: I am not sure that we are so many miles away from Tiptree. The inspector acknowledged that the proposed site was 'part of an attractive and prominent ridgeline', particularly when viewed from the west, because of the contrast with the surrounding area. For him it was not 'unremarkable' as a landscape, as your landscape expert claimed. It must also be borne in mind that the inspector does not refer simply to the visual, nor to the Nimby case. He backs the local people's feeling that such a would-be designated LLA could serve as a countryside gap to separate the Medway towns from the outskirts of Sittingbourne itself.[13] The earlier LLA inspector judged too that Tiptree Hill was a prominent physical feature within its local landscape providing a backdrop to the North Kent Marshes, whilst the Saxon Shore Way, passing beneath it, provided good views over the sea.[14] All of this in the context of your debate leads me to conclude that the idea of the aesthetic appears in one of three ways in what you say. The first of these is whether Tiptree Hill is 'unremarkable from a scenic quality point of view. The guiding principle here would be taking the site to be a landscape viewed as a picture, as something framed, as it were. A second possibility is offered by the idea that the aesthetic is to be sought in relation to something seen as existing physically in the world; a prominent ridgeline. A third criterion which could be used to justify the label aesthetic – a route taken by both inspectors – recognises the legitimacy of designating a countryside area whose occupants identify the hill as constituting something in a countryside gap, generating a different quality of life from what can be enjoyed in the towns.

CPO: Since our original discussion has now metamorphosed into such a highfalutin discourse, may I suggest that the general thrust of your last remarks is to distinguish the expressive from the indexical and both from the interpretative; the first in relation to scenic quality, the second relating to a physical existent and the third giving a sense of the meaning and significance a landscape might hold for its inhabitants. Accordingly we can ask our philosopher which of these three criteria he regards as essential in addressing the issue as to what is central to the aesthetic. After all, it would appear that on the evidence so far he has only added or developed suggestions we have made in our conversation.

VIP: It emerges then from the proceeding discussion that there is an important sense in which we have the necessary tools to hand with your three categories: the expressive, the indexical and the interpretative. These three are, of course, identified by the thinker Charles Peirce in his theory of signs: the first indicates 'a mere quality', the second an actual existent' and the third represents 'a general law', For an interpreter a sign can signify a 'possibility', or refer to a 'sign of fact', or instantiate 'a sign of reason'. The question then arises, which you put to me, as to whether we should follow the 'Italian school' in focusing on the expressive, the Dutch Renaissance artists in emphasising existing objects or some intermediate view relating both requirements to each other interpretatively. Following the idealistic bent of Peirce's philosophy – that the evolving and spontaneously created universe is to be regarded as living mind – I would go for the first, seeing nature as a picture. To make sense of the aesthetic is to understand the activities of someone creating art, since the chief thing for the artist 'is the qualities of feelings' and of course Peirce would admit that to emphasise 'immediacy as feeling' is to identify that with subjectivity.[15]

BCR: The mind boggles at the thought that our universe is haunted, for isn't that the conclusion of your argument. If the universe is a creation like a work of art, then aesthetic appreciation must lead to the idea of a designer. It is difficult to see how that claim can be justified, even if I do know a cyclist who likes to ride over Tiptree Hill at dawn which, he says, gives him the feeling of being enveloped in some kind of reality much vaster than his usual hum-drum experience. But, in saying that, I think he is emphasising not the first possibility which, through our senses of seeing and hearing, separates us from what we experience but rather the third possibility. Because this area excites all his senses, especially the smell and taste of the sea air whose movements strike his face, he comes to be drawn into the landscape in some fashion. In either case, by the way, the aesthetic then has value because it offers a kind of spiritual uplift but why should that entail religious belief?

VIP: It must be admitted some of Peirce's remarks have led to reading him in line with your first possibility. It has to be contended that aesthetics concern what he calls Firstness, feelings evoked in freshness, freedom and spontaneity – 'Firstness is what is present to the artist's eye' – encouraging at least one commentator to assume that the best example of experiencing Firstness is to sit back and enjoy a piece of music,[16] in the sense, presumably, that listening is a contemplative activity enabling us to give our full attention to a work of art. But it would be most acceptable for me to say with you that if such feelings are to be enjoyed in regard to nature, then part of the experience of its aesthetic quality is that, within such feelings, there is a sense of what Peirce called the reasonable ensuring contact with 'a sort of intellectual sympathy'. Indeed if he were asked what part such feeling qualities played 'in the economy of the Universe', he would say that the universe itself 'is a vast representatum, a great symbol of God's purpose, working out its conclusions in living realities'. But he would not regard such aesthetic awarenesses as proving God's existence![17]

LAC: This fundamental lack of understanding I fail to comprehend: our dependence on the specular image is *not* a matter of choice. It is central to modernism's advance in our culture because it is carried by the development of the new technology.[18] This fact in itself is sufficient for us to admire what you say about Peirce's identification of the aesthetic with vision. But we should not skirt around his nineteenth-century obsessions. I think it more sensible to take the advice of his student John Dewey and look for the roots of the aesthetic perspective in seeing man as part of nature so as to account for the fact that 'his impulsions and ideas are enacted by nature within him' as Dewey puts it[19] rather than taking on a comprehensive study of religious matters, since God-talk is only used to deal with what we don't know. That is why Appleton's work is so important: scientific explanations can substitute for religious mysticism!

CPO: I am of the view that little of what has been said is of much relevance to interests of acknowledged importance in our culture. Throughout the long life of the Tiptree debate one thing is clear since it is demonstrated in the inspector's reports: the sheer weight of public opinion![20] In what has become known as our 'notorious report' of 12 September 1995[21] the full extent of that opinion was overlooked. As a general matter of policy, aesthetic matters are subjective and where numbers of persons register their views, the authorities have to react accordingly. There is my view. What is our philosopher's, when he has finished quoting Peirce?

BCR: I think it is a very cynical view to say that what matters aesthetically simply comes down to counting heads. What you said earlier could be lost if not grasped quickly: your distinctions between the expressive, the indexical and the interpretative. Our philosopher seems to have gone for the first, me for the third, but what about the indexical. Surely knowledge about what exists is a scientific matter and it must be borne in mind that science is a form of human inquiry. To take aesthetic consideration in that direction, or in either of the other two, would cast aesthetics in an anthropocentric direction, wouldn't it? Not much room for an ecological approach here!

VIP: Ecological philosophy is a demanding *mistress*, extracting her price so that, as she ages, her demands become correspondingly high. The question arises as to what my real position is in regard to these matters. Come to my lecture and we will be able to see the degree to which our discussion has helped to sharpen our insights and made us more critical as to the details of the formulations I am about to offer this audience, rendering the place of what has been called the 'specular image' within them!

The value or values we ascribe to nature has been the central concern of this book. In particular the focus of our attention has been upon a particular natural area in Kent called Tiptree Hill. Can an argument be sustained for developing it for human purposes or for not doing so – in this case a waste disposal site – as opposed to some other area, whether that be a site on the Isle of Grain, Dargate, Oare Marshes, Richborough or somewhere else since, as was pointed out in chapter XI, the Tiptree inspector in his report stressed that additional facilities were required for waste disposal by landfill in Kent. Two arguments have not been faced directly within this book: a practical argument and a more philosophic one. The former was pressed at Dargate in view of the fact that no further progress had been made in relation to developing more incineration facilities since the close of the Tiptree Inquiry: more recycling of waste should and will take place within the county. Normally it is assumed that whereas the county council is responsible for generating policy as to where and how waste is to be disposed, the local borough council is concerned with the way it is collected.[22] And it is true that local boroughs have increasingly become involved in recycling waste – tins, cardboard, paper and glass – but are there sufficient directives in this area and even if there are, how might these be enforced? In any case, any such recycling processes, even if already in hand, would not ameliorate the present situation, namely that a landfill site of some kind is required if Kent is not to continue exporting its waste out of the county.

The second argument is philosophic, bearing on man's relationship with nature more generally conceived. We can't go on regarding nature commodifically or instrumentally, a resource to absorb items of no use to us, or force nature to undergo changes consequent to human activity, since such activity involves depleting natural resources – for either mineral extraction or waste filling – as well as the destruction of species or pollution. The natural world is essential to the survival of humankind as a species. We are creatures of nature dependent upon natural resources and, as was indicated in chapter VII, these are significant for their recuperative capacities for human beings. In addition there are the basic needs which nature satisfies: our need for air, clean water, food etc. At the Dargate inquiry, for example, the issue was raised as to whether material in the air spreading from the proposed tip site would affect fruit growing in the area![23] Indeed, in the light of our increasing population size, our interests in these kinds of provision have increased rather than lessened during the twentieth century. Now this argument is so obviously the case, why has it not been used in focusing upon the chief value nature has for us: its value if we are to survive as a species?

A number of reasons can be given for not cashing fully this survival case within this book, though in a piecemeal fashion its claims have been addressed. After all the inspector's report did refer to the pollution issue, the ornithological and natural conservation arguments and the possible effects upon clean water etc. besides the need for the mineral clay and a waste tip in the area. One obvious reason is that putting the survival-case argument so straightforwardly is simply to forward the case that no more human intrusions of any kind should be considered in relation to the natural world, not even in any part of Kent! Desirable as that goal might be, it simply isn't practicable in relation to the present situation. Arrangements for further waste disposal are required in the short term so that time can be saved to allow for better plans to be agreed upon to prevent waste tipping in future.

The second reason is that this human survival argument is too general to prove relevant in deciding the site of a tip whether on the Isle of Grain, Tiptree, Dargate or some other locality. That is why specific arguments had to be considered. Interestingly, most of them were instrumental: either there was or there was not a need for a mineral or a waste disposal site; the need for good agricultural land would be undermined or that pollution effects would harm humans.

A third reason for not dealing with the survival need case directly is that it seems to carry little weight in the public domain. There is very little support for it – taking the human species survival case that seriously – from a political point of view on either side of the Atlantic or the English channel, save perhaps in Germany. Of course, it has already been implied both herein and more widely in this book that it ought to be taken more seriously, but that is beside the point; clearly it isn't! Indeed a remark by the visiting idealist philosopher reflects that public sentiment, at least in our millennium: as the strategic value of a perspective in nature ceases to be 'crucial for survival, so its value emerges in an aesthetic experience'. Clearly the warrant for that claim requires examination.

Fourthly and finally, another reason for not stressing the survival need argument is that in its formation it presupposes one way of regarding nature: instrumentally, rather than scientifically or aesthetically. What exactly legitimates that attitude? Indeed, it is an implication in the writing of this book that a case for instrumental value – the value of nature consisting in its capacity to sustain the survival of humankind – does not provide sufficient reason for ascribing value to nature, since in its terms we can make no sense of the locution 'viewing nature for its own sake', the burden of the investigation in chapter XII. Rather, if we are to get at the latter, the concerns of the aesthetic dimension have to be taken seriously. It may be the only way we can move from an anthropocentric – which is what the human instrumental argument constitutes – towards an ecocentric point of view. And the participants in our imaginative discourse at the start of this chapter have to take the aesthetic dimension seriously because, as the landscape architect consultant pointed out, the Tiptree inspector concentrated on that perspective particularly in his report.

This skirmish, then, with two possible objections to the line of argument sustaining this book – the recycling possibility and the survival-need case – can be related to issues surfacing in the discussion between the four disputants: the move away from a concern with natural survival and the grounding for an aesthetics of nature. In relation to the first issue, it is clear that most of us don't grow our own food; we buy it, if we are lucky, at our local fruit and vegetable shop. We don't hunt and kill animals for food; we buy meat from a store. We don't walk or run as our main form of transport; we drive everywhere. We rarely view a sunset or sit by an open fire outside; we turn on the TV set. Now what affect does all this have on the second issue, the grounding of an aesthetic perspective? Let us return to the discourse of our participants. Note by the way, that each of them does illustrate one of the four possible cognitive standpoints delineated earlier in chapter X: the county planning officer (CPO) represents the planner s/he who puts into practice recommendations after evaluating perspectives from other sources than his or her own. Enured seems an appropriate label since the CPO's original recommendation on 12 September 1995[24] derived from his assessment of the arguments and cognitive claims put to him from various sources. In Iwade Hall during the inquiry he reversed that judgement. The landscape architect consultant can be regarded as representing the experimentalist in the sense that he takes the scientific paradigm as his authority. The borough council representative is more the practical man concerned with the argument as to how far nature is useful for humans in raising their spiritual awareness. He could be described as an activist in the way he put theoretical studies, for example the findings on noise studies, to good use at the inquiry. Finally the visiting idealist philosopher seems more concerned with aesthetic experience and the role of artistic creation in enlivening what might be described as a more humanistic point of view. But what are we to make of his claim regarding the shift in human concerns away from mere survival and his ground for aesthetic experiencing?[25]

Hegel first distinguished natural man as 'a natural being' from man as a social individuality shaped by a culture. This social creature thereby became capable of holding different attitudes rather than one of total dependence towards nature, as this creature became increasingly capable of 'setting aside' its former 'natural self'.[26] The latter Lukacs labelled man's 'second nature' – 'the nature of man-made structures' – 'devoid of any sensuous valency of existence'. Now the human being is estranged from nature. His attitude towards it is at most one of sentimentality. Such an emotion represents a compensation for his new experience within a 'self-made environment' realised 'as a prison instead of as a parental home'.[27] In chapter V we quoted Peirce in 1902 who could be regarded as making sense of this transformation:

> But *fortunately* (I say it advisedly) man is not so *happy* as to be provided with a full stock of instincts to meet all occasions, and so is forced upon the adventurous business of reasoning, where the many meet shipwreck and the few find, not old fashioned happiness, but its splendid substitute success. When one's purpose lies in the line of novelty, invention, generalized theory – in a word, improvement of the situation – by the side of which happiness appears a shabby old dud – instinct and the rule of thumb manifestly cease to be applicable' (Peirce: vol. 2, para. 178).

The rationale for such a social 'transformation of the situation' disappears for those living after such a social transition. This transformation makes a life, once meaningful to their forbears, no longer so for them, especially in Western culture as that has developed following the industrial revolution. Such a 'second nature', generated historically, comes to be regarded as natural, unproblematic within a culture to which it is now attuned. Whilst 'first nature' has to be interpreted biologically, 'second nature' – emerging from it – has to be seen historically, now socially as more and more sophisticated technologies transformed the relationship of human beings to their world. That is partly what is meant by Benjamin's claim that 'a different nature opens itself to the camera than opens to the naked eye'.[28] But to appreciate that remark fully we need to return to the landscape architect consultant (LAC)'s case for *specularism. The visiting idealist philosopher (VIP) too, seems charmed by this approach to aesthetics, that the aesthetic attitude is to be captured by Peirce's phrase 'seeing nature as a picture', so that in vision, nature's qualities can be appreciated to incite pleasure in the viewer.[29] So we see the VIP taking, as Peirce did, the artist, particularly his or her immediacy of feeling in a subjective experience in creating a picture, as a paradigm for the aesthetic perspective. Landscape viewing would then amount to viewing the natural world as if it were a work of art. But here the VIP departs from the LAC, the former seeking a route to religious awareness from aesthetic experiencing: the latter, a different route to sustain specularism more clearly, a route spelt out in chapter VIII. There, landscape painting was compared to cartography: whereas the cartographer fixed space, the landscape painter froze time. Kenneth Olwig writes: 'Where cartography made possible an urban and regional planning in which the earth was bounded and reformed according to the quadratic net of the world, landscape painting provided a means of visualising the world which the landscape architect used to reshape the topography according to his designs.'[30]

Five important features of specularism are elicited by this comparison of the activities of the landscape painter and cartographer. The *static* becomes central to aesthetics. The landscape painter fixes things as they appear to a spectator, just as the cartographer by means of a grid provides a window from which to construct a survey or map some area. Secondly both activities imply *distancing* – sustained by the viewing spectator. Thirdly, both activities privilege seeing or *vision* over the senses, so sustaining Foucault's claim that we live in a 'specular civilization'.[31] Fourthly, both activities are associated with *the rise of science*, the latter again emphasising physical seeing, inquiry – where questions requiring an answer are put to nature – and the enframing of the environment to ensure that inquiry proceeds satisfactorily. The specular image appears in all kinds of scientific inquiry: the photograph of a planet's surface, the microscopic plate, the aerial photograph, the satellite image, the distribution map etc.[32] Finally, works of art depicting landscapes, like the areas they image are both subject to *commodification*: they acquire exchange value in the market place.

Well before Foucault's speculations on the specular image, Benjamin in *The Work of Art in the Age of Mechanical Reproduction* of 1936 anticipated the specular culture we enjoy every day. That specular culture is shaped by the technically mediated perceptions offered by the photographic image, visual creations manipulable on a computer screen or the TV presentation which together generate different standards for what is regarded as important in vision. These standards supplant more traditional ways of experiencing.[33] Whereas these new standards become legitimated within public practice to inform what it is now natural to perceive and do, technologically unmediated perception becomes relegated merely to the private dimension. Consider the remark of someone viewing beautiful scenery in Montana at Roger's Pass – a prairie against the backdrop of the Highwood Mountains: 'It looks unreal, like the scenery painted in the background of a TV program.'[34] A specular image of what landscape looks like is used as a paradigm against which more 'normal' viewing is to be evaluated! Here, without the viewer being conscious of their own perceptual activity, the properties of the specular image – enframing, isolating, freezing, distancing – are imposed upon, if not intimated within, a person's life-world.

The landscape architect consultant (LAC) legitimates this new way of seeing as natural by turning to Jay Appleton's prospect–refuge theory. What has come about as a social transformation in the way human beings view their environment is now to be passed off as natural! Indeed Appleton makes use of John Dewey's approach to aesthetics to sustain a biological theory of aesthetic experiencing cast hedonistically, as what brings pleasure to a spectator visualising natural scenery. Of course it is a matter of debate in the literature as to how far Dewey would accept the reduction of cultural levels of experiencing to the biological.[35] In this way the LAC can separate his view from that of the visiting idealist philosopher: present aesthetic charm represents the veneer of past instinctive behaviour!

In chapters VII and XII a notion of the aesthetic was intimated, conceived as an experience where objects or entities are regarded as standing outside their more usual everyday context as placed within our instrumental concerns. Under such conditions both entities or objects experienced and their experiencers, with their subjective appraisals, become transformed.[36] An advocate of specularism might enjoin this account of the aesthetic but a crucial difference emerges on examining certain kinds of phenomena. Consider such witnessing the glow from a burning Berlin, as British pilots did on their return from a bombing raid; the image of an exploding hydrogen bomb or the depiction in the film *Apocalypse Now* of Wagner's music accompanying the machine-gunning of peasants from American helicopters. These phenomena characterise an aestheticism not the aesthetic perspective at all.

In the latter case a distance is placed between the experiencer and the perceived objects or entities, as these constitute sources of our normal affections and everyday perceptions. In the case of aestheticism, the distance is placed between the experiencer and any affections s/he may have which affects him or her bodily or spiritually.[37] Now the enjoyment of shock, not an aesthetic experience, is the central feature of experience to be characterised as aestheticism, a central feature of specularism itself. And as a glance over the three previous examples already cited confirms, specularism celebrates the four 'ps' which now typify our 'new' humanity: power, pleasure, purposefulness and projection.

Appleton's prospect–refuge theory posits our enjoyment of power in being able to enjoy views over an area taken as a whole, without necessarily being seen, as central to aesthetic experience. This is the source of our pleasure in a prospect. But once hedonism is slipped into an account of aesthetic experiencing the latter can come to cover such recreational activities as the racing of power-driven vehicles or water-skiing.[38] And the celebration of any sense of power in an account of the aesthetic confirms the place of purposefulness within such experiencing so that 'in the experience of landscape there is no fundament difference between the perception of the foxhunter, the deer-stalker, the mountaineer, the fell-walker, the poet, the painter'.[39] Finally, objects or entities are defaced in relation to any place they may have in the origination of the aesthetic dimension, since the latter is regarded as something projected by humans onto such objects or entities out of their own subjective experience. So, for Appleton, if 'a forest is beautiful, this is not because of some quality, "beauty", which resides in it, but because, in his strategic relationship with the environment, a creature, including man, feels that he needs a place where he can hide, and this need is supplied among the trees and in the shadow of their foliage'.[40] Specularism undoes an account of aesthetic experiencing which tries to do justice to such notions as experiencing the fragility of the natural world in a way which may not bring pleasure necessarily. It undoes an account of aesthetic experiencing which focuses on an unintentional form of awareness, enabling the human subject to attend fully to an object or entity's qualitative features. Hence we must turn elsewhere in our search for a way of grounding an aesthetics of nature!

An alternative route is suggested by not focusing on the subjective pole of human experiencing but rather on the object itself. This move is suggested by the borough council representative (BCR)'s suggestion in the four disputant's fictional discourse: aesthetical experiencing has to take account of what is known about what is experienced from a scientific point of view. If science serves in our world as a paradigm for how knowledge can be acquired, surely it might serve too as a way of grounding aesthetic awareness. After all, in chapter XII, it was suggested it was needed to interpret something's internal value so as to make sense of the natural events causing what is experienced to be what it is. In addition, too, science may offer an account of how an innate sense of value can be achieved as living things gain their ends from a teleological point of view.

Now you may point out that this scientific route faces two immediate problems. Firstly, we may know people who have plenty of cognitive awarenesses – whether ecological, hydrological or geological – but for whom a piece of countryside only seems suitable for waste tipping, clay extraction or logging! Secondly, we may know people too who can experience non-humanly created entities aesthetically with little knowledge of the science which might account for the nature of the existence of such things!

These two problems might be quickly side-stepped: we are not concerned, as we were with specularism, with the attitudes, dispositions or appraisals of the human being but rather with the nature of the object perceived. So why not attend to the design and/or orderings in what is experienced.[41] In relation to the first possibility, as the borough council representative pointed out in our fictional discourse, the notion of design forwards the idea of a designer, so that the aesthetic provides an entry point for religious awareness! Clearly Simone Weil employs this move but she does not use it to establish God's existence, any more than Peirce does when he employs his 'A Neglected Argument for the Reality of God'. Rather she grounds that proof ontologically: for Weil, given that God does exist, his actuality can be grasped aesthetically.[42] Why not ground God's existence aesthetically? The reason is obvious: to picture God, say, as an artist would be to project onto divine existence an all too human conception! But what about the second possibility: focusing upon the appreciation of order in nature as the basis for aesthetic experiencing'? Now how are we to account for this feature? Through a religious or scientific rationale or by an appeal to myth?

Let's try again. Surely an appeal was made to knowledge concerning the nature of the orderings we experience in appreciating natural objects aesthetically so as to fix those 'aspects that are relevant' to aesthetic experiencing. Not only is a 'cognitive requirement' in aesthetic experiencing thereby satisfied but a rationale can be provided for what it is about an object which allows human beings to be 'moved by nature' as Carroll puts it.[43] In other words, whereas in appreciating works of art aesthetically, our appraisal of them can be grounded upon appropriate ways of interpreting them in the light of requisite categories required for their appreciation, no such categories exist for appreciating nature aesthetically, save those offered by the natural sciences.[44]

The trouble with this whole approach is that whilst it can be shown that grasping something cognitively in a scientific sense is compatible with experiencing it aesthetically, we have not established that the latter requires the former. Can't an aesthetic experience be enjoyed without any kind of scientific cognition at all? In that case 'being moved by nature' becomes disconnected from any relationship it might enjoy with scientific cognizing. So we are still left with the problem of the role the latter is supposed to play in relation to the former. Noel Carroll suggests we focus upon whether or not a particular state of 'being moved by nature' can be regarded as aesthetic. To account for the appropriateness of that state he invokes Appleton's prospect–refuge theory. Present aesthetic charm is a residue from past instinctive behaviour.[45] The adequacy of that theory has already been explored!

Rather than focus upon either the subjective or the objective poles in making sense of aesthetic experiencing, we have a third possibility. This surfaces in our fictional discussion at the beginning of this chapter and in chapter X in a discussion of Christopher Tilley's *A Phenomenology of Landscape*. But rather than support Heidegger's possible desire to return to some form of medievalism, an alternative route needs to be taken whilst ensuring, at the same time, that aesthetic experience is understood as what occurs between the human subject and an object, in this case, one which is not humanly created. That route, however, must grant the prior significance of the object in such experiencing if we are to avoid the pitfalls of specularism. Simone Weil indicates our direction when she asks why it is we find something in nature beautiful. To leave the question unanswered delivers us an aesthetic mysticism.[46] Already an account of aesthetic experiencing – employing Bullough's notion of psychical distance in a new way – has been emphasised to make sense of these two necessary features in aesthetic experiencing: the preponderance of natural objects – nature's objective expression – and the mystical nature of this feature. Indeed at the end of chapter XII, the notion of intrinsic value was cashed holistically in terms of the play between the inherent, the innate and the internal senses of valuing within aesthetic experiencing. In a concentrated attention upon some entity or element in experience, where the experiencer becomes aware of something transcending the sheer contingency of what exists, it makes no sense to separate what the subject brings to this experience from what is regarded as belonging to its object. But how are we to make sense of what it is that transcends this moment's particularity?

Perhaps a clue to answering this question is provided by Appleton's advocacy of his prospect–refuge theory, despite our earlier attack upon it. Rather than focus upon primitive man's *viewing* of his environment to ensure human survival, perhaps we need to examine the *status of the object* in that viewing, 'the echo of the real supremacy of nature in the souls of primitive man' constituting nature's otherness, its 'moving spirit', its mana for human beings.[47] This idea represents Adorno's attempt to make sense of Benjamin's claims for an object having an aura within an aesthetic perspective. Benjamin accounts for this feature in human experiencing in two ways. The first approach emphasises the *existence* of the object in a way which need not run counter to scientific cognizing. The aura of natural objects is defined as follows: 'A strange weave of space and time: the unique appearance or semblance of distance, no matter how close the object may be. While resting on a summer's noon, to trace a range of mountains on the horizon, or a branch that throws its shadow on the observer, until the moment or the hour become part of their appearance – that is what it means to breathe the aura of those mountains, that branch'.[48] Benjamin employs a second sense of aura, which emphasises more the *actuality* of the object within aesthetic experiencing to use Hartshorne's helpful distinction in a new way.[49] In emphasising again the significance of attentiveness in perception, Benjamin remarks: 'Experience of the aura thus rests on the transportation of a response common in human relationships to the relationship between the inanimate or natural object and man... To perceive the aura of an object we look at means to invest it with the ability to look at us in return'.[50] Here we are released from the anthropocentric dimension since what is revealed by this 'look' lies beyond intentionality as consciousness is overtaken by what is unanticipated. This is the reason why Bergson's *memoire involuntaire* is so important for Benjamin, since this sense of aura recalls what can occur in unintentional scenting, tasting – in the case of Marcel Proust – or feeling. In such cases the object can't be held at a distance, even if it is initially in entering the aesthetic dimension. Our experience of it absorbs all our capacities and all our senses. Note the borough council representative's remarks about the bodily experiences of the cyclist.[51] Such a conception of aura seems akin to a sense of 'being moved by nature' in that it may be enjoyed independently of any sense of scientific cognizing. And the intensity of how this actuality is enjoyed or endured depends – as Benjamin and Weil in their different ways appreciated – on the human subject's capacity for attending to the object, appreciating it as a particularly unique focus for aesthetic experiencing.

This chapter began with a fictional discussion between representatives taking part in the Tiptree Inquiry grappling with the landscape architect consultant (LAC)'s defence of specularism in chapter VIII. Three ways have been indicated for grounding an aesthetics of nature. Firstly, there is specularism, which emphasises vision in delivering a sense of the aesthetic but overlooks the transformation of our ways of seeing by technological means. Secondly, there is dependence on a scientific point of view to ground categories in experiencing nature by pointing to the degree of orderliness to be found in nature's expressive activity, a view which can't account for a possible aesthetic experience which is not rooted in such scientific cognizing necessarily. Thirdly, there is a more 'holistic' approach which roots the aesthetic in the felt relations between subject and object.

This third approach, characterised elsewhere as perspectivalism,[52] picks up the second approach's stress upon something being a natural entity in aesthetic experience. Neither the wo/man in the street nor the philosopher would deny the possibility of an object's existence making an aesthetic experience possible in the way Benjamin's first sense of aura would suggest. Argument issues – as the debate between Brecht and Benjamin illustrates[53] – around Benjamin's sense of aura emphasise the actuality of this kind of experiencing. This is hardly surprising in view of the contingent and involuntary aspects of such experiencing which focuses upon unique particularities as opposed to the manipulative effects generated in the monocultural consequences of commodification. But without an emphasis placed upon the actuality as well as the existence of this form of experiencing, the poet's activity would remain unintelligible; it would then become impossible within an aesthetic experience of natural objects 'to read what was never written',[54] something required in the artist's activity, a capacity constituting what it is to be a human being. If it is no longer possible to preserve such submission to natural objects, then their authority, necessary to fulfil the actuality of aesthetic experiencing, can't be grounded. And without that grounding, not only would our normative model for ethical discourse be undone but I can't see how a real concern for the existence of a world created independently of, even if shaped by, human beings could be secured. Can you?

NOTES

[1]'Words of Comfort', Avoncliff, Wilts, 28 February 1995.

[2]The quotation is defended in Peirce's 'The Law of Mind' in *Philosophical Writings of Peirce* selected, edited and introduced by Justus Buchler, New York, Dover 1955, chapt. 25, p.353 and is to be found in 'Man's Glassy Essence', *Collected Papers of Charles Sanders Peirce*, vols, 1–8, vol. 6 paras. 102–163, 6.238–271, cf. 6.268.

[3]Town and Country Planning Act 1990; Section 79 Appeals by South Eastern Minerals Limited, Land at Tiptree Farm, High Oak Hill, Funton, Sittingbourne, Kent File nos. APP/W2275/A/95/ 259578 APP/W2275/A/96/271958 and APP/V2255/A/96/27236O, 12 October 1998, para. 9.8.3, p.144.

[4]Peirce, 'The Law of Mind' op. cit., letter authorised by the Secretary of State to sign on behalf, 12 October 1998; Planning Directorate, Minerals and Waste Division; Department of the Environment, Transport and the Regions para. 6.

[5]*Landscape Assessment: Principles and Practice*, a report by Carys Swanwick of Land Use Consultants to the Countryside Commission for Scotland, June 1991, p.247.5(v).

[6]Jay Appleton, *The Experience of Landscape* (1975), pp.62–68.

[7]Christopher Tilley, *A Phenomenology of Landscape*, pp.11, 18.

[8]Tilley quotes Heidegger as follows: 'Let us think for a while of a farmhouse in the Black Forest, which was built some two hundred years ago by the dwelling of peasants. Here the self-sufficiency of the power to let earth and heaven, divinities and mortals, enter in simple oneness into things, ordered the house. It placed the farm on the wind-sheltered mountain slope looking south, among the meadows close to the spring. It gave it the wide overhanging shingle roof whose proper slope bears up under the burden of snow, and which, reaching deep down, shields the chambers against the storms of the long winter nights. It did not forget the altar corner behind the community table; it made room in its chamber for the hallowed places of childbed and the "tree of the dead" – for that is what they call a coffin there: the Totenbaum – and in this way it is designed for the different generations under one roof the character of their journey through time. A craft which, itself sprung from dwelling, still uses its tools and frames as things, built the farmhouse'.

'Building Dwelling Thinking' (1951) in *Poetry, Language, Thought*, chapt. IV, pp.143–161 trans. A Hofstadter, New York: Harper Colophen Books, 1971, quoted in Christopher Tilley, *A Phenomenology of Landscape* p.12.

[9]Martin Heidegger *Being and Time* (1927), para. 416, pp.468–469.

[10]See the five-key steps argument of the landscape architect consultant in chapt. VIII.

[11]Svetlana Alpers, *The Art of Describing: Dutch Art in the Seventeenth Century* (1983), New York, Penguin, 1989 especially the 'Introduction', pp.41–45 and pp.124–126. I am indebted to Pat Turner for pointing out this source to me.

[12]Svetlana Alpers, *The Art of Describing*, op. cit., pp.xix, xx, xxiii, xxvii, 43.

[13]Town and Country Planning Act 1990, op. cit., para. 9.8.3, p.144.

[14]Paras. 1–17 of the inspector's Conclusions in the *Report of a Public Inquiry into Objections to the Swale Local Borough Plan*, Date of inquiry 3 December 1996 to 17 February 1997. The Planning Inspectorate, Tollgate House, Hamilton Street, Bristol B52 7DJ, 23 July 1998, pp.100–104, paras. 11, 12, p.103.

[15]For Peirce's theory of signs see chapt. 7 of *Philosophical Writings of Peirce*, Peirce, 'The Law of Mind' op. cit., 2.243; 1.43, 2.165 where the first number represents a paragraph not a page number.

[16]J J Zeman, 'The Esthetic Sign in Peirce's Semiotic', *Semiotica* 19, pp.241–258, p.243.

[17]Peirce, 'The Law of Mind', 1.418; 5.44; 5.113 and 5.119, cf. too 'A Neglected Argument for the Reality of God', 6.452–93.

[18]See the 'landscape architect consultant's case' in chapt. VIII.

[19]'The fact than science tends to show that man is part of nature has an effect that is favourable rather than unfavourable to art when its intrinsic significance is realised and when its meaning is no longer interpreted by contrast with beliefs that come to us from the past. For the closer man is brought to the physical world, the clearer it becomes that his impulsions and ideas are enacted by nature within him. Humanity in its vital operations has always acted upon this principle. Science gives this action intellectual support. The sense of relation between nature and man in some form has always been the actuating spirit of art.' John Dewey, *Art as Experience*, pp.338–339.

[20]Inspector's Conclusions, Report of a Public Inquiry into Objections to the Swale Local Borough Plan, op. cit., para. 11, p.103: 'there is considerable local support for the LLA designation of the line of hills between Iwade, Newington and Lower Halstow.' In relation to the actual appeal the later inspector wrote 'I have read all of the written representations and, while I have taken them into account in my conclusions, it is not generally possible to refer to them individually. There is substantial public opposition to the appeal proposals. At the application stages, some 360 letters and petitions were submitted about the proposals, almost all of which were against them. Also, there were about 500 letters at the appeal stage, including some duplication where more than one copy of the same letter found its way to me. The Tiptree Action Group – TAG – represents six local parish councils which object to the appeals. To those who have written in, on both sides, I would advise that the level of support or opposition to a proposal is not in itself a ground

for making a planning decision, unless the representations are founded on valid planning reasons which can be substantiated.' Town and Country Planning Act, 1990, op. cit., para. 9.1.2 p.135. But consider 9.8.3 p.144 'Having spent a number of weeks in Iwade and the area of the appeal site, and despite the concession by KKC that the site is "unremarkable" as landscape, I can well understand the affection that locals feel for this important local landscape feature in the rural gap between the Medway towns and the settlements west of Sittingbourne.'

[21]Report of the county planning officers to the Planning Sub-Committee, 12 September 1995, from the county planning officer. Subject: 'To Extend An Existing Clay Quarry, Restoration to Agriculture by Infilling; Provision of Site Access Roads and Improvements to Local Roads, Tiptree Farm, Funton, Lower Halstow – South Eastern Minerals Ltd (MR 878.669)'.

[22]County targets are set out in the Kent Waste Local Plan, p.43, para. 4.1.11, namely 'by 2011 an 80% reduction in the volume of the non-inert waste needing to be disposed of in Kent, or the waste having been first processed (e.g. sorted for materials recovery, and the residue then reduced in volume)'. If achieved, this would give a residual stream at the end of the Plan period of 500,000' cubic metres. Para. 31, p.10 Closing Submissions of Kent County Council, Application for Land Raise Disposal Facilities at Lamberhurst Farm, Dargate, Kent (SW/96/931; CA/96/1031) Monday, 22 February 1999. Incidentally, during the time of this book's writing, increasingly KCC has encouraged the borough council's involvement in decisions reagarding waste disposal.

[23]'If you or I were given the choice between buying fruit from a site adjacent to a landfill site, or from an orchard deep in undisturbed countryside, which would you choose? Not one in a thousand would choose the former. The county council is currently seeking further independent clarification from retailers in an effort to obtain a policy statement from them on the issue.' See endnote 22 above p.51 para. 129.

[24]Report of the County Planning Officers to the Planning Sub-Committee, op. cit.

[25]Much of what follows derives from Boulting, N E 'The Aesthetics of Nature' in *Philosophy in the Contemporary World*, vol. 6, nos. 3–4, Fall–Winter, pp.21–34.

[26]Hegel, G W F, *Phenomenology of Spirit*, sect. 489.

[27]Lukacs, George, *The Theory of the Novel*, pp.63–64.

[28]Walter Benjamin, 'The Work of Art in the Age of Mechanical Reproduction' in *Illuminations*, pp.217–251, p.236.

[29]That is how Peirce saw these matters – hedonistically – until after 1883 (Peirce, CP. 5.111).

[30]Kenneth R Olwig 'Sexual Cosmology' in *Landscape: Politics and Perspectives* ed. by Barbara Bender, pp.307–343 p.328.

[31]J Thomas, 'The Politics of Vision' in B Bender, see endnote 30 above, p.22.

[32]J Thomas, Kenneth Olwig, *Landscape: Politics and Perspectives*, op. cit., above p.25.

[33]Walter Benjamin, *Illuminations*, op. cit., especially sect. XIII.

[34]This example is Jennifer Smith's in a letter of 24 February 1999.

[35]S C Bourassa, *The Aesthetics of Landscape*, New York, Belhaven Press, 1991, p.47.

[36]E Bullough, *Aesthetics: Lectures and Essays*, pp.93–96.

---

[37]N E Boulting, 'Edward Bullough's Aesthetics and Aestheticism: Features of Reality to be Experienced', *Ultimate Reality and Meaning*, vol. 13, no. 3, September 1990, pp.201–221, especially sect. 5.

[38]Jay Appleton, *The Experience of Landscape* (1975), rev. ed. 1996 New York, John Wiley.

[39]Jay Appleton, *The Experience of Landscape*, op. cit. p.151.

[40]Jay Appleton, *The Experience of Landscape*, op. cit. p.216.

[41]Allen Carlson, 'The Aesthetics of Art and Nature' in *Landscape, Natural Beauty and the Arts*, pp.228–243.

[42]Boulting N E, 'Necessity, Transparency and Fragility in Simone Weil's Conception of Ultimate Reality and Meaning' in *Ultimate Reality and Meaning*, vol. 22, no. 3, September 1999; 'Grounding the Notion of Ecological Responsibility: Peircian Perspectives' in D A Crosby and C M Hardwick (eds.) *Religious Experience and Ecological Responsibility*, New York, Peter Lang, sect. IV.

[43]Noel Carroll, 'On Being Moved by Nature: Between Religion and Natural History' in *Landscape, Natural Beauty and the Arts*, Allen Carson, op. cit., pp.254, 244–266; p.253.

[44]Noel Carroll, op. cit., pp.255–256.

[45]Noel Carroll, op. cit., pp.262, 263–264.

[46]Simone Weil, *Gateway to God* UK: Collins, Fontana Books., 1982 p.101: 'for the beautiful gives us such a vivid sense of the presence of something good that we look for some purpose there, without even finding one.' Those who continue asking 'Why?' are given an answer: 'Silence is the answer the word of God.' In that case aesthetic mysticism becomes converted into an aesthetic theism!

[47]Max Horkheimer and Theodor W Adorno, *Dialectic of Enlightenment*, p.15.

[48]Walter Benjamin, 'A Small History of Photography' in *One-Way Street and Other Writings*, p.250.

[49]'Actuality is how or in what concrete thing or things, an essence is actualised; mere existence is only that the essence is somehow or in something concretely actualised,' *The Zero Fallacy & Other Essays in Neoclassical Philosophy*, Charles Hartshorne, p.81; 'To exist is to be somehow actualized in some individual – by 'actuality' is meant the how, the state of actualization', *Creative Synthesis and Philosophic Method*, p.254.

[50]Walter Benjamin 'Some Motifs in Baudelaire' in *Charles Baudelaire: A Lyric Poet in the Era of High Capitalism*, p.148.

[51]Cf. a remark by Paul Cezanne: 'The landscape thinks itself in me and I am its consciousness.' So he characterised his artistic activity as making 'visible how the world touches us' bodily. (Quoted in Maurice Merleau-Ponty's 'Cezanne's Doubt' in *Sense and Nonsense* trans. by H L Dreyfus, Northwestern UP 1964 pp.17, 19.

[52]N E Boulting, 'The Aesthetics of Nature' in *Philosophy in the Contemporary World*, vol. 6, nos. 3–4, Fall–Winter, pp.21–34.

[53]For Benjamin's friend Brecht, this second sense of aura would go too far in the direction of ecocentrism. Benjamin's theory was 'rather ghastly'; it was 'all mysticism'. cf. Susan Buck-Morss's, *The Origin of Negative Dialectics*, pp.78, 149.
[54]Walter Benjamin 'On the Mimetic Faculty' in *One-Way Street and Other Writings*, p.162.

Appendix I
GLOSSARY OF TERMS

The terms listed below give a sense of their meaning as they are used in some philosophic circles, but the main function of this list is to give a sense of how they are being employed within the text you are reading. Some of the terms are defined as Peirce would, since the writer wishes to convey his meaning at a particular point in the text.[1] Some of the terms listed, however, might not be regarded as philosophical terms at all, but that issue would generate as much argument as would the issue as to what is to count as philosophy in this, the new century.

abduction: This term refers to a form of inference Peirce later called *retroduction. Hence abduction is used to describe the nature of inquiry: retroduction initiates inquiry where a likely *hypothesis is formed (see *Firstness) its consequences are drawn out (*Thirdness) in *deduction and these are then put to the test (*Secondness) in an *induction in experience.

action: What brings causal effects into the world; different from *conduct since Peirce speaks of action and reaction in the same breath: 'Whatever exists, *ex-sists*, that is really acts upon other existents, so obtains a self-identity and is definitely individual' (Peirce: vol. 5, para. 428);

[1]Consider, for example, the term 'common sense'; a definition is given and the number 5.369 appears. That notation stands for *Collected Papers of Charles Sanders Peirce* ed. by C Hartshorne, P Weiss and A. W Burks, volumes 1–8, 1931–1958, Cambridge, MA: Harvard UP where 5 stands for the volume number and 369 stands not for the page but the paragraph number.

'there can be no resistance where there is nothing of the nature of struggle or forceful action. By struggle I must explain that I mean mutual action between two things, regardless of any sort of third or medium, and in particular regardless of any law of action'. ⋆Secondness 'is the predominant character of what has been done' (Peirce: vol. 1, paras. 322 and 343).

activist: A practical person who applies knowledge in life-situations.

activity: For the purposes of this book an activity can be described as being in a state of attending to something, being aware of something whatever that may be in ⋆contemplating it, directly experiencing by feeling it or 'experiencing' it as a result of its mediation through the new technology, using it for some purpose, or thinking/theorising about it. The activity may well make use of all these dimensions in relation to what it is to be involved as a human being.

actuality: How the existence of something is manifested.

aesthetic: A word used to refer to what is appreciated for itself alone usually in a heightened state of receptivity. An aesthetic experience of something is one where attention is given to the thing itself for its own sake. Aesthetic value refers to what is prized for itself and not for any other reason.

aestival: The dictionary points out this term relates to 'belonging to' or 'appearing in summer'. Aestivation refers to a petal arrangement in a flower bud before its expansion, presumably in sunshine. The term is used herein as a noun and refers to what is produced in an ⋆aesthetic moment when ⋆consciousness alights by chance upon something. Lending it complete attention can yield an aestival, something emerging out of the object's inter-reaction with a subject. Each chapter, except the first, begins with one such, composed at some particular time and place.

agapism: The 'mere proposition' that 'the law of love' is 'operative in the cosmos'. *Agapasm* is a mode of evolution, 'evolution by creative love', 'whose nature is divined before the mind posesses it, by the power of sympathy, that is, by virtue of the continuity of mind' (Peirce: vol. 6, paras. 302 and 307).

anancasm: Mechanical necessity as an operative principle in the universe; a degenerate form of *agapasm (Peirce: vol. 6, para. 303).

analogical reasoning: Given that a small collection of objects agree in various ways, it can be inferred that they are similar in other respects too. Given the development in his day of astronomy, Peirce argued that since Earth, Mars, Jupiter and Saturn are known to revolve on their axes, other planets probably did so too (Peirce: vol. 2, para. 733).

anthropocentric: Human centred; referring in some way to the human state in given conditions to realise its possibility.

a priori: Claims which are wholly independent of experience; necessary and universal truths presupposed in any kind of thinking about the world e.g. 'Circles can't be squared' despite *Hobbes's efforts!

a priori method: Peirce's label for a way of arriving at the truth which promises to deliver 'opinions from their accidental and capricious elements' (Peirce: vol. 5, para. 383) because they are 'agreeable to reason' (Peirce: vol. 5, para. 382).

argument: The activity or the act (outcome) of inferring a conclusion from a set of premises in a rule-governed process by means of reasoning. Peirce identified three types of such reasoning:

*retroduction, *induction and *deduction.

assumption: A premise of an argument; a claim or proposition taken for granted.

authority: An institution, or a person representing such, whose opinion is accepted because s/he may be regarded as an expert on some subject-matter.

axiomatic: An assumption not subject to doubt. Any claim can be doubted theoretically. But an axiomatic claim may be regarded as indubitable for practical purposes and so appears to be self-evident. Upon it other presuppositions might rest. If, for example, it were accepted as axiomatic that human beings 'live to earn' then the focal point of learning would be to equip students, through training, to do so.

being aware: Open to experience, so as to be able subsequently to give one's complete attention to something for its own sake in order to ascertain the truth about it, as opposed to seeking to use it for *commodific or *instrumental purposes.

bifurcation: A division into two distinct sorts of thing. Consider *Plato's separation of being and becoming; Descartes' Mind and Body; Hume's Fact and Value. If a philosopher is to follow in the footsteps of these thinkers then the problem arises as to how the two distinct sorts of thing delineated relate to each other.

causal: That which brings about an effect (cf. *Secondness).

casuistry: A way of dealing with problems in *ethics particularly as these arise in medical ethics, by focusing on a particular individual case and relating its circumstances to relevant laws or rules to which it can be applied.

cognitive: A word referring to what is taken to be known or understood as opposed to what may be believed, responded to or directly experienced.

cognitive interest: The cognitive aspect of an interest deriving both from nature and from humanity's cultural break with nature. Habermas argues that there can be three such cognitive interests: one in technically controlling the environment; a second, a communicative one, gaining understanding of someone's point of view; a third, an emancipatory one, exploring critically the nature, status and relationship between the former two so as to enhance a reflection upon our condition within society. This third represents ⋆philosophy's heart's core.

cognizing (cognition): The act of considering what is taken to be known or understood (cf. ⋆cognitive).

commodific: To be bought or sold in the market place; exchangeable; concerned with profit.

common sense: '…thought as it first emerges above the level of the narrowly practical' (5.369); 'instinctive beliefs' e.g. 'there is an element of order in the universe' (6.495).

commonsensism (critical): The 'critical acceptance of a sifted common sense of mankind regarding mental phenomena' (5.494).

community: This term can be restricted to the idea of an actual or indeed potential number of persons: 'This community, again, must not be limited, but must extend to all races of beings with whom we can come into immediate or mediate intellectual relation. It must reach, however vaguely, beyond this geological epoch, beyond all bounds' (2.654). ⋆Aldo Leopold uses the term to include what lies beyond man's cognition: ⋆Ecology 'is the science of communities, and the ecological conscience is therefore the ethics of community life' (*The River of the Mother of God and other Essays* ed. by S L

Flader and J B Callicott, Wisconsin P, 1991 p.340).

conceptual imperialism: Any kind of social reality can be described in many ways as a result of the kinds of concepts employed. Consider the activity of schooling. It could be described by using a set of endogenous concepts – those making sense of the activities of the students in developing their knowledge and understanding – or by a set of exogenous concepts explaining how such activity serves to slot youngsters into their place in a technologically organised society. To claim that either set of concepts is the set that ultimately explains the nature of schooling is to commit conceptual, even perhaps cultural imperialism.

conduct: 'Action is second, but conduct is third' (1.337); 'future conduct is the only conduct that is subject to self-control' (5.427); '...conduct controlled by ethical reason tends towards fixing certain habits, ...the nature of which does not depend upon accidental circumstances'; '*in that sense*' is '*destined*' (5.430).

consciousness: A word which goes beyond the term 'mind' because it refers to the totality of *thoughts and *feelings, both *intentional and *unintentional to which a person can be subject. As Sartre argued in *Being and Nothingness*, one can be conscious in the sense of being aware* of something or someone without bringing that awareness to an articulated thought (cf. N E Boulting 'Sartre's Existential Consciousness: Implications for Subjectivity' *Philosophy in the Contemporary World* vol. 5, no. 4, Winter 1998 pp.11–23).

contemplation/ contemplating: Meditation; giving one's whole attention to something, to regard it for its own sake alone (cf. *aesthetic and *intrinsic). For aesthetic purposes, this arises unintentionally: suddenly one finds oneself contemplating something. For most people this occurs in an art gallery. Is that why Peirce regarded the aesthetic experience of nature as seeing it as a picture? To regard contemplation as significant would be rather curious in contemporary life where sheer busyness ensures our lives are filled with stimuli, desires and problems to be solved!

content of an argument: What is being established, the substantial conclusion proved.

credibilism: In order to begin enquiry we have 'to start with propositions' considered to be 'perfectly free from all actual doubt' but we 'have to acknowledge that doubts may spring up later' about them (Peirce: vol. 5, para. 376, no. 6).

cultural: The imposition of the *cultured upon the natural world according to a design, creating a culturescape, cultivated or culturable landscapes.

cultural relativism: 'Time', 'moral', 'cause', effect', 'number', 'aesthetic harmony' etc. are concepts we employ to make sense of our world. Their successful use, it might be argued, settles for us the forms of experience we have of the world. To say they are merely warranted for the conceptual schemes applied in our own culture alone is to advocate cultural relativism since another culture may use different conceptual schemes making these two schemes *incommensurable with each other. To say that one culture's conceptual schemes are the ones needed to make sense of experience is to commit *conceptual imperialism.

cultured: Intentionally made subject to and involved in the expressive activities of human beings, whether artistic, cultural or technological.

decisionism: Given a range of possible actions, each one is critically examined in the light of its probable consequences and the one judged to be most efficacious in the pursuit of some end, with the least damaging consequences, is adopted.

de facto: Customary practice; according to what takes place factually though not legally; not recognised in a ⋆*de jure* sense.

de jure: Legal or according to law.

demarcation: Separating one kind of thing from another by drawing a line between them. Can astronomy be demarcated from astrology? The philosopher ⋆Popper thought so, since he considered the former to be regulated by evidence or experimental activity in order to test its ⋆scientific theories.

deduction: A ⋆form of reasoning displayed in mathematics (Peirce: vol. 2, para. 778). It is validated in that 'the facts presented in the premises could not under any imaginable circumstances be true without involving the truth of the conclusion' (Peirce: vol. 2, para. 778).

describing: The activity of setting out in discursive form the nature of something, event, state of affairs or whatever.

dichotomy: Division into two mutually exclusive parts. Consider the claim 'Everything must be green or not green'.

distracted: A passive condition where consciousness cannot concentrate upon or attend to some specific state of affairs. You may be trying to speak to someone and notice that his or her attention has, unintentionally perhaps, been taken by an aircraft flying overhead or by an item of news on television.

distractive: An active condition where consciousness is intentionally focused upon something in an ⋆activity preventing attention being given to some other state of affairs. Someone may be unable or unwilling to speak to you because, say, they are completing some business on their computer, having to speak to someone on their mobile phone or have to finish the hoovering of a room.

dualism: Another word for bifurcation; two basic principles or entities, only, give rise to everything that is taken to exist.

ecocentric: That standpoint was adopted in the UN's 1982 World Charter for Nature: 'Every form of life is unique, warranting respect regardless of its worth to man, and, to accord other organisms such recognition, man must be guided by a moral code of action' (cf. L W Wood, Jnr., 'The United Nations World Charter for Nature' *Ecology Law Quarterly* (1985) pp.977–98). Such an *ecological sensibility implies adopting a view of the natural environment, taken as a whole, generating principles illuminative of the human condition. Recognising human beings as part of the natural world would then provide the key for sustaining any form of cultural life.

ecological: Pertaining to the habitats of living things, the modes of life with respect both to them and of other species as well as the relations of these to the physical conditions of the environment.

ecology: Pertaining to the habitats of living organisms and the influence of the environment upon them (cf. *ecological). An ecological problem can be distinguished from a problem in ecology. The latter is often regarded as a scientific problem. Consider how rat poison may find its way into the food chains of other living creatures. An ecological problem is usually thought of as more *normative since it arises out of man's dealings with nature and the resulting consequences.

egocentric: Literally the ego or the self at the centre; self-centred. Seeing everything in terms of one's self and one's own self-interest.

emblematic: Given a commodity which once had a function and value – a football programme costing one shilling – in a different cultural context its function changes to become a 'souvenir from deceased experience' whilst its value depends on its rarity as well as its significance as a souvenir. *Commodities, then, can return as emblems. (cf. 'The emblems return as commodities.' *W Benjamin 'Central Park' sect. 32a *New German Critique* Winter 1985, vol. 34 pp.32–58).

empirical: Drawing from what is experienced or observed. A proposition which is capable of being tested. 'John Major won the 1997 British election' is an empirical statement; it makes a claim about what can be put to the test. It happens to be false and thereby can't be regarded as a factual statement.

encultured: Unintentionally made, subject to the activities of human beings as a result of the consequences of such activities within a culture.

enured: The word 'inured' refers to what is habitual, done without thought. The term 'inured' then can be taken to refer to action, to what is done. I am using the word enured to refer to how we think, better, to our reaction at the prospect of thinking, which is now not to think at all but, rather, accept some thinking done for us: 'After all what can I say; some authorities think this, others that!'

epistemology: The theory or study of knowledge.

ethics: The theory or the study of theories accounting for what makes something *moral. The *meta-narrative for the 'narrative' acting morally!

existentially (existence):	Pertaining to what can be taken to endure *ontologically.
Fact/Value dichotomy:	There are factual statements, regarded as value-free and value statements. Strictly speaking we can only speak of a Fact/Value distinction not a *dichotomy here, since this issue makes no reference to other kinds of statement, e.g. analytic statements (true or false in virtue of the words used) and *metaphysical statements. However, it might be argued that describing the facts of a case is a quite distinct activity from evaluating them. And because there is a logical gap between facts and values – one cannot argue from the one to the other – no *ought statements can be derived from factual claims.
fairness:	Seeking to be equitable or just.
fallacy:	An error in reasoning; a misleading argument; an invalid *argument. If a false claim follows from a set of true premises, there must be a fallacy in the argument.
fallibility:	'...we cannot in any way reach perfect certitude or exactitude. We never can be absolutely sure of anything' (Peirce: vol. 1, para. 147). Here Peirce points out our fallibility, our capacity to err. Indeed he uses the term *fallibilism* to denote his doctrine 'that people cannot attain absolute certainty concerning questions of fact' (Peirce: vol. 1, para. 149); 'that our knowledge is never absolute but always swims, as it were, in a continuum of uncertainty and indeterminacy' (Peirce: vol. 1, para. 171). The fact that what we think we know is capable of revision means that we don't have to be dogmatic on the one hand nor need be entirely sceptical on the other.
feeling:	'...is a mere tone of consciousness' (Peirce: vol. 7, para. 530); 'consciousness of a moment as it is in its singleness, without regard to its relations

whether to its own element or to anything else' (Peirce: vol. 7, para. 540); 'the consciousness which can be included within an instant of time, passive consciousness of quality, without recognition or analysis' (Peirce: vol. 1, para. 377); 'an instance of consciousness which is all that it is positively, in itself regardless of anything else' (Peirce: vol. 1, para. 306). 'To be conscious is nothing else than to feel' (Peirce: vol. 1, para. 318) (cf. ⋆Firstness).

Firstness: The quality, complete in itself, in or of what can be experienced: redness, bitterness, hardness etc. What brings life, a sense of freshness and freedom. Such feeling-qualities may not be realised in experience, so Peirce refers to these suchnesses as may-bes; they are possibilities (Peirce: vol. 1, para. 304) in experience and defy description since they are unique, immediate, spontaneous and vivid.

form of argument: Kind or process of reasoning – of which Peirce claims there are three: ⋆retroduction, ⋆deduction and ⋆induction – as opposed to the ⋆content or ⋆substantive meaning of the premises in that reasoning (cf. ⋆procedural).

functionalism (teleological): Since ⋆teleological means the logic of the goal or end, functionalism can be understood as the idea that there is an end or purpose for every living thing. ⋆Aristotle explained the existence of things in terms of their 'final' or ultimate goal, for the sake of which a thing exists, reacts to or acts on something else.

functionally: In relation to a purpose, which is to be fulfilled. To say that man has a function is to presuppose some form of functionalism, ⋆teleologically understood.

geocentric: Earth-centred; in relation to the earth. Aristarchus of Samos, living around 270 BC, may have been the first astronomer to reject a geocentric universe in favour of an *heliocentric one.

heliocentric: Sun-centred; in relation to the sun. Copernicus (1473–1543) tried to prove in his *De Revolutionibus* of 1530 that the sun was the centre of the universe.

hypothesis: Peirce used this term initially to delineate a form of inference later called *retroduction. An outcome of that form of inference; a conjecture about something; a statement of a possibility whose steadiness 'enables us to think about it – and to mentally manipulate it – which is a perfectly correct expression, because our thinking about the hypothesis really consists in making experiments upon it.' Without its steadiness 'we could not think about and ask whether there was an object having any positive suchness' (Peirce: vol. 1, para. 322). Something 'which looks as if it might be true, and which is capable of verification or refutation by comparison with facts' (Peirce: vol. 1, para. 120).

icon (image): 'A sign which would possess the character' or quality 'which renders it significant even though its object had no existence' (Peirce: vol. 1, para. 304). The sign corresponding to *Firstness, which asserts nothing. Its mood is supposal or potential in character (vol. 2, para. 291).

iconic: Likeness; what 'may represent its object mainly by its similarity, no matter what its mode of being' (Peirce: vol. 2, para.276).

idea: '...an event in an individual consciousness' (Peirce: vol. 6, para. 105) towards which *thought can be directed; a living feeling which affects others and carries an insistence into the future bringing other ideas with it (Peirce: vol. 6, para. 135) in forming an *interpretant.

idealism: '...the idealist in metaphysics, attacks the category of reaction' (see *Secondness) (Peirce: vol. 5, para. 85). If then only ideas exist they 'tend to spread continuously and to affect certain others which stand to them in a peculiar relation of affectibility. In this spreading they lose intensity, and especially the power of affecting others, but gain generality and become welded with other ideas'. (Peirce: vol. 6, para.104).

ideological: What is referred to as *rationalisations at the level of individual human action – reasons justifying human actions substituting for real ones – serves as ideology at the *cultural level. In both cases *consciousness is held, despite the illusion of autonomy, by an interest (cf. J Habermas *Knowledge and Human Interests* trans. by J J Shapiro, London: Heinemann 1976/8 p.311)

impartiality: Even-handedness; characterised by being unprejudiced; not taking sides or being partisan (yet see *incommensurable).

Imperative (Right/Ought) fallacy: Consider a number of actions that could be regarded as *moral: 'Concern for my mother is a priority'; 'Joining a rally to protest about the wrongs my government is committing'; 'Writing a letter of condolence to someone whose best friend has died is something I should do'. What ought I to do now? To jump from a claim about what it is *right to do to one recommending what I *ought to do i.e. give preference to one of the former claims, is to propagate this fallacy.

incommensurable: Non-comparable or non-measurable by a common standard. For example, there may be no impartial, neutral, independent standpoint by which two theories can be weighed against each other. Nor can the claims of one theory be 'translated' accurately or impartially into the language of the other.

index: A pointer; its mood is indicative or declarative in character (Peirce: vol. 2, para. 291).

indubitables: Acritical propositions. In inquiring into something there is no point 'going behind' (Peirce: vol. 5, para. 440) them, though one might do so on some other occasion. So there isn't a 'definite and fixed collection of opinions that are indubitable', at least theoretically, even if there are practically, e.g. 'fire burns'.

induction: A process of reasoning by which a general proposition (an 'all' statement) is inferred from premises based on experience ('some' statement). For Peirce if 'there is anything real ...it follows necessarily that a sufficiently long succession of inferences from parts to whole lead men to knowledge of it' (Peirce: vol. 5, para. 351). Its validity is sustained by 'the necessary relation between the general and the singular' (Peirce: vol. 5, para. 170).

inherent: Pertaining to the cultural or historical in relation to something's creation.

innate: Characterising what belongs to something at its origins and which never ceases to inform its development *teleologically speaking.

inquiry: What can lead to the settlement of opinion (Peirce: vol. 5, para. 375) or the attaining of a state of belief (Peirce: vol. 5, para. 374) (cf. *abduction and *scientific method).

insight: A sudden, clear, penetrating perception into some complex situation, resulting from what Peirce referred to as man's rational *instinct.

instinctual: Habitual as a result of genetic or cultural acquisitions. Animals have 'the more active functions', 'adaptive characters calculated to insure the continuance of nature' (Peirce: vol. 8, para. 21). Humans are provided with an instinctual ability to form *hypotheses about the natural world – the basis of Galileo's Il Lume Naturale – since unless man has 'a natural bent in accordance with nature's, he has no chance of understanding nature at all' (Peirce: vol. 6, para. 477). Such instinctual insights still have to be tested to ascertain their truth!

instrumental: A means for something else. Instruments can be means for making music.

intelligibility: Comprehensibility; being clear; understandability.

intentional: Relating to what is deliberate, purposed or planned (see *Thirdness).

internal: Pertaining to something's particular existence regarded *objectively and what, in evolutionary terms, accounts for its present *actuality.

interpretant: A *thought, which gives meaning to or translates a sign into something else (Peirce: vol. 5, para. 283), The significant outcome of a sign (Peirce: vol. 5, para. 473). What is created in the mind of a person as the result of a sign's impact, which is another equivalent or more developed sign (Peirce: vol. 2, para. 228).

intrinsic value: Aimed at, desired or regarded as an end in itself and not as a means to something else within experience, as that arises from the inter-relation between object and subject. Here there is a play between the *inherent, *innate and the *internal dimensions of what it is to be an object.

logical socialism: 'He who would not sacrifice his own soul to save the whole world is illogical in all his inferences, collectively. So the social principle is rooted intrinsically in logic' (Peirce: vol. 5, para. 354). Even if he has not that capacity himself, he 'will perceive that only the inferences of that man who has it are logical, and so views his own inferences as being valid only so far as they would be accepted by that man. But so far as he has this belief, he becomes identified with that man' (Peirce: vol. 5, para. 356)

logistical: Determined by logic. For Peirce 'logicality inexorably requires that our interests shall not be limited' (Peirce: vol. 2, para. 654).

meta-narrative: A theory about the nature of an account, story or narrative. If someone said that the reason you suffered from stress was that you were trying to reconcile what you really desired to do – given in your dreams or through the slips you make in discourse – with what it seemed you ought to do as a duty, since that is what your parents would expect you to do, such a narrative might be justified theoretically by Freudian theory. Freud's meta-narrative, involving the Id, Ego and Super-Ego, could be used to make sense of your stress story.

metaphysical: Non-*empirical and non-analytic. Such a claim offers its advocate a picture of the world and his or her place in it, a picture which may be undermined by factual claims but never knocked out by a single one of them. Examples: 'God exists'; 'Both what is becoming and what is, consists of processes of change'; 'Only minds and bodies ultimately exist'. These claims transcend our everyday experience of the world. Such presuppositions lie at the base of and make possible much of what counts as scientific theorising.

method of authority: As for the *method of tenacity but where 'the will of the state' acts 'instead of that of the individual' (Peirce: vol. 5, para. 379).

method of tenacity: Going 'through life, systematically keeping out of view all that might cause a change' in his or her arbitrary opinions (Peirce: vol. 5, para. 377).

methodological solipsism: Philosophising or indeed thinking about a critical communication from a standpoint outside it i.e. without necessarily being committed to the rules which make such a communication possible. The fact that a *solipsist says anything need not imply that s/he stands under or sustains a commitment to an intelligible communication with others. Discourse might be regarded merely as behaviour designed to manipulate others into thinking s/he is forwarding a point of view, whilst, in reality, s/he is standing outside it.

methodology: The general study of methods. These inform the activities undertaken in different disciplines in their enquiries. The philosopher's interest lies in exploring the connections between a discipline's methodology and what is claimed for or by its use.

modernist tradition: A phrase used to identify belief in the Enlightenment's tradition plus the importance of establishing some kind of rational society in the light of it.

moral: Refers to action which is right or wrong. Acting morally carries the idea of acting under some kind of obligation or duty with respect to others. 'I ought or ought not to do this at the level of my *conduct!'

Musement: Speculative activity of the mind focusing on the nature of *Firstness, *Secondness and *Thirdness and the relations between them as a result of a purposeless encounter one has with something in experience.

nationalism: Love of one's nation for the glory which would seek its prolonged existence in time and space; stronger than *patriotism.

naturalism (ethical): The natural world or indeed the natural dispositions of human beings account for or serve to define the way we are to think about *moral discourse.

non-anthropocentric: Not human-centred; not referring in any way to the human state in given conditions to realise its possibility.

normative: Setting a standard or establishing a value as a guide.

objective: Belonging to what is presented, often through *scientific inquiry, as the object. What is considered to belong to the realm of facts (see *present-at-hand).

ontology: The study of what can be taken to exist (cf. *metaphysical).

ought to do: An action, indeed the only action which should be done by a specific person in a particular situation; a duty-bound action binding on someone morally. For Kant, for example, it was not only *right to tell the truth but one always ought to tell the truth on any and every occasion! (cf. the *Imperative fallacy).

paradigm: An ideal model or pattern often used as a set of beliefs or unquestioned theory to serve as a justification for some practice.

patriotism: A negative sense of *nationalism. *Simone Weil defined it as a love of nation, A nation being 'earthly, can be destroyed, and is all the more precious on that account'; a feeling generating warmth, 'which the sentiment of nationalism lacks', for what 'is incomplete, frail and fleeting' (S Weil *The Need for Roots* Routledge and K Paul, 1978 pp.164–165).

perspectivalism: Aesthetic awareness is not gained merely through viewing something but by experiencing it bodily, sharing an intimacy with it.

personifying: Attributing or projecting what is something human upon something non-human;

representing something in terms of an animal or human form.

phaneron: Whatever can be present to *consciousness 'in any sense or in any way whatever, regardless of whether it be fact or figment' (Peirce: vol. 8, para. 213).

phaneroscopy: Peirce's term for *phenomenology. Unlike *Hegel, whom he saw as too occupied with *Thirdness, he did not seek to confine philosophical enquiries simply to what we think about in relation to experience. And unlike Husserl, he tried to distinguish quite clearly between three sorts of thing about which humans can be conscious of: *feelings (Firstness); 'two-sided consciousness of effort and resistance' in *actuality/existence (*Secondness) (Peirce: vol. 1, para. 24); *thoughts (*Thirdness) which are 'neither qualities nor facts' (Peirce: vol. 1, para. 420).

phenomenology: The study of what presents itself to *consciousness. *Hegel used the term in relation to different forms of consciousness enabling a sense of self-consciousness to emerge. The term is used to characterise Husserl's approach to *philosophy, which was much less speculative than that of Hegel. It focused upon descriptions that can be made of what we experience.

philosophy: An activity where how we feel, act and think can be explored to ascertain the reasons we have for so feeling, acting and thinking. Hence the kind of justification given for these reasons can be examined. That rationale informs this present book, though there may be other ways of approaching such activity; the adequacy of some alternatives are explored in this book's Preface (cf. *second-order activity). This activity can be divided roughly into three significant areas:

thinking about existential possibilities – *metaphysics – which may assist in clarifying why we feel the way we do in relation to experience; considering our actions may reveal what we value either *aesthetically or *ethically; thinking about how we gain knowledge of the natural and social world, *epistemology, and what makes thinking valid (logic).

postmodernist tradition: A term characterising the rejection of any kind of *meta-narrative in relation to the way the *modernist tradition identifies itself. It thereby initiates the end of what might be regarded as the presumptive role of *philosophy, seen as an ultimate, privileged discourse concerned with truth seeking. This latter claim is often associated with the idea that there is no necessary relation between systems of signs and the external world.

practical: As opposed to theoretical, which focuses upon *ideas, this term refers to action where knowledge is applied; carrying the implication of being useful.

pragmaticism: So as to be clear about the meaning of terms, Peirce invented this doctrine, intended not to be applied either to human actions or to questions about the truth but to ideas so that 'our idea of anything is our idea of its sensible effects' (Peirce: vol. 5, para. 401): 'Consider what effects, that might have conceivable bearings, we conceive the object of our conception to have. Then our conception of these effects is the whole of our conception of the object' (Peirce: vol. 5, para. 402). '…a *conception*, that is, the rational purport of a word or other expression, lies exclusively in its conceivable bearing upon the conduct of life' (Peirce: vol. 5, para. 412).

precept: A rule or a principle to guide or direct human *conduct.

prescribing: Advising, recommending what should be done so as to guide action.

present-at-hand: Relating to the world through thought, an attitude necessary for scientific explanation where things are regarded independently of any use they may have.

procedural rationality (critical rationality): Knowledge grows through the formulation of theories, conjectures or opinions which are made subject to criticism and testing so as to decide on their provisional acceptance, as opposed to simply assuming that such theories, conjectures or opinions are true.

procedural: Relating to the way something is done or decided upon, not to that thing's ⋆content (cf. ⋆substantive). If a procedure is thought to be irregular, its outcome might be judged invalid simply on that ground.

proposition: Something which 'need not be asserted or judged. It may be contemplated as a sign capable of being asserted or denied.' It 'professes to be really affected by the actual existence or the real law to which it refers' (Peirce: vol. 2, para. 252). A sign 'connected with its object by an association of general ideas' (Peirce: vol. 2, para. 262).

prospect: A view; what makes viewing other things, set at a distance from us, possible.

rational: Manifesting reason in making sense of some acknowledged activity such as deciding what is true, good or consciousness-raising.

rationale: A set of reasons for feeling about or acting for or reflecting on something in a particular way.

rationalisation: An invented reason, which legitimates or makes an outcome – what has been said or done – appear reasonable. Like an excuse it offers a ground for that outcome. Unlike an excuse it sets itself up as a valid ground for that outcome.

rationality: Something governing the way we are to act and think in a rule-governed way. But how are these rules to be justified? To ask that question is to raise the problem of grounding rationality itself (cf. W W Bartley III, *The Retreat to Commitment*, La Salle, Open court 1984).

ready-to-hand: Available for use and not considered at the level of thinking about something as an existing thing.

realism: '...a realist is simply one who knows no more recondite reality than that which is represented in a true representation' (Peirce: vol. 5, para. 312); 'to say that of the real, objective facts some general character can be predicated, is to assert the reality of a general' (Peirce: vol. 6, para. 100). There has to be existing individual entities to verify the reality of such generals if such generals are to be determined to be true, thereby making Peirce a *verificational realist, (cf. K Thompson 'Peirce's Verificational Realism' *Review of Metaphysics* (l978, 32 pp.76–98, p.1).

reasoning: The activity of reaching conclusions by drawing out inferences expressed by premises; thinking out the implications of an assumption in the form of an *argument. Gaining access from what we know to what we do not yet know by a series of inferential steps. This account of reasoning might make it appear *deductive in character; there is also *inductive and *retroductive reasoning.

refuge: What makes hiding or sheltering possible.

reify: Convert something into something else, e.g. regarding a possible heir to the throne as 'the people's princess', the 'Queen of Hearts' or projecting divine powers on to a ruler.

retroduction: A form of reasoning Peirce sometimes called

*abduction or even *hypothesis. It focuses upon how theories or *ideas arise (Peirce: vol. 5, para. 590) supplying new ones (Peirce: vol. 2, para. 777). It is based on what Peirce called a rational *instinct to guess something as to the truth, even if such guesses turn out to be false so that it 'merely suggests that something may be' (Peirce: vol. 5, para. 171).

reverie: A flow of *thought or fancies triggered by meditating upon something in experience; an activity which wanders away from that something; a waking dream.

right to do: A correct, defensible form of action in a situation for which reasons can be given.

scientific: Relating to what can be inquired into by the logic of science (cf. *abduction).

scientism: Disciplines or forms of awareness can only be counted as knowledge to the degree to which they adopt the methods of the natural sciences. Someone who said, for example, that 'Epistemologically all there is is physics and stamp-collecting' would be advocating scientism.

Secondness: Refers to exertion in response to something of some kind; what is brute and thereby has been done or come into existence so that it is unalterable and resistant to our will. *Causation and constraint embody Secondness since that refers to what 'insists upon forcing its way to recognition as something other than the mind's creation' (Peirce: vol. 1, para. 325); the facts of the case; unalterable.

second-order activity: If the ideas we have in exploring the world in which we live are taken for granted in such first order *activity, then second-order activity refers to enquiring into the validity of these ideas themselves.

sham reasoning: A process of reasoning predetermined by some known end or outcome, so that the *reasoning merely decorates or covers up its real purpose or intent. Unlike reasoning proper, it is not the process of inferring which determines what the conclusion will be, but rather a pre-thought conclusion.

sincere: Trying to be honest or genuine or frank.

solipsism: The idea that only one's own experience and oneself ultimately exist resulting from the notion that knowledge can only be grounded within one's own personal experiential states enclosed upon themselves. The external world is a dream!

specular view: A perspective on something which is enframed, as is the case in looking through a windscreen, a window, a mirror or seeing something created for the screen, as in the case of the computer image or television. The term 'specular image' refers to what is created by this perspective.

specularism: Seeing other things, human beings, indeed nature itself, as in a picture so that nature's qualities in vision can be appreciated because they incite pleasure in the viewer as they might be by a picture. The resulting representation appears to place emphasis upon what is fixed, seen at a distance, capable of visualisation, useful for scientific investigation in relation to its representative qualities.

subjective: Belonging to the *consciousness of the subject or the person undergoing an experience. Whereas a reverie is completely subjective, an *aestival arises in the interaction between the object and the subject.

substantive: Relating to the content of what is done or decided as an outcome or result.

symbolic: Pertaining to the law-like or habitual in terms of what is 'a general type' (Peirce: vol. 2, para. 249). It acts through or applies to instances or replicas of itself, though it is inclusive of *iconic and *indexical elements (Peirce: vol. 2, para. 294–295) whilst requiring an *interpretant.

teleological: Subject to a purpose or design served by an entity which may or may not be regarded as part of some larger predetermined plan. An evolutionary adaptation might serve a function i.e. given a teleological explanation without supporting the idea of some super-designer.

theocentric: God-centred; in relation to God. Of course belief in God might not invoke the idea of a God-centred universe. In Whitehead's cosmology it might be argued that the existence of God can be derived from the other categories his *metaphysics employs. This issue is one of dispute in that scholarship.

Thirdness: The rule-governed, the law-like which typifies not feeling (*Firstness) nor action and reaction (see *Secondness) but *thought. It makes predictions possible since it ensures that 'future facts of Secondness will take on a determinate general character' (Peirce: vol. 1, para. 26). Whilst an unthought reaction is specific (Secondness), *conduct illustrates Thirdness because of its general character, its susceptibility to reason, to what can be learnt.

thought: 'A thread of melody running through the succession of our sensations' (Peirce: vol. 5, para. 395). 'It should rather be understood as covering all rational life, so that an experiment shall be an operation of thought' (Peirce: vol. 5, para. 420). A person's 'thoughts are what he is 'saying to himself', that is, is saying to that other self that is just coming into life in the flow of time' (Peirce: vol. 5, para. 421). Continuity 'like generality, and more than generality, is an affair of thought, and

is the essence of thought' (Peirce: vol. 5, para. 436).

truth: 'That which any true proposition asserts is real, in the sense of being as it is regardless of what you or I may think about it. Let this proposition be a general conditional proposition as to the future, and it is a real general such as is calculated really to influence human conduct' (Peirce: vol. 5, para. 432); 'truth's independence of individual opinions is due (so far as there is any 'truth') to its being the predetermined result to which sufficient inquiry would ultimately lead' (Peirce: vol. 5, para. 494); 'the ideas of truth and falsehood, in their full development, appertain exclusively to the experiential method of settling of opinion' (Peirce: vol. 5, para. 406).

truthful: A truthful person is s/he who wishes his or her 'opinions to coincide with the fact.' Such a person 'who confesses that there is such a thing as truth, which is distinguished from falsehood simply by this, that if acted on it should, on full consideration carry us to the point we aim at and not astray' (Peirce: vol. 5, para. 387).

tychism: The ⋆proposition that absolute chance/fortuitous variation is an operative principle in the universe; a 'way of thinking' which gives 'birth to an evolutionary cosmology' (Peirce: vol. 6, para. 102). *Tychasm* is an evolutionary mode, 'evolution by fortuitous variation'. Tychasm is a degenerate form of ⋆agapasm, (an evolutionary mode suggested by the proposition ⋆agapism) (Peirce: vol. 6, paras. 302–303).

unintentional: By chance or accident; not pre-thought (cf. ⋆Firstness).

unity:	An integration so that what is brought together is always a many ensuring that the whole and its parts complete and complement one another.
utilitarian:	Useful to human beings. As an *ethical doctrine forwarded by such thinkers as Bentham and Mill, an action is right to the degree that its consequences maximised human happiness, utility or profit.
verificationist:	Making sense of what counts as *empirical by making that issue dependent on the way its claim can be confirmed by scientific inquiry.

Appendix II

AN INDEX OF THINKERS

The thinkers listed below, other than Charles Sanders Peirce, are referred to in the book's text. The term 'thinker' extends beyond philosopher, as some of them, for example Charles Elton and Aldo Leopold, would not be regarded as philosophers in the traditional sense. Rather, they engaged in philosophical thought about their activities elsewhere, in their cases in zoology and ecology respectively, just as Peirce did in relation to the natural sciences. What is said, in the case of each thinker, is not to be taken as exhaustive in terms of each thinker's contribution to human thought but, rather, how their ideas have contributed something with respect to what is discussed within the text. Consequently, whatever the reader finds about a particular thinker in what follows below can hardly escape the charge of distortion. Terms used in this and the former glossary are in writing marked with an asterisk.

ADORNO, THEODOR (1903–1969)

A musicologist and a leading member, along with Max Horkheimer, of the Frankfurt school of Critical Theory. Though his best-known book is in social psychology – The *Authoritarian Personality* (1950) – he is best known philosophically for his joint work with Horkheimer, *The Dialectic of Enlightenment* (1947) which presented not only a new conception of history but a critique of *modernism through an attack on *procedural rationality, a dispute he continued with *Karl Popper and Hans Albert in the 1960s (cf. *The Positivistic Dispute in Sociology* T W Adorno (et al.) (1969)). Max Horkheimer also credited Adorno for his assistance with Horkheimer's book *Eclipse of Reason* (1947), the best introduction to this school of philosophy. Adorno's most famous book is his aphoristic *Minima Moralia* (1951), a work haunted by the loss of his close friend and colleague *Walter Benjamin in 1939, which Adorno claims concerns 'the teaching of the good life' now distorted by *commodific values. If that work can be approached without knowledge of Hegel's philosophy, the latter is necessary for an appreciation of *Negative Dialectics* (1966) which serves as a critique of *Hegel's totalisation approach to *philosophy. Adorno is concerned to show how ordinary everyday concepts, meant to inform us about existing things in the world, continually misrepresent their referents; the latter's non-identification with ideas. In *Aesthetic Theory* (1970) he tries to explore how the non-conceptual, how that gap between the concept and the thing, can best be expressed, and it is to modern art that he bestows this particular role. Art initiates us into other ways of experiencing, outside the usual concepts meant to settle for us our everyday experience of the world.

APEL, KARL-OTTO (1922–)

German professor of philosophy at Frankfurt-am-Main university (1972–1990) and then at the Johann Wolfgang Goethe University, Frankfurt. He has followed the bent of Peirce's pragmaticism in Peirce's *On the Doctrine of Chances,* as he makes clear in his *From Pragmatism to Pragmaticism* (1967/trans. '81) – published in German in 1967, translated in 1981. He focuses on the possibility of employing in *rational discourse the notion of an ideal communication act, which is subject to rules flowing from the very nature of reason itself. In this way he was able, in *Transformation of Philosophy* (1973/6), to take up Marx's challenge in his *Theses on Feuerbach* of 1845 – 'Philosophers have only interpreted the world in various ways; the point is to change it – whilst retaining his philosophy's Kantian or transcendental character. Using such a methodology he was able to attack the school of late Wittgensteinian philosophy in his *Analytic Philosophy of Language and Geisteswissenschaften* (1965/7). In *Understanding and Explanation* (1979/84) he attempts to clarify the 'explanation versus understanding' debate in terms of his 'transcendental-pragmatic' approach. He does this by employing Habermas's theory of three *cognitive interests to distinguish three kinds of knowledge: the deductive-nomological, the historical-hermeneutic and the critical reconstructive sciences. He published *Towards a Transcendental Semiotics* in 1994 and *Ethics and the theory of Rationality* in 1997. Following Peirce, then he argues that it is possible 'both to thematize the processes of natural evolution and those of human inquiry into, and (into an) interpretation of, natural evolution, and to conceive of the latter as continuations of the former.' Such a *metaphysics of evolution may risk 'going far beyond the paradigmatical suppositions of 'normal science' through speculative designs that are at the same time bold and conceptually vague.' As he admits, this is the direction of Peirce's own speculations in the early 1890s.

APPLETON, JAY (1919–)

An English geographer and author of *The Experience of Landscape* which suggested a fresh theoretical approach to the *aesthetics of landscape following some of *John Dewey's ideas, in his *Art as Experience*, relating these to habitat and his own *'prospect–refuge' theories. This approach has had important consequences for the activities of landscape assessment and evaluation (cf. G Swanwick's *Land Use Consultants Landscape Assessment:* Principles and Practice for the Countryside Commission for Scotland 1991.)

ARISTOTLE (384–322 BC)

Picking up a suggestion from the beginning of his teacher's work, Plato's *Meno*, this Greek philosopher and scientist, in his *Politics*, suggests three ways in which human beings can become good: through nature, habit and *reason. The first is understood as a certain character or set of qualities of body and soul bestowed by *nature* or by some divine dispensation and which through *habit* can be turned to good or bad. However, human beings can reject both habit and natural impulses if *reason* convinces that another course of action is better whereas nature's role is outside human control. Whilst arguments and teaching may not be efficacious in making men good, it is the cultivation of good habits, which nourish the seeds of a moral life, enabling the human being to love what is noble and hate what is bad. But the training in good habits must be subject to what is the end of our nature: *reason* and the intellect.

This trichotomy is echoed in his examination of what leads to securing happiness. He considers three candidates: the life of enjoyment or pleasure, the political life and, thirdly, the contemplative life. It is re-emphasised when he considers the three features which guarantee the effective functioning of the state: human productive functions in agriculture and arts and crafts (*poesis*), public service (*praxis*) whether in guarding the city, jurisdiction and statesmanship or religious activities, and thirdly *contemplative activity to secure those values central to the life of reason sustaining the other two spheres. Three kinds of knowledge can be distinguished by focusing on these three features: technical expertise employed by craftsmen to produce objects of use (*techne*); practical knowledge or practical wisdom (*phronesis*) – where *thought is expressed within *action and controls it, rather than being outside it, as in *Hobbes – and speculative knowledge, a kind of philosophic wisdom (*sophia*). S/he enjoys wisdom to the degree s/he can combine knowing (*episteme*) with the demonstration of why things are as they are, exercising insight or vision (*nous*). To do the latter is to exercise every nerve to live in accordance with the best thing, as he puts it in his *Nicomachean Ethics*. Though Aristotle argues that in the polis technically proper arrangements are necessitated to ensure the appropriate institutional reproduction of society, he also shows that those arrangements, unlike our own situation, do not have to determine the contents, purposes and moral goals pursued within the state. In other words, there is more to human existence than being preoccupied merely with conditions of survival. Rather we should be asking 'What is it to live well?' What is the nature of the good life for humans? There is considerable discussion in the literature of what is called the question of the two lives. Does the best form of life presuppose the exercise of man's highest disposition, that of contemplation, or is it to be understood in terms of some kind of complete, active, political life? (cf. L Nannery 'The Problem of the Two Lives in Aristotle's ethics' I*nt. Philos. Quart*. (1981) vol. 21, no. 3, pp.277–293). But what is clear is that you can't have intellectual virtue without moral virtue and once the latter is properly entrenched the former can flower. This examination of his ideas central to his ethics and politics illuminates his differences from Plato; Aristotle rejected the Theory of Forms. Whereas Plato's attention was focused upon the realm of Being to make sense of the physical and social realm, Aristotle proceeds from the latter in order to advance to the former. Hence his interest in physics: *On the Heavens, Physics, On Generation and Corruption*; in psychology: *On the Soul*; in Natural History: *On the Parts of Animals*, *On the Movement of Animals*, *On the Generation of Animals* where he employs *teleological explanations to delineate the nature and behaviour of living organisms; in Logic where he invented the syllogism in such works as *On Interpretation, Categories, Prior and Posterior Analytics* besides inquiring into the nature of being in his *Metaphysics*, required to underpin his logic in accounting for essences, substances and species. Many of the ideas, then, to be found in his works still haunt us today: we only have to consider, for example, his theory of tragedy as that is presented in his *Poetics*.

BARTHES, ROLAND (1915–1980)

A French literary critic, a structuralist – someone who holds that it is only through grasping the interrelations between the phenomena of human life that the latter can be made understandable – and a semiologist, the latter term referring to someone who inquires into signs, going beyond the surface actualities of particular speech acts, to examine a range of hidden systems of meanings. In his book *Elements of Semiology* (1964) he claimed to apply his pursuit of semiology to any system of signs: 'images gestures, inimical sounds, objects and the complete association of these which form the content of ritual, convention or public entertainment: these constitute if not language, at least systems of signification' to which it can be replied Peirce grounded that insight first. Barthes's contribution to what is referred to as the structuralist debate was to analyse signs to be found in popular media culture (*Mythologies* 1957).

BAUDELAIRE, CHARLES PIERRE (1821–1867)

A Parisian poet famous for employing symbolism, better, allegory in his work so that Paris becomes the focus of his lyrical poetry, itself a celebration of his alienation from that city. A solitary man whose only companion was a woman of mixed race – Jeanne Duval – who not only became his mistress but a kind of inspiration for his *Flowers of Evil* (1857). In that work Baudelaire celebrates his sense of isolation in the crowds of Paris, a city he sometimes casts as a landscape, sometimes as a room in which he lived. One way of explaining the title of his famous book of poetry is to appreciate that in his writings images of death and of woman haunt the streets of Paris. After all Baudelaire's only sexual relationship was with a prostitute. Hence the rejection of his work as perverted, horrid, even satanic (cf. *W Benjamin *Charles Baudelaire* (1935–9 (trans. 1973)).

BENJAMIN, WALTER (1892–1940)

Admired as a foremost literary critic of the twentieth century, an historian of Jewish mysticism and member of the Frankfurt School. Recognised in European intellectual circles, but not in English philosophy departments, as offering suggestions towards a philosophy of history, a philosophy of language, a critical *epistemological project and pioneering ideas in *aesthetics, particularly in relation to the way artistic and literary production is governed by social and technological conditions in society (cf. G Smith (ed.) *Benjamin: Philosophy, Aesthetics, History*, Chicago UP 1983). Perhaps this ignoring of his philosophic contribution is due to Benjamin's distrust, as Adorno puts it, of 'all limitations placed on the realm of possible knowledge, the pride of modern philosophy in its illusionless maturity'. On the one hand, Benjamin advocated his friend Brecht's revolutionary approach to theatre whilst forwarding what he called 'immanent criticism': principles for interpreting and criticising a work of art must not be brought to it from outside but be drawn from within the work itself. As he put it in *Author as Producer* (1937): 'the tendency of a literary work can only be politically correct if it is also literarily correct'. Yet he agreed with his other friend *Adorno that nothing in culture escapes functionalisation. In response, Fascism aestheticises politics: 'Communism responds by politicising art' as he concludes *The Work of Art in the Age of Mechanical Reproduction* (1936), but Adorno rejected his claim that photography and cinema can be defended as ways of 'politicising aesthetics' because of his, Adorno's, own hostility to the mind-numbing and dominating effects of the mass media. Yet Adorno passed the highest of praises on his former friend: 'He never wavered in his fundamental conviction that the smallest cell of observed reality offsets the rest of the world' (cf. Adorno's *Prisms* Cambridge: MIT 1982).

BERLIN, ISAIAH (1912–1997)

Russian-born historian and philosopher at Oxford University who also served as a British diplomat. He is famous for his distinction between negative and positive freedom in his *Two Concepts of Liberty* (1959). The former focuses on a conception of freedom from constraint which haunts, for example, Hobbes's philosophy. The latter emphasises one where the individual is empowered to do things, a conception defended for example in Marx's philosophy. His 1959 work characterises his interests in the political philosophy of his day.

BRECHT, BERTOLT (1895–1956)

Poet and perhaps Germany's greatest dramatist of the twentieth century. A Marxist prepared to regard his works as experiments with social phenomena played to an audience required to view them in a detached manner. His epic theatre, where situations are portrayed rather than plots, required the audience to see the actors as actors, props as props and the stage as a stage so as to prevent any kind of identification with what is experienced in his, Brecht's, 'dramatic laboratory', a place where every technical means is used to raise the audience's consciousness to the issues explored. That technique is well to the fore in *Mother Courage* (1941) a play which, for *Adorno, distorts the historical social reality of the Thirty Years War – presented to forward an up-to-date communist message – thereby undoing the whole point and structure of Brecht's epic drama.

BULLOUGH, EDWARD (1880–1934)

English professor of literature at Cambridge University and author of the classic article 'Psychical Distance' as a Factor in Art and an Aesthetic Principle' (1912). Subsequently published various articles on the relationship between psychology and aesthetic appreciation. Even though he is critical of *Schiller's* 'illusion theory', Bullough was influenced by his conception of ⋆aesthetics. This comes out in his view that anything can be appreciated aesthetically and also in his last article 'Recent Work in Experimental Aesthetics' (1921–1922) where his distinctions between four kinds of attitude formation – the ⋆practical, the ⋆scientific, the ⋆ethically valuable and the ⋆aesthetic – mirror those drawn by Schiller between the physical, the cognitive or logical, the moral and the aesthetic.

DARWIN, CHARLES (1809–1892)

English naturalist and originator of the theory of evolution first set out in *The Origin of Species by Means of Natural Selection* (1859) though he had access to it twelve years earlier after his journey around the world aboard HMS *Beagle*. Believing his theory was in need of ⋆verification, his book illustrates his huge efforts to amass cases confirming his theory. In *The Descent of Man and Selection in Relation to Sex* (1871) he sought to trace the origins of the human race as well as explore theories in relation to sexual selection. Reference is still made to *The Expression of the Emotions in Man and Animals* (1873).

DEWEY, JOHN (1869–1952)

American educational theorist and philosopher. He was a philosophy professor 1904 until 1930 but was concerned to show how philosophy could be put to good use within a culture. So he tried to test his theories of education in his laboratory schools at Chicago University. He was very influenced by his teacher Peirce in relation to considering the stages of an adequate Logic of Scientific Investigation to provide validity for his conception of truth as a 'warranted assertion' having developed James's pragmatism: the true idea is what is useful or works in life-situations. His ⋆moral theory is outlined in *Human, Nature and Conduct* of 1922. Dewey's problem-solving approach to the philosophy of education is placed in the context of a general social philosophy. The titles of his books give a sense of the importance of a child-centred education in a society committed to democracy:

The Child and the Curriculum (1902), *Democracy and Education* (1917) and *Experience and Education* (1938). It is thanks to *Jay Appleton's work that more interest is shown in Dewey's charming book *Art as Experience* (1934) where Dewey attempts to explore the following thesis: 'Because experience is the fulfilment of an organism in its struggles and achievements in a world of things, it is art in germ.'

ELTON, CHARLES (1900–1991)

An English ecologist who made four Arctic expeditions, yielding material for his books on animal ecology. His ecological *'community concept', found in his *Animal Ecology* (1927), served to change the ecological *paradigm of 'nature-as-organism' to 'nature-as-society' which may have influenced the writings of his friend Aldo Leopold who reviewed his *Exploring the Animal World* of 1933.

FISCH, MAX H (1900–1995)

American historian of philosophy who did much to save Peirce's ideas from obscurity (cf. his essays in *Peirce, Semeiotic and Pragmatism* (1986)). A scholar in the history of American ideas and in Vico's philosophy.

FOUCAULT, MICHEL (1926–1984)

French historian and philosopher. His thought is indebted to *Hegel insofar as there is firstly a concern to relate the history of thought to general history. Secondly, there is a concern for how far human beings can be regarded as *cognitive creatures and the degree to which such cognitions can be related to the power of social practices forming them, as these creatures engage in such activities. At the same time he rejects the idea that there ever can be a specific science focusing upon the human being as such. These concerns yield the insight that our everyday concepts bearing on sexuality, identity and the *moral point of view have emerged from previous historically formed constructs. So, in his *Discipline and Punish* (1975), he shows that with the passage of time the meaning and significance of punishment, for example, has changed fundamentally whilst his earlier *Madness and*

Civilization (1965) mapped out previous Western attitudes towards the insane.

FRANKFURT SCHOOL OF PHILOSOPHY

Came into existence in 1923 with the founding of the Institute of Social Research at that University. Inspired by *Marxism, it generated critiques of historical and sociological inquiries. Soon its leading members were Horkheimer, *Adorno, *Benjamin and Marcuse whose works referred to Freud's psycho-analytic discoveries as well as existentialism to generate an approach called 'critical theory'. The latter focuses on political philosophy, cultural critique, *aesthetics and debates about the nature of *modernism. The members of the School, except Benjamin, emigrated to America during the rise of Hitler because of their strict opposition to Fascism and Stalinism, returning to Frankfurt after the Second World War. Its leading figures then became *Habermas and *Apel.

GALILEI, GALILEO

Usually known as *Galileo* (1564–1642) Italian astronomer, philosopher and mathematician. He came to reject the philosophical approach of *Aristotle as a result of his earlier medical studies. As a mathematics professor, he forwarded the theory that all falling bodies descend with equal velocity. His astronomical observations convinced him of the truth of Copernicus's *heliocentric claims that, along with his atomistic theory of matter, got him into trouble with the church authorities. Not only is he credited with the distinction between primary and secondary qualities – the latter, tastes, colours, smells etc. – but also the relativity of motion, as well as inventing the idea of the ideal experiment, a thought experiment, where the ideal case of a body moving without impediment could be used to legitimate a new system of mechanics. This imaginative possibility may have influenced *Hobbes's conjectures e.g. the state of nature, as did Galileo's use of the resolutive-composite method: given some experience, all elements of it, which can't be quantified, e.g. secondary qualities, can be stripped away 'resolving' the situation by enabling an analysis to be made of these quantifiable elements

– the primary qualities; shape, size, quantity and motion. Variables are then related to each other to provide a formula for what is occurring. In this way a 'composition' can take place reconstructing what originally happened in mathematical terms by drawing out consequences in relation to the natural laws thought to apply. So we find Hobbes saying, at the start of *De Corpore*: 'Philosophy is such knowledge of effects and appearances as we acquire by true ratiocination from the knowledge we have first of their causes or generation: And again, of such causes or generations as may be from knowing first their effects'.

HABERMAS, JURGEN (1929–)

Like Adorno, his former teacher, and Marcuse much of this German philosopher's work has focused on a critique of critical or *procedural rationality, whose advocates, such as *Popper and Hans Albert, do not recognise its intimate connection with instrumentality, with a technical cognitive interest pursuing scientific knowledge so that the latter can be used to advance the human control of nature and other natural processes. The lack of such recognition is often associated with *scientism, the view that science is not just one form of human knowledge but rather that any kind of cognitive activity must be able to satisfy the canons of scientific methodology to be counted as knowledge at all. Such scientisation, when applied to the public domain by politicians, reduces public debate about what ought to be done to problems of technically controlling or manipulating the masses in order to satisfy predetermined goals. To meet this reductive challenge, in his *Knowledge and Human Interests* (1968/trans.1972), Habermas distinguishes three *cognitive interests: the technical, manifested in keeping mass production in society developing; the communicative, ensuring a measure of understanding between members of a society; the emancipatory to guarantee the possibility of an awareness of where a culture is developing and how best it might achieve the values of an open democratic society. The latter interest serves to remove possible distortions in our self-understanding as well as those in a contemporary culture. Out of a concern with this third cognitive interest, Habermas has focused upon how it is possible for authentic communication acts to take place between human beings, so as to make clear what value commitments must be sustained if genuine linguistic communication between persons is to occur. This endeavour is often summarised in terms of his theory of 'transcendental pragmatics' through which he faces attempts by *procedural pragmatists – for example *Rorty – to sustain the equating of truth with what happens to be efficacious in the way of belief. In his pursuit of this endeavour Habermas seeks, as a *modernist, to continue the enlightenment critique project through examining 'the philosophical discourse of modernity' (cf. *Philosophic Discourse on Modernity* (1996)).

HARTSHORNE, CHARLES (1897–2000)

Author of some twenty books beginning with his *The Philosophy and Psychology of Sensation* (1934), an American process philosopher and metaphysician strongly influenced by Whitehead and Peirce adopting the view that being has to be defined in terms of becoming. The latter is accounted for in the emergence of actual entities or 'drops of experience', to use James's phrase, growing out of each other, each one constituting a potential for the becoming of the next. He is famous for his panentheism which sustains the idea of God transcending the world whilst including it, – so mediating between theism on the one hand and pantheism on the other – an idea employed in a book he co-authored with W L Reese *Philosophers Speak of God*; for his discovery in 1944 of Anselm's second proof for God's existence and for his original book on ornithology *Born To Sing* (1973). His selected essays in his book *Whitehead's Philosophy* contains critical pieces on the contemporary fashionable philosophy of the day whilst his last work, published in the year of his hundredth birthday – *The Zero Fallacy and other essays in Neoclassical Philosophy* (1997) – includes articles on the philosophy of science, the philosophy of mind, *ecological ethics and *aesthetics.

HEGEL, GEORG WILHELM FRIEDRICH (1770–1831)

The last of the great German idealists who saw himself completing the development of Kant's philosophy. His most influential work was *Phenomenology of Spirit* of 1807, an attempt to advance the mind from its immersion in sense-awareness through a number of phases so as to achieve the more adequate stage of a comprehensive philosophy. In dealing with one of these phases he delineates what can be called a *phenomenology of the master–slave relationship, still worthy of attention today. Much of the writings for example of *Marx and *Heidegger are indebted to this work as is Sartre's *Being and Nothingness*, though Hegel's most well-known treatise is his *Philosophy of Right* (1821) containing in its Preface the following controversial lines: 'When philosophy paints its grey in grey, then has a shape of life grown old. By philosophy's grey in grey it cannot be rejuvenated but only understood. The owl of Minerva spreads its wings only with the falling of the dusk.' In both these famous works he seeks to show how universal notions are evinced in individual experience, individual facts and within history and how these are related to each other. Once trained in his methodology, the students of his phenomenology come to form a universal spirit in appreciating the logical development of these universals, to gain complete knowledge of the absolute itself. He is also to be credited with putting the philosophy of history and the history of philosophy on the philosophical studies map. These activities were important to him since historical facts provided the data from which philosophical inquiry proceeded.

HEIDEGGER, MARTIN (1889–1976)

A German philosopher who was professor at Marburg (1923–1928) before succeeding Edmund Husserl at Freiburg University (1929–1945). His most significant work was *Being and Time* (1927). In this work he examines different modes of 'Being' in the light of his fundamental *ontology – underpinning his explorations in history and his reflections on nature – so as to analyse the different human modes of being in the world, the authentic mode called *Dasein*. Since the latter is prior, how we relate to the world is dependent not so much on our *cognitive awareness, as in Kant, but how we operate in that world as living, practical human beings. He was notorious for his support of Hitler, the rationale for which he sustains in his *Introduction to Metaphysics*, a series of lectures given at Freiburg in 1935 rendered in the following words still to be found in reprints of this work in 1959 and 1976. Here he refers to publications of the day on the subject of values: 'All these works call themselves philosophy. The works that are being peddled about nowadays as the philosophy of National Socialism but have nothing whatever to do with the inner truth and greatness of this movement (namely the encounter between global technology and modern man) – have all been written by men fishing in the troubled waters of "values" and "totalities".' One way of entering his thought initially is by examining his very interesting essay 'The Origin of the Work of Art' (1935) in *Poetry, Language and Thought* and his *The Question Concerning Technology* (1954).

HOBBES, THOMAS (1588–1679)

The Father of Western Philosophy, if not psychology, a title more usually attributed to his French rival Descartes, since the French have traditionally been much more interested in philosophy than the English! Indeed that lack of interest in his ideas, compared to those of someone like *Wittgenstein, exists even within English philosophical circles whereas in America there is virtually a Hobbes industry. Hobbes was Francis Bacon's secretary between 1621 and 1626 where he may have been initiated into a rejection of *Aristotle's philosophy, visited *Galileo in 1636 and, at about that time he published his *De Corpore* in 1655, he became embroiled in an unfortunate argument with John Wallis, a geometrician, concerning the possibility of squaring the circle. The 'Malmesbury Monster', as he was known, thought of the world as a mechanical system whose bodies are driven entirely by attraction and repulsion. This view legitimated Hobbes's psychology which, with his *metaphysics, *epistemology and political philosophy, is explored in his *Leviathan* of 1651, though his new science of the state is more obviously presented in his *De Cive* of 1642. These are his two well-known works in a large corpus. Human beings are presented for the most part as wholly isolated and self-interested so that if in a state of nature, they would be at war with each other in a condition where each man's life would be 'solitary, poor, nasty, brutish and short' (*Leviathan* chapt. 13). Hence the need for a sovereign to establish a social contract, where each man would forgo his right to all, thereby providing a means for each person to overcome his fear of living with others.

HUXLEY, JULIAN (1887–1975)

English humanist, biologist and brother of the novelist Aldous; political and social problems were informed by his knowledge of science to form his 'evolutionary humanism', an *ethical theory grounded on natural selection.

JAMES, HENRY (1843–1916)

Brother of William, he was an American novelist. He regarded himself as a pragmatist (cf. R A Hocks, *Henry James and Pragmatistic Thought*, North Carolina, UP 1974). He took his brother's distinction between the ambulatory and the saltatory, which William applied to different kinds of knowledge, seriously but applied it to writing. In the latter case the act of writing is to proceed from one position, claim or idea to another in accordance with a predetermined end; formula-writing for example. Henry's art was to illustrate in his novels the possibility of enjoying 'intervening parts of experience through which we ambulate in succession'. This means that the reader is engaged from the start of such a story as *The Turn of the Screw*, trying to make sense of the significances rendered from the possibilities suggested in the narrative. His great novels, such as *The Ambassadors* (1903), *The Golden Bowl* (1904) and *Spoils of Poynton* (1897) illustrate this technique to the full. But whereas Henry was full of admiration for William, William regarded his brother's novels as far too long!

JAMES, WILLIAM (1842–1910)

Born in New York and Henry's brother, as well as being a benefactor and close friend of Charles Peirce, crediting him with the doctrine which he, William James is most associated: pragmatism. This doctrine was unpacked in a book of that name in 1907 and as well as in his *Will To Believe* (1897). However, whereas for Peirce this doctrine was about the meaning of a conception, which he later called *pragmaticism, for James it meant hooking the gaining of truth to successful human action.

What tends to make something true then is what one can come to believe in and what is efficacious in leading one through the thickets of everyday life. This doctrine originated in his *Principles of Psychology* (1890) where human consciousness is represented as an instrument for dealing with practical problems in the world. His book *Essays in Radical Empiricism* (1912) advocated the view that knowable reality at least was to be conceived of as pure experience, a doctrine very influential upon ★Whitehead's ★metaphysics (cf. Craig R Eisendrath *The Unifying Moment* Harvard UP 1971).

KANT, IMMANUEL (1724–1804)

A German philosopher who predicted the existence of the planet Uranus before it was discovered by William Herschel in 1781. His three most important works deal with human awareness – the *Critque of Pure Reason* (1781) followed by the second edition of 1787 – human ★action in the *Critique of Practical Reason* (1788) and ★aesthetics, the *Critique of Aesthetic Judgement* (1790). In his *Perpetual Peace* (1795) he forwards the idea of a world system of autonomous states. His works advocate what is called 'the Copernican revolution': it is not the case that our knowledge conforms to the nature of existing things but if these things are taken to exist, they must conform to our ways of understanding them. We have no access to things in themselves in experience; *noumena*, accessed through ★thought. We have access merely to objects as they appear to us; Phenomena. This doctrine is central to Kant's transcendental ★idealism. In his first critique Kant argues that our grasp of the immediate objects in experience depends not only upon sensations but what is imposed upon experience ★*a priori* by the mind, such a disposition inbuilt into our sensibility. This was Kant's way of reconciling the rationalistic claims of Leibnitz – that all truths can be attained through valid ★reasoning – to Hume's empiricism, that all knowledge is gained through the senses. Indeed, it is only that *a priori* element which enables the ordering and structuring of our sensations into intelligible unities. Our awareness of such categories as 'cause and effect' or 'relation' for example and the application of such schema as 'time' and 'space' – subjective forms of human sensibility through which the manifold of sense has access to mind – these can't be acquired from experience but belong rather to our cognitional capacities for understanding the world. The second critique develops themes found in his *Groundwork to a Metaphysic of Morals* (1785) where the well-known categorical imperative is presented: 'Act on only that maxim which you can at the same time will to become a universal law'. The third critique deals with aesthetic judgments which he relates to reflection on what he calls 'purposiveness in nature' so that he comes to define beauty in terms of a feeling one can have along with other human beings in relation to a 'form of purposiveness insofar as it is perceived apart from the presentation of a purpose.'

KIERKEGAARD, SOREN (1813–1855)

Danish religious thinker and philosopher thought of as one of the founders of existentialism: a philosophy emphasising the importance of individuals and their unique choices in a universe devoid of meaning thereby generating a sense of the absurd in human life. As manifested in *Philosophical Fragments* (1844) and *Concluding Unscientific Postscript* of 1846, his philosophy is a reaction to *Hegel's in that the individual once more comes to the fore along with his or her choices which form future selves. He specifically rejected the idea of a science of the human spirit so as to offer a place for Christian faith. Emphasis is placed on the will which in order to be directed towards an *ethical mode of life, requires a religious orientation. His *Fear and Trembling* (1843) explores the clash between ethical and divine duties, the former emerging as shared social norms whose status is merely contingent and historical in nature, whereas the latter proceed from a completely unconditioned transcendental origin. In that case the former then can't be regarded in Hegelian fashion as following or mapping out the path of some developing spiritual absolute.

KLEIN, YVES (1928–1962)

A French artist, judo expert and leading member of the European neo-Dada movement after the Second World War. His *Anthropometries* involved naked, young women painted possibly in his International Klein Blue patented colour, pressing the front part of their bodies upon canvases. At the same time a twenty piece classical orchestra played his *Monotone Symphony* for twenty minutes followed by twenty minutes' silence whilst the painting with such 'living brushes' continued. Later he worked with a flame-thrower and water to create his *Fire Paintings*.

LEOPOLD, ALDO (1887–1948)

Born in Burlington, Iowa, he was always interested in natural history. After working for the US Forest Service, he became interested in *Game Management* as his famous book of 1933 is called. He is best known for his *Sand County Almanac* of 1948, which made him the foremost conservationist of the twentieth century forging, in that work, a conception of *ecological sensibility to lead a human being to adopt appropriate conduct in relation to caring for the land!

LUKACS, GEORG (1885–1971)

Hungarian Marxist philosopher whose *The Theory of the Novel* (1916) and *History of Class Consciousness* (1923) proved so significant to the development of the *Frankfurt School of Philosophy. These works may have developed out of his study of *Marx's Economic and Philosophical Manuscripts of 1845*. Adorno never forgave Lukacs for disowning his 1923 work after it was banned by the Russian Communist Party. That work delineates the notion of *reification, extending Marx's work on the idea of fetishism attached to *commodities being extended into culture as a whole. His 1916 work was important because of the development of the distinction, drawn by *Hegel, between what is socially and historically produced in humans – 'second nature' and what is supposed to be 'natural' in humans, an animal or 'first nature'.

LYOTARD, JEAN-FRANÇOIS (1924–)

A French *postmodernist who draws attention to the collapse of the philosophies such as Kantianism, Hegelianism or Marxism, philosophies which have sought to offer a *meta-narrative postulating, for those engaged in inquiry, some truthful outcome or panacea which can forward justice in the modern world (cf. *The Postmodern Condition: A Reflection on Knowledge* (1979)). Instead, what the pursuit of ⋆philosophy offers is a number of language games or interpretative discourses possessing their own internal criteria, which have evolved historically, for deciding what are or are not appropriate moves to be made within a particular narrative purporting to forward philosophical activity. These different interpretative discourses may simply be ⋆incommensurable with each other. Anyone who rejects this point of view – say someone upholding the idea of rational consensus, the values of the Enlightenment or the importance of critique as ⋆Habermas does in his *Philosophical Discourse of Modernity* (1986) – such a thinker must be a dogmatist.

MARX, KARL (1918–1983)

How many well-known subjects fall under the name Marx? Four you may reply: Groucho, Chico, Harpo and Zappo! In the case of Karl Marx we may have four subjects too! As a sensuous materialist in such works as *The Holy Family* (1844), *Towards the Critique of Hegel's Philosophy of Right* (1844) and *Economic and Philosophic Manuscripts of 1844* where he works as a ⋆philosopher in ⋆Hegelian fashion seeing the existence of the human being in terms of social relations as humans work upon nature, so that whilst man's nature is realised in his activity, material nature becomes 'saturated with spirit' to use Merleau-Ponty's phrase. This unity of man with nature, according to this humanistic or romantic Marxism, serves as 'the true resurrection of nature'. But such an ideal is undone in a world of 'estranged' labour, raising the problem of alienation. This kind of Marxism proved very influential upon the ⋆Frankfurt School of Philosophy: ⋆Adorno, ⋆Horkheimer and ⋆Benjamin were influenced by ⋆Georg Lukacs's *History and Class Consciousness* (1922) which forwarded it despite the author's renunciation of that book later. After 1845 Marx writes more as a social historian mapping out his materialistic conception of history offering a philosophy of history with 'practical intent', as ⋆Habermas puts it. Now Marx shows how relations of production generate social relations determined by the significance of the shift from use to exchange-value in such works as *The German Ideology* (1845–1846), *Manifesto of the Communist Party* (1848) up to his Preface to *A Contribution to the Critique of Political Economy* of 1859. As an historical materialist he delineates forces of production, understood in terms of tools and available technology, social relations and both from forms of social consciousness. It is after 1859 that his 'economism', as Gramsci called it, comes to the fore. Now, as a scientific materialist, Marx claims that the productive forces determine social relations and then in their turn social consciousness. In *Capital* (1867) his determinism generates the idea that the laws underpinning the capitalist mode of production – founded on capital and the creation of surplus value – originate 'tendencies working with iron necessity towards inevitable results'; their own undoing. Now human beings are merely personifications of particular class-relations and class interests. Finally, in his post-capital years Marx, in his *Letters*, appears more as an ironical prophet throwing doubt on his economic determinism, considering whether, in his Preface to the Russian edition of the 1848 *Communist Manifesto*, a country could pass from a rural, agrarian mode of production to communism without passing through capitalism. As a more sophisticated materialist, he seems to distance himself from capital's iron-law necessity thesis, whilst he weakens its predictive role especially in relation to events in Russia and Germany. So when we are now told that Marxism has failed or been falsified it is not at all clear to what strand in his thinking that judgement applies. And even if it can be shown that all four strands influenced his writings at different times in his life, complicating this rather simplistic account rendered to you, it still remains the case that as the validity of the third, deterministic strand is undone so the relevance of the first dimension, interpreted as a cultural critique of capitalism, has sustained!

PASSMORE, JOHN (1914–)

An Australian philosopher who published his classic *A Hundred Years of Philosophy* in 1957 and its sequel *Recent Philosophers* in 1985.

PETERS, RICHARD STANLEY (1919–)

An English philosopher who, as an admirer of Kant's philosophy worked in political philosophy; his book on *Hobbes was published in 1956 and *Social Principles and the Democratic State* with S I Benn in 1959; in philosophical psychology; his book on *The Concept of Motivation* was published in 1958 and *Brett's History of Psychology* in 1956; in ethics; *Ethics and Education* was published in 1966. His greatest contribution was to establish, no matter how temporarily, the significance of the philosophy of education as an academic discipline which began in the latter work, developing themes from his *Authority, Responsibility and Education* (1959). Education is the initiation into worthwhile activities to those who become committed to it, so developing knowledge and understanding, which are not inert, but involve cognitive perspective in the learner who takes part willingly and wittingly in processes of learning, in a morally acceptable manner (cf. *Education and the Education of Teachers* (1977) and *Moral Development & Moral Education* (1981)).

PIAGET, JEAN (1896–1980)

Just as *Dewey forwarded child-centred education in the philosophy of education so did Piaget in the psychology of education. As a Swiss psychologist, his work focused on the stages of *cognitive development through which the maturing child passes as s/he learns to perceive, cope with, think about and apprehend his or her everyday world. A good way of entering into his psychology is through his *Language and Thought of the Child* (1926) and *The Moral Judgment of the Child* (1932). The latter work has proved influential in Lawrence Kohlberg's efforts to map the development of moral consciousness in human beings (cf. 'Stages of Moral Development as a basis for Moral Education' parts 1 and 5 in *Moral Education* ed., by C M Beck (et al.) Toronto UP 1971).

PLATO (428–347)

The best-known of Greek philosophers because of the dialogues he wrote about *Socrates. The standard view of their relationship is that Plato presents Socrates veridically, as primarily concerned with *ethical questions, in such early dialogues as *Laches*, *Lysis*, or the *Hippias Minor* and *Major*, transforming him into more of a fictional character in his middle dialogues such as the *Republic*, the

Symposium and *Phaedo* where the Platonic doctrine of forms and a strict *dualism between being and becoming are clearly enunciated. So in Book V of the *Republic*, in the image of the Sun, we find a *bifurcation between an intelligible realm, illuminated by the chief Form – the Good, cause of knowledge and truth – and the physical world illuminated by the sun which makes light and vision possible. Both in the *Meno* and in the *Republic* – through the Divided Line Analogy and the Cave Allegory – possible stages in the passage upwards from the physical to the intelligible world are indicated. First, the unquestionable trust in images; secondly common sense beliefs and opinions, which inform the practical world of the character Anytus in the *Meno*, leading to what *Weil called 'knowledge of the third kind': causal, deductive, theoretical thinking typifying mathematical reasoning. These levels seem akin to Peirce's distinctions between *Firstness, *Secondness, and *Thirdness. But there is no Fourthness in Peirce's philosophy, that is an invisible realm of Forms characterising an unchanging atemporal realm, itself ordered by the Good, as Plato has it, causal of what exists in the physical world of experience, as is made clear in the *Phaedo*. This standard view depends on the way Plato's texts are ordered chronologically. Julius Tomin has refuted this view in a series of articles (cf. for example 'Plato's First Dialogue' *Ancient Philosophy 17* (1997) pp.31–45) and his case remains unchallenged. Though his critique would not reject the above interpretation of Platonism, his placing of the *Phaedrus* as the first dialogue casts Socrates more as a genuine historical character, the real Socrates. Those stressing the importance of the dialogue form see Socrates more dramatically, an expression of how Plato conceives the pursuit of philosophy ought to be forwarded by a figure 'Socrates'. Hence, in a dialogue, the reader is shown certain things, as Leo Strauss has it (cf. *The City and the Man*, Chicago, UP 1964 chapt. 2); perhaps something about Plato's characters, the way they employ particular arguments and features of those arguments themselves. In this way the reader is encouraged to do philosophy with Plato through the written word much as the historical Socrates did in relation to the young men of Athens through the spoken word. But this expressive, rather than historical treatment of Socrates, can be developed further. Not only can we never know Plato's own authentic teaching, save for what he tells us in his *Seventh Letter*, if it is genuine, but because Plato, by means of the dialogue form, can distance himself from the activities of his fictional character Socrates, he not only honours him – the usual way of crediting Plato's activity – but he can be ironic towards 'Socrates', even critical of his activity (cf. Stanley Rosen *Plato's Symposium*, Yale, UP 1969/87). Such an approach, implied by this 'Strauss–Rosen' thesis, may help to deal with the strange beginning of the *Republic*, which opens in dialogue form, and the contrast with what follows in Books II–X, which seem much more didactic in character.

POPPER, KARL (1902–1994)

An Austrian philosopher who settled in England; he has consistently attacked English analytic philosophy and ⋆verificationism in the name of falsificationism: as many observations as you like might confirm a universal theory but only one is required to refute it. The susceptibility of theories to such a procedure demarcates science from pseudo-science. For Popper, in *Conjectures and Refutations* (1963), since claims for Marxism are never deflected by refutation of them and Freudian psychoanalysis never generates empirical predictions, both are to be discredited. His falsificationism extended beyond the philosophy of science to issue in a defence of ⋆procedural or critical ⋆rationality more generally conceived: any claim can be made subject if not to critical testing then at least to critical discussion guided by the ideal of approaching the truth. There can be nothing more rational than taking any decision and being critical of it, so that whereas factual statements about states of affairs in the world are accepted, we create standards or ⋆values, emerging from social decision-making, are created by us within an open democratic society. This approach was used in his most popular book *The Open Society and its Enemies* (2 vols. 1945) to attack the social philosophies of *Plato*, *Hegel* and *Marx* and to advocate piecemeal social engineering in response to social problems in *The Poverty of Historicism* (1957).

PROUST, MARCEL (1871–1922)

A French novelist who wrote thirteen volumes – *Remembrance of Times Past* – which expresses the significance of his own experiences as they relate to his felt life, revealing realities beyond or below superficial *consciousness. He was very influenced by the process philosopher Henri Bergson, using a distinction of his within his novel to distinguish voluntary from involuntary memory: the first refers to the intentional recall of experiences, the latter to the accidental flooding of the mind by past impressions due to the fact that they have been associated previously with something now felt or experienced in the present.

RORTY, RICHARD (1931–)

An American rhetorician and philosopher, former student of Hartshorne but not a convert to his unfashionable Process Philosophy. Rather, in his *Philosophy and the Mirror of Nature* (1979), he regards *metaphysical statements, following *Wittgenstein, as attempts at 'escaping from time and history' and, following *Hegel, he is concerned as much with reading the history of philosophy as with the nature of philosophical activity itself. Here he follows the *postmodernists in seeing philosophic discourse as starting and proceeding from 'floating, ungrounded, conversations' characterising 'the contingent starting points' from which the participants in such discourse are placed. There is no *meta-narrative which can better that condition, though Rorty does follow Dewey and Popper in advocating a *procedural rationality, thereby regarding science as informing us about the natural world. The importance traditionally attached to the philosophic enterprise, then, needs minimising given that, as Foucault might have said, its discourses 'are contaminated with the needs of a repudiated community'. Indeed religious belief becomes redundant in a culture where the individual can regard him/herself as part of 'a body of public opinion capable of making a difference to the public fate' as opposed to appealing to some transcendental power beyond that community to give one's life some greater significance than it already has. These themes are continued in his *Contingency, Irony and Solidarity* (1989).

ROUSSEAU (JEAN-JACQUES) (1712–1778)

French political philosopher, composer, playwright, novelist, educationist and autobiographer who was born in Geneva. His 1750 essay *Discourse on the Arts and Sciences* presents an attack on *modernism through a critique of his own culture, Here he expresses a scepticism towards science and the advance of technology, contrasts the relative innocence of a life close to nature against city corruption, lampoons the idea of social progress and condemns the arts of his time as a source of corruption. In 1762 he published *On The Social Contract* since a critique of 'What is' implies 'What ought to be'! He now explicates principles necessary to redeem the social realm forwarding the idea of Sovereign Authority: 'Each of us puts his goods, his person, his life, and all his power in common under the supreme direction of the general will, and we as a body accept each member as a part indivisible from the whole'. But for an individual to become a citizen, Rousseau has to show what kind of education is required. In addition he provides a 'genealogy' to explain humankind's corruption which makes the idea of sovereign authority necessary. The latter he does in his Second Discourse *On the Origins of Inequality* (1754) and his *Essay on the Origins of Languages* written at about the same time. With respect to the former he decides to educate a young person as a person rather than as a citizen first, since both enterprises can't be done at the same time. *Emile or Education* is the classic work on child-centred education rivalling Plato's *Republic* as a masterwork in the philosophy of education.

RUSSELL, BERTRAND (1872–1970)

Welsh philosopher and controversial polemicist who was a mathematician – his *Principle of Mathematics* was published in 1903 – and author of many other books. His most famous essay *On Denoting* (1905) considered the problem of how it was possible to speak of a non-existing object. He suggested employing descriptive phrases so that what is being referred to no longer appears as if it were a name. Consider an atheist reflecting on the question 'Does God exist?' This can be translated into 'Is there an entity which is all powerful, all good and all-knowing?' Now a debate can ensue about the possibilities of these predicates referring to something without employing the term God at all! He worked with Whitehead on *Principia Mathematica* (1910–1913) but his most popular work is *The Problems of Philosophy* (1912) along with his *History of Western Philosophy* (1945).

SANTAYANA, GEORGE (1863–1952)

Though he was born in Madrid, much of his activity as a philosopher, after being a student of William James, took place at Harvard from 1889 where he was a professor from 1907 to 1912. His philosophy could be described as naturalistic and materialistic as he exercised a devastating wit against the claims of German idealism and the transcendental claims of religion, though two of his best pieces were tributes to *Spinoza, one in the Preface to the 1910 *Everyman* edition of Spinoza's *Ethics* and in his *Ultimate Religion* essay in his *Obiter Scripta* of 1936. Though he was sceptical about knowledge of the external world, he argued that it was an act of 'animal faith' which ensured our belief in it, a position he set out most brilliantly in *Scepticism & Animal Faith* (1896). His doctrine of essences, introduced in this book as pure possibilities, came more to the fore in *The Realm of Essence* (1929). Controversy surrounds his 1911 address to an audience at California University – 'The Genteel Tradition in American Philosophy' (in his book *Winds of Doctrine* 1926) – where he fires a parting shot at the culture which sustained him before returning to Europe: does he in that essay pioneer an ecocentric view of the world and man's place in it?

SCHELLING, FRIEDRICH WILHELM JOSEPH VON (1775–1854)

German *idealist philosopher famous for forwarding his romanticism, which was championed by the English poet Coleridge, and for his 'Identity Philosophy', derided in *Hegel's *Phenomenology of Spirit* of 1807. We start with Kant's approach to *aesthetics: on viewing objects aesthetically, in a 'disinterested' manner, they are seen as if they manifested a purpose though we do not know if they manifest such a purpose in themselves. This is the root of Kant's 'purposiveness without purpose': a *teleological harmony within nature may be required in appreciating it aesthetically, but we have no guarantee nature is so ordered. In his 'Treatise Explicatory of the Idealism' in the *Science of Knowledge* (1797), Schelling rejects the mechanistic view of nature on which Kant's stance is based. Secondly, how can we account, he asks, for the nature of the phenomena we enjoy and how is it related to its object, the thing-in-itself? Schelling's answer is to say that the latter is generated by mind, by some kind of spiritual activity, just as the former originates in our mental activity occurring in ordinary experiencing. Kant's noumena, then, are the creations or products of spirit. As *Apel puts it: 'Nature is visible mind: mind is invisible nature.' These two realms – 'the objective aspect of the absolute' and the human intellectual realm, actuality and potentiality – are realised in one unified system – the absolute – just as they are by humans in their everyday activity. The former embodies ideality in reality whilst the latter embodies reality in ideality as he puts it in his *On University Studies* (1803). This claim enables him, in *Ideas Towards a Philosophy of Nature* (1797), to develop an insight of *Spinoza: the Absolute is an eternal act of producing – natura naturans – whilst its products are to be seen as 'the mere body or symbol' of the Absolute – natura naturata. In relation to the former, productive activity begins unconsciously, fulfilled in the emergence of consciousness. But the artist's activity reverses this process: the consciousness of the artist initiates the act of creation and what is produced carries unconscious implications outstripping his or her *intentions producing the work. As he states in his *System of Transcendental Idealism* (1800): 'it is art alone which can succeed in objectifying with universal validity what the philosopher is able to present in merely a subjective fashion.' *Philosophy, then, is to the ideal what art is to the real, and whilst the former is a higher reflection of the latter it is the latter which provides access for the former. Is this formulation the origin of *Adorno's claim in his *Aesthetic Theory*: 'philosophy says what art cannot say, although it is art alone which is able to say it by not saying it'? Later, he undid the position in his earlier work, developing a more 'positive philosophy' which sought to sustain a philosophically defensible religious position on the basis of reinterpreting Christianity's historical development. But it is during his 'Identity Philosophy' period when he maps out the distinctions between 'different modes of living' – the scientific, the instrumental and the aesthetic – in his address *Concerning the Relation of the Plastic Arts to Nature* (1807). Here he puts the concern for values at the centre of philosophical inquiry and, in *The Philosophy of Art* (1807), he anticipates the arguments of Lynn White Jr's. *The Historical Roots of Our Ecologic Crisis* (1967) by claiming that man's estrangement from nature, embodied in modernism's development, has its roots in the Christian religion.

SCHILLER, JOHANN CHRISTOPH FRIEDRICH VON (1759–1805)

German poet whose poem 'Ode to Joy' was set to music by Beethoven in his ninth symphony. He was also an historian and author of the play *Don Carlos* (1787). In his *Letters on the Aesthetic Education of Man* (1795) he developed Kant's lead in exploring the relationship between ⋆aesthetics and ⋆ethics, the former a potential feature of human nature. It was this work which had an enormous influence on Peirce. Schiller saw the possibility both for individual human beings and in the history of man for a development to occur from man's physical state, where s/he is ruled by a sensuous nature, through the aesthetic dimension to a state where s/he can achieve ⋆moral ideals. The significance of beauty lies in its capacity to reconcile the battle between man's sensuous nature on the one hand and his/her rational will on the other.

SOCRATES (469–399 BC)

Born in Athens and remained in Athens. He wrote nothing, founded no school and had no disciples yet was venerated by Plato and Aristotle, the former casting him as a character in his dialogues. He turned philosophy away from speculation about the universe to focus initially on ethical problems, employing his 'Socratic Method' in analysing what terms meant in Greek discourse. He is famous for his midwifery: the idea of drawing from the person he questioned knowledge they already possessed or to demonstrate their ignorance. Hence the charge of corrupting the youth of his day and his sentence to death by drinking hemlock.

SPENCER, HERBERT (1820–1903)

English philosopher and prophet of social progress who wrote an evolutionary tract – *Principles of Psychology* (1855) – before Darwin's *The Origin of Species*. Because he applied evolutionary theory both to *ethics and sociology he became regarded as a 'Social Darwinian'. Indeed he coined the phrase 'survival of the fittest': societies evolve in a natural competition for resources and the fittest survive. This thesis seems to have *moral overtones; it is not just the result of *empirical investigation, so that ethics comes to have a natural basis. Peirce objected to the idea of making evolution depend 'upon one chance principle' (Peirce: vol. 6, para. 4) (cf. *anancasm). As can be seen in his *System of Synthetic Philosophy* (1862–1893). Spencer regarded *philosophy as 'the science of the sciences' so that when he published his *Education*, of 1861, he advocated an important role for science in the natural development of the child's intelligence as well as the creation of pleasurable interests since the latter encourages tendencies which will have survival value.

SPINOZA (BARUCH OR BENEDICTUS DE) (1632–1677)

Born in Amsterdam into a Jewish émigré family from Portugal to avoid Catholic persecution. He was famous for his interests in optics and astronomy. His new ideas both in philosophy and theology are brilliantly illustrated in his *Tractatus Theologico-Politicus*, the first important historical treatment of the scriptures, published anonymously. His only book to be published with his name on the title page in his own lifetime was his *Short Treatise on the Correction of the Understanding* of 1662. In his most well-known book, his *Ethics*, published after his death, he delineates, in Book I prop. 29, natura naturans – 'eternal or infinite essence' or God 'conceived as a free cause – and natura naturata – what follows 'from the necessity of God's nature' – so that the creative act of producing is distinguished from what is produced. In this work he seeks to build a complete metaphysical system deduced with a kind of mathematical exactness from certain axioms, theorems and definitions so as to demonstrate what is good for human beings. In that sense his philosophy could be regarded as a precursor to Whitehead's world view in sustaining the idea that *ethical enquiry presupposes *metaphysics. At least this is the programme of that book as it developed between 1665 and1670. The rest, beginning with Book III – 'Concerning the Origins and Nature of the Emotions' – as that developed between 1670 and 1675 – is not so much deductive or 'top-down' in character as 'bottom-up'! Now Spinoza tries to deal with *Hobbes's problematic – 'For men are not born for citizenship, but must be made so'; 'none of the objects of my fears contained in themselves anything either good or bad, except insofar as the mind is affected by them'; 'we call that good which we certainly know to be useful to us' – but it is not clear how his efforts to do so in the remainder of his *Ethics* can be reconciled with Books I and II (cf. Anthony Negri *The Savage Anomaly* Minnesota UP 1991). Controversy also surrounds the issue as to whether, following the later strand, he should be seen as the father of humanism or, in taking the former strand, as revitalising a conception of the divine.

THOREAU, HENRY DAVID (1817–1862)

American poet and essayist who met the American philosopher Emerson, who in his turn met Wordsworth. Indeed, for a time, Thoreau was regarded as a follower of Emerson. In 1845 he built a shanty house near Walden Pond where he composed the classic work *Walden, or Life in the Woods* (1854) and it is this work which shows his uniqueness. His 1851 statement 'In wildness is the preservation of the world' is now heralded as the ground for ecocentrism.

WARNOCK, SIR GEOFFREY, J (1923–)

An Oxford philosopher committed to the defence of analytic philosophy in such works as his essay on *Berkeley* (1953), in his *Contemporary Moral Philosophy* (1967) and in his *The Object of Morality* (1971).

WEIL, SIMONE (1909–1943)

A French mystic and philosopher. A modern day Platonist whose development passed through four phases mirroring *Plato's allegory of the cave, if we are to understand her philosophic quest in terms of understanding the occurrences in her life and vice-versa. In 1934 she was released from her teaching appointment and worked in a factory in Paris and then at a Renault factory. In *Factory Work* (1934–1935) she explains how conditions of 'forced labour' are endured where human beings are treated as objects. Here they are made subject to Plato's first level: images or fantasies enabling the mind to escape oppression by filling the void or 'sheer emptiness' of experience, such a void caused not merely because of God's absence – or the Good in Plato's terms – but because of a lack of awareness about the nature of that sheer emptiness. Such oppression, endured under conditions of social necessity, generates a more intense form of suffering than that endured under conditions of natural necessity, confronted in physical labour generating beliefs about what exists, mirroring Plato's second level. Here she follows *Marx, despite her highly original criticisms of his philosophy in *Oppression and Liberty* (1933–1943), in regarding productive labour on nature as generating value. Natural labour evidences her belief in determinism since, as she puts it later in *Gravity and Grace* (1942): 'All the natural movements of the soul are controlled by laws analogous to those of physical gravity. Grace is the only exception.' But s/he who endures 'the peasant's toil' is 'necessarily obedient to the world's rhythm' under the sun and stars, as she was as a farm labourer. Now it is possible to appreciate that the 'distance between the necessary and the good is the distance between the creature and the creator.' Under conditions of natural necessity it is possible to experience affliction, the non-presence of God: indeed the world 'must be regarded as containing something of a void in order that it may have need of God.' But if we are now able to 'read necessity beyond sensation we may be able to read 'order behind necessity'. Given appropriate tools and training, creative work can strike a proper balance between the human will and necessity, as she puts it in *Intimations of Christianity amongst the Ancient Greeks* (1941–1943) so that necessity can be conditioned through methodical action informed by knowledge of the third kind, proper reasoning or thought, as she points out in her *Notebooks* (1940–1942). The final and fourth stage of this *cognitive ascent is characterised by the idea of reading 'God behind order' so that we are brought up against 'the veritable mysteries, the veritable undemonstrables, which constitute reality' in *contemplative activity. She characterises the difference between the last two stages as follows: 'The discoursive intelligence, which grasps relationships, the one that presides over mathematical knowledge, lies in the boundary between matter and spirit. It is intuition alone which is purely spiritual.' And it is from this realm the intuitive, that the other three are derived. In Platonic fashion she claims ordinary human existence 'is not real, it is a mere setting' whilst it is reality which makes us subject to God: 'The only good which is not subject to chance is outside the world.' Stars and blossoming fruit trees realise a sense of values in two ways. The former provides us with an image of the impersonal and utter permanent quality characterising reality: the latter, extreme fragility typifying earthly existence. But why do blossoming fruit trees give rise to the beautiful? The reply: 'Real presence' endows her with mysticism. Once interpreted as the 'Silence of God', such mysticism leads to religion, to an aesthetic theism where the self is decreated in its unification with divine love. She sustains this claim by offering a kind of ontological argument – her 'experimental proof' – for God's existence. Her last work – *The Need For Roots* (1942–1943) – brings these themes together in an important work in social philosophy.

WHITEHEAD, ALFRED, NORTH (1861–1947)

An English mathematician and logician who worked with his former student Bertrand Russell on *Principia Mathematica* (1910–1913). After the First World War he went to America because the intellectual climate in England was unsuited to his increasingly *metaphysical concerns. His popular works – *Science and the Modern World and Religion in the Making* – were written between 1925–1926 but his most worked out cosmology makes its appearance in *Process and Reality* (1929). In *Adventures of Ideas* he attempts to apply his more theoretical insights to historical and sociological enquiries. Unfortunately his work has been side-stepped in philosophy because the movement Process Studies has been preoccupied with theological matters, merely a strand in Whitehead's own thought, and because his work is regarded, with the exception of Charles Hartshorne, more and more religiously by his followers rather than as an opportunity to generate critical thinking with respect to recent developments in philosophy outside *ecological ethics and theology.

WITTGENSTEIN, LUDWIG (1889–1951)

The philosopher, whose ideas stand between the philosophy student and a degree in England at least, though not necessarily in America. Austrian born, he forwarded two distinct approaches with respect to the relationship between language and the world. The first is manifested in his *Tractatus Logico-Philosophicus* (*TLP*) written from jottings set down in notebooks he kept as a First World War soldier, a book for which *Bertrand Russell wrote the Introduction. *TLP* consists of a series of numbered remarks dealing with a conception of meaningful language seen as made up of propositions which serve as 'pictures' mirroring the facts constituting the nature of the world. In this way the limits of language can be explored in order to make clear what can be spoken meaningfully; 'whereof one cannot speak, thereof one must be silent'. But, as Paul Engleman shows in his *Letters from Ludwig Wittgenstein* (1967) this latter formulation was misunderstood: it wasn't that non-*empirical claims such as those made in *ethics, *aesthetics and religion were unimportant. Rather, aesthetic expression, ethical acts and religious belief were made manifest in people's lives whereas ethical propositions do not exist. Moreover, what could be spoken succinctly and clearly was not the most important aspect of human life: there is a higher sense to our experience from which value claims are derived, derivation then which lies outside our world, something manifested as that is evoked in ordinary discourse. From 1929, teaching at Cambridge University, he took an entirely different approach, presented in such works as *Philosophical Investigations* (1953) and *On Certainty* (1969). Now discourse is seen as much more open-ended and subtle in its everyday use, manifested in different kinds of language use or 'language-games' where there is no straight correspondence between what is said and some factual case in the real world; rather the distinctions between the real and the unreal are manifested within language use itself. And the grammar within language use casts its shadow upon thought to give us a sense of an apparent harmony between our ordinary discourse and reality. The way this occurs in aesthetics, the psychological use of language and religious discourse is explored in *Lectures and Conversations in Aesthetics, Psychology and Religion* (1966). Once the god of analytic philosophy, where criticisms of this new-found form of *idealism by Ernest Gellner, John Watkins – disciple of *Karl Popper – *Karl-Otto Apel and *Charles Hartshorne were for the most part simply ignored, he is once more in fashion as a *postmodernist in advocating a much reduced significance for philosophy. Now its task is taken to be the dissolving of philosophical questions by scrutinising the way words are employed according to rules befitting appropriate use. The significance of *metaphysical statements is reduced, since these are no more than linguistic expressions projected onto the world by human beings, the force of which they come to regard as revealing *existential necessities.

WORDSWORTH, WILLIAM (1770–1850)

Romantic lakeland poet. A close friend of the poet Coleridge, who was the first person to hear the famous poem 'The Prelude' of 1805. It is the first form – rather than the tampered version – of the poem, never published in his own lifetime, which inspired Whitehead's classic treatment of romanticism as well as aspects of his process philosophy in *Science and the Modern World* (1926)

Appendix III
A TRIBUTE TO FATHERS?

An issue has haunted the writing of this book's contents: the issue of *cognition.[1] Each of this book's chapters began with an image, narrated something to the reader – considered to be factual – and was completed with reflections of a *cognitive nature meant to relate the former two elements to each other. Regarding such reflections, imagine a man with his hands on his hips, his bodily weight carried by one side of his posture asking 'How do you know?' or 'What makes you think that?' Now why invoke the image of that questioning man? Because the kind of *philosophy into which an examination of this book may have initiated its reader is one which issues from the sort of questions he asked. And if that man was this author's legal parent then Charles Sanders Peirce is his intellectual father, as far as the pursuit of philosophy is concerned. Indeed, had the latter thinker met the former, he might have returned a question to the questioner: 'Is knowledge all-of-a-piece, or are there different ways of knowing?' We, in our turn, might wish to ask the philosopher where an answer to that question fits in his conception of philosophy as a whole. After all, consider the many areas of intellectual activity Peirce developed: logic, semiotics (his theory of signs), *scientific methodology, *phenomenology, *epistemology, *metaphysics, *aesthetics, religion and so on. Given this vast array of intellectual pursuits, Peirce can be regarded as an ideal thinker to refer to, since his thinking can be applied to the many interdisciplinary fields of enquiry this book seeks to employ. But how are we to make sense of his philosophical contribution taken as a whole?

Those who have sought to deal with that issue – the way Peirce's many intellectual pursuits are to be related to one another – have generated a host of golden opportunities for interpreting his philosophy. These can be indicated by asking a number of questions which may appear somewhat technical at first, before a proposal can be considered reflecting the author's present stance. So we can ask whether his philosophy is to be regarded as a kind of unified project resulting in a *realistic metaphysics.[2] Or did Peirce develop a sequence of systems[3] in falling back into a speculative *idealism through misunderstanding Kant's philosophy? Or did these systems, rather, reflect stages of development enabling us to see Peirce's philosophy in a more unified architectonic way?[4] Or is that philosophy to be understood as strands of a system issuing from different kinds of premises Peirce used to make sense of what can be gathered from experience, strands which also reflect the intermittent manner in which he wrote?[5] Or was Peirce first and foremost a logician, who, in, defining logic broadly, came to equate it with his theory of signs (semiotics) thereby generating, through his deep interest in science, a logic of scientific discovery?[6] Because of his confession in 1897 that for 'about forty years, I have been diligently and incessantly occupied with the study of methods of inquiry, both those which have been pursued and those which ought to be pursued', that he was 'saturated, through and through, with the spirit of the physical sciences' as well as being 'a great student of logic' (Peirce: vol. 1, para. 3)[7] and perhaps, too, because of his somewhat startling remark in his more well-known 1877 paper 'The Fixation of Belief' that 'each chief step in science has been a lesson in logic' (Peirce: vol. 5, para. 363), I have interpreted Peirce's philosophy in the light of his advocacy of taking a logical point of view, casting that in terms of the logic of the sciences understood semiotically.[8] This stance has been challenged[9] but, perhaps as a result of my defence,[10] I have been carried in a direction that has resulted in the writing of this book Let's make that stance a little clearer.

This chapter began with an image of a father questioning his son. Now consider the following narrative in which a rather different father listens to a story his four-year-old son tells him. 'I have a friend. His name is Digger. He wanders in Epping Forest. He lives on one of the islands in the lake surrounded by the forest's trees. He chops off the heads of the lions that prowl in the forest at night. So it is quite safe for me to wander in the forest too.' To this story the father replies: 'Can Digger do me a favour?' 'What is that?' his son asked. 'Do you think Digger could let me have one of those lions' heads?' His son's reply was again quite short but to the point: 'What do you need one of those for; you've got one of your own?' This narrative may prove useful to illuminate aspects of Peirce's philosophy as the stages of his philosophical development are set out, told in the light of a biographical sketch[11] covering four periods in his life: the first stage until 1870 saw his preoccupation with his reaction to Kant's philosophy; the second (1870–1884) marked the birth of his doctrine pragmatism; the third (1885–1902) celebrated his interest in metaphysics whilst the fourth (1902–1914) opened his defence of what he came to call pragmaticism and its application to wider cognitive concerns.

Charles Sanders Pierce was the son of a Harvard professor of mathematics and astronomy – a leading mathematician of his day – Benjamin Peirce. Benjamin Peirce's friends were Emerson, Longfellow and other literary figures. His son Charles was born 10 September 1839. At the age of eleven, assisted by an aunt and an uncle, Charles wrote a history of chemistry. A year later he read Whateley's *Elements of Logic*, which encouraged him to think that any intellectual pursuit – including chemistry – could be regarded as so many exercises in logic. At the age of sixteen he studied Schiller's *Aesthetic Letters* and came to know much of Kant's *Critique of Pure Reason* by heart. Those wishing to emphasise the importance of metaphysics to Peirce focus on these texts assisting him in formulating his doctrine of categories,[12] but we are already getting ahead of ourselves; let's stay within the terms of Peirce's own biography.

After graduating from Harvard in 1859 he gained a position in the United States Coast and Geodetic Survey. Whilst at Harvard he had studied under Agassiz, acquiring his method of classification and so knew of the latter's dispute with Darwin; the *Origin of Species* had just been published. In 1863 Peirce published his *The Chemical Theory of Interpretation* and later he almost anticipated Mendeleev's work on the periodic table.[13] Now his former position was made permanent. In 1867 he was elected to the American Academy of Arts and Sciences. Such early intellectual development, however, he confessed, was dominated by his being 'a passionate devotee of Kant'.[14]

Already you must be gaining the impression of a young man who was always 'epistemologically minded'.[15] And that mindedness may have arisen through his debates with his father. Indeed, the issue in one of their debates might provide a way of entering the thickets of Peirce's own philosophy: the status of mathematics itself. For Peirce's father, mathematics seemed to be a branch of logic, a view taken later by the mathematician Dedekind. Charles 'argued strenuously against' this view[16] so that later the son was able to write that Dedekind's position would not result from holding his father's true position, namely that mathematics 'is the science which draws necessary conclusions' (Peirce: vol. 4, para. 239), a claim with which Benjamin Peirce began his *Linear Associative Algebra* of 1870.[17] This claim made mathematics ⋆deductive in character. In their debates, the father – as a mathematician – and the son – as a logician – were both struck 'with the contrary nature' of their interest 'in the same proposition'. The son did 'not care particularly about this or that hypothesis or its consequences, except so far as these things may throw a light upon the nature of ⋆reasoning.' The father was intensely interested in efficient methods of reasoning, with a view to their possible extension to new problems'; but he was not, '*qua* mathematician' troubled 'minutely to dissect those parts of the method whose correctness is a matter of course'. Therefore, whereas the father studied 'the science which draws necessary conclusions' the son was interested in logic, 'the science of drawing conclusions' (Peirce: vol. 4, para. 239), in activity which is explored in chapter V of this book.

The deductive nature of mathematics can soon be illustrated. For something to be regarded as an equilateral triangle for example, it can be demonstrated what features – equal sidedness, three sidedness – are necessary and then it can be shown what can be deduced from such feature: equal angledness, that each angle must be sixty degrees for example. Such inferences are true whether or not there are such entities as equilateral triangles in the world.

Now consider the following diagram as such:

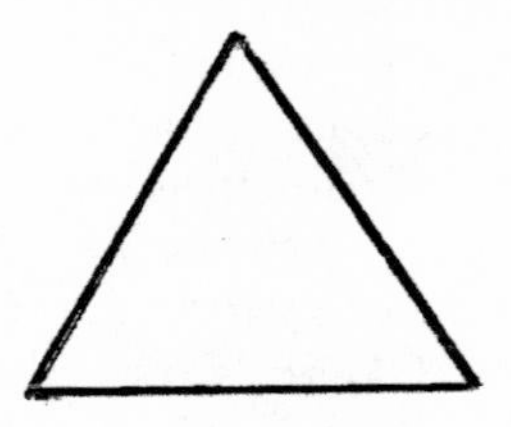

Don't consider anything else. Just attend to that diagram. You are treating it monadically, as a monad. Since we are dealing only with its shape – one element – it can be regarded as non-relative, since we are not considering it with respect to anything else. Now consider it in relation to the following:

equilateral triangle

We now have a dyad established, that is a relation between two sorts of entity which require each other so as to make sense of one another.[18] We may want to know why the diagram represents an equilateral triangle. We can cite the rule: three equal sided figure. We now have a triad, a relationship between three sorts of entity – a diagram, a label and now a rule – whose connections would be impossible without all three of these elements. At a glance you can see that triads require dyads and monads whilst dyads need only monads. So in logical order we have first, monads; second, dyads; third, triads. Such elements make up the abstract and formal categories and our thinking takes place with respect to these basic relations since any other kind of relation is reducible to these. Such mathematical relations underpin the thinking behind Charles Peirce's article 'New List of Categories' of 1867 allowing him to distinguish quality from relation and both from representation, which mediates between the first two categories. These three categories underpin his classification of signs into icons, indices and symbols. But once more we are in danger of running ahead of our story, so let us pause a moment to consider that other conversation between father and son in relation to the son's 'Digger story'!

Of course, the story could, perhaps should have been considered monadically, that is treated as simply a narrative in its own right. Clearly we can make sense of the story in it own terms. Like a painting, it draws us in; it seems to have a qualitative tone to it and it feels as if there might be somewhere a possible figure – Digger. Indeed, according to Peirce, we can compare it to our geometrical diagram, the equilateral triangle. The latter can't be a pure icon because it serves as a token for what is a type, that is, in this case, for what it is to be triangular 'but in the middle part of our reasonings we forget that abstractness in great measure, and the diagram is for us the very thing. So, in contemplating a painting, there is a moment when we lose the consciousness that it is not the thing, the distinction of the real and the copy disappears, and it is for the moment a pure dream – not any particular existence, and yet not general. At that moment we are contemplating an Icon' (Peirce: vol. 3, para. 362).

The father in his questioning was not so much interested, perhaps unfortunately, in the pictorial features of his son's story, that is to say its monadic singularity in producing iconic effects. He'd obviously got the sense of it, but he was after something else; was there a referent here? Was his son using a name to point to some living, existing creature? His concern went beyond what Peirce called in his *On a New List of Categories* 'quality', 'the ground' of the story, and more towards something dyadic: its 'relation' to something existing, its 'correlate'. Given that in this forest, during the early 1970s, men were known to expose themselves to children, perhaps his additional concern may be forgiven.[19]

You might have a more interesting question to raise about the son's narrative. Your concern might not lie in making sense of it as a monadic entity in and for itself alone, nor in its dyadic suggestiveness with respect to some referent, but rather the meaning, the point of the story. How is the story to be interpreted? Such an interpretation is a triadic issue because it mediates between the sense of the story and its possible relation to an existing state of affairs. Is the story to be interpreted as a cry for help with respect to what can't be told about someone in the forest? Is it to be interpreted as the way a child compensates for moments when s/he has no friend available? Is it to be interpreted as a wish on the child's part for a forest akin to accounts to be found in Maurice Sendak's famous yarn *Where The Wild Things Are*, a favourite tale of this particular boy. These suggestions are but a few of the possible ⋆interpretants that can be adopted for this story, ways of representation.

You may want to know why so much time has been given to interpreting this 1867 article of Peirce. The answer is that Peirce regarded his theory of the categories as his 'one contribution to philosophy, the gift I make to the world' following 'three years of almost insanely concentrated thought hardly interrupted by sleep'.[20] Later he replaced the terms 'quality' by the name ⋆Firstness, 'relation' by ⋆Secondness and 'representation' by ⋆Thirdness. These are what can be 'before the mind or in consciousness' in relation to any kind of appearance and the study of the description of these forms of consciousness Peirce called ⋆phaneroscopy at the turn of the century (Peirce: vol. 8, para. 303), his word for ⋆phenomenology. And the writing of this book brings this aspect of Peirce's philosophy to the fore!

Firstness, Secondness and Thirdness are terms Peirce used then to indicate the essential elements of any phenomenon, any happening or occurrence to be found anywhere, elements which are taken or abstracted from any specific manifestation. No matter what kind of phenomenon we face, these elements will be present somehow. Now it is a case of making these three categories clearer and we can once more refer to our 'Digger story'. If the father had given himself up completely to the story without any other kind of end in view, a certain feeling would have or may have been induced in him. Such Firstness refers to 'passive consciousness of quality without recognition or analysis' (Peirce: vol. 1, para. 377), the kind of ⋆consciousness 'which involves no analysis, comparison or any process whatsoever, nor consists in while or in part of any act by which one stretch of consciousness is distinguished from any other, which has its own positive quality which consists in nothing else, and which is of itself all that is, however it may have been brought about' (Peirce: vol. 1, para. 306). The 'Digger story' then, considered as it was told by the son without any further questioning or remark, would have elicited that sense of Firstness. You, the reader, have no access to that dimension since you were not present to hear the way it was actually rendered at the time. But that is not the only phenomenological difficulty we will have to face, as you will see!

Peirce spells out Secondness: 'Besides Feelings, we have sensations of reaction; as when a person blindfold suddenly runs against a post, when we make a muscular effort, or when any feeling gives way to a new feeling' (Peirce: vol. 6, para. 19). One example he gives anticipates Whitehead's notion of *causal efficacy in relation to the issue of causation. Consider the following case: 'In the dark, the electric light is suddenly turned on and the man's eyes blink'. 'He feels that the experience of the *eye* in the matter of the flash are causal of the blink. The man himself will have no doubt about it'.[21] Now Peirce's version: 'While I am seated calmly in the dark, the lights are suddenly turned on, and at that instant I am conscious, not of a process of change, but yet of something more than can be contained in an instant. I have a sense of saltus, of there being two sides to that instant. A consciousness of polarity would be a tolerably good phrase to describe what occurs' (1.380).

In the 'Digger story' seen under this form of awareness, we are not interested in the sense the story has or the meanings rendered within it, but what it points to: does 'Digger' refer to an existing person or something quite fictional? Such an orientation seems in a sense to deface the story itself, since our attention is now directed away from it to something it may indicate! Indeed it can be and has been argued[22] that this *indexical significance of things is what dominates contemporary consciousness where no thing is done for nothing; there must be a cause for something. 'Constraint is Secondness' (Peirce: vol. 1, para. 325).

Peirce illustrates the difference between Thirdness and Secondness perhaps with his own experience in mind as 'a fugitive from Pennsylvania law in New York City'[23] between 1895 and 1898. He pointed out in 1902 that even if 'order and legislation' refer to Thirdness (Peirce: vol. 1, para. 337) and even if, as he puts it in 1902, a court may issue *injunctions* and *judgments* against me' he need 'not care a snap of my fingers for them – but when I feel the sheriff's hand on my shoulder, I shall begin to have a sense of actuality. Actuality is something *brute'* (Peirce: vol. 1, para. 24). So, whereas 'action is second', when we have the rule-governed nature of conduct, since the latter refers to the future, we have a 'third' (Peirce: vol. 1, para. 337). In the case of the 'Digger story', our uncertainty in regarding the status of Secondness makes the prospect of relating it to the sense of Firstness problematic. So, at the level of making a judgement or enabling an understanding of what has been narrated, a number of interpretations seem plausible. Now to that phenomenological difficulty referred to earlier.

The Secondness of Thirdness has already been drawn to your attention: in the case of a genuine Third, conduct – the sheriff's hand on the shoulder – actualises Thirdness through Secondness. Otherwise we would simply have Thirdness but without any practical effect; imagine living in a country where there were laws but the police, say, simply chose not to enforce them. But what about the Firstness of Thirdness? In this case the particular sensation of someone's hand touching one's shoulder, sensed independently of its coercive nature suggests itself as a possibility. Imagine a child in a courtroom, for example, feeling something making contact with his shoulder but not registering its significance in any way. For the child it would present 'mere possibility' as to its significance, thereby something quite 'vague' (Peirce: vol.1, para. 537) since for him there is 'no conveyance or embodiment of thought at all' (Peirce: vol. 1, para. 538) as to what that grip might mean. Given now that you have a sense of the Firstness of Thirdness you may want to raise that phenomenological difficulty: is it possible to experience Pure Firstness? After all, the way we have considered objects of possible experience so far – a child's story, diagrams, paintings, published narratives, questions, remarks, a post, electric lights, the structure of a courtroom – all these are embodiments of Thirds, of what humans have constructed, the creative outcomes of human thought. Is it possible to experience Firstness independently of anything human beings have created?

I once asked Charles Hartshorne this question. His answer was immediate: 'a pure intuition of the divine'. What could he have meant. But you may be more interested in furthering two other questions in relation to the possibility of experiencing Pure Firstness: 'How is an answer to that question relevant to the writing of this book?' and 'Does this book then finish with a religious conclusion?'

In relation to the first question, throughout this book the notion of Firstness is connected to the aesthetic, particularly in the case of chapter XIII which focused on the issue of the aesthetics of nature. Indeed, in that chapter Peirce's 1908 article 'A Neglected Argument for the Reality of God' – not his existence, incidentally – was referred to and the connection between aesthetics and a spiritual way of viewing the world was explored. Of course, someone could contemplate a mountain within the Rocky Mountain range in Montana or a smaller one within the English Lake District, as Wordsworth reported in the first book of his, *The Prelude*, and thereby experience 'the haunting presences of nature' as Whitehead put it.[24] And if natural laws were to be regarded as 'habits of the universe' as Peirce, in his advocacy of idealism, intimates (Peirce: vol. 6, para. 25; Peirce: vol. 1, para. 415)[25] then wouldn't the appreciator of natural scenery still be enjoying a sense of the Firstness of Thirdness? Moreover, if such habits were not disconnected from divine activity, then once more a route might be opened to God's existence. But, as we shall see, Peirce's possible advocacy of such metaphysical claims are not nearly as central to his philosophy as is the case for the philosophies of Spinoza or Whitehead. Furthermore, Peirce provides neither a design nor an ontological argument for God's existence. His possible reality is suggested as deriving from other kinds of considerations within Peirce's philosophy itself.[26]

All this, however, is really beside the point. Pure Firstness is a Pure Monad. It can only be realised through sheer intuition because, in grasping it no reference is made either to analysing anything or inferring something. To be concerned with the former would be to deal with a plurality of elements: to be concerned with the latter would be to focus on something in relation to something else. To grasp a Monad is to directly intuit something non-relative, unique and quite 'independent of anything else' (Peirce: vol. 6, para. 32). As an example, Peirce cites a consciousness 'whose whole life, alike when wide awake and when drowsy or dreaming, should consist in nothing at all but a violet colour or a stink of rotten cabbage'. His focus here is not upon 'what psychological laws permit' but simply what can be imagined (Peirce: vol. 1, para. 304). Given such a case, we can now make sense of this notion of a pure sense of Firstness. It is abstracted from experience. It is a possibility. It is a kind of essence in that no actual feeling, manifested quality or form of sensory enjoyment can be a pure First.

What has just been said is still a bit misleading since Firstness, really, should not be regarded purely as a feeling. After all, when we feel we always feel something. Nor should Firstness be regarded simply as a quality alone, since the only way to experience quality is to feel it; we experience redness or blueness as distinguishable felt unities.[27] So Firstness is best regarded as a feeling-quality or 'Quality of Feeling' (5.44).

Yet reconsider Peirce's own examples: 'a violet colour' or 'a stink of rotten cabbage'. These are possible feeling-qualities not associated with any specific body at all, whereas actual violetness, for example, is in relation to something existing. So we arrive then at the non-relative, just as we do for the possible, by abstracting from relations for any kind of experienced concreteness. Whereas a Second is actual and concrete, a First is potential and abstract. So Pure Firstness 'is the most abstract, and is pure possibility'.[28] At its extreme it is 'the primordial continuum of quality as such'. This kind of 'primordial potency' which Hartshorne equates with the 'pure power of God'.[29] Now you have the rationale for his remark that experiencing Pure Firstness is to have access through intuition to the divine.

Whether or not we follow Charles Hartshorne's direction here is once more beside the point, for what is really being intimated is the problematic status of Firstness itself, especially in a culture orientated more and more around existing material things and thoughts about them: Secondness and Thirdness respectively. Moreover, it is important to make Peirce's doctrine of the three categories, underpinning his phenomenology, clear not only because it was introduced to the reader in chapter I but because it provides the necessary anchorage for his semiotic or theory of signs[30] to which we can now turn.

We now need to make a transition from Peirce's phenomenology to his theory of signs. That can be done by considering his remark that 'the ideas in which Thirdness is predominant are, as might be expected, more complicated' (Peirce: vol. 1, para. 338). The easiest of these, however, is the idea of a sign or representation itself: 'A sign, or representamen, is something which stands to somebody for something in some aspect or capacity' (Peirce: vol. 2, para. 228). Indeed, in his article 'Consequences of Four Incapacities' of 1868, the notion of the sign is defined as a triadic relation: '1st, it is a sign to some thought which interprets it; 2d it is a sign *for* some object to which in that thought it is equivalent, 3d, it is a sign, *in* some respect or quality, which brings it into connection with its object' (Peirce: vol. 5, para. 283). Such a sign or representamen can be one of three kinds, either a general sign or symbol, an index or an icon (Peirce: vol. 5, para. 73) but whereas the latter two 'assert nothing' (Peirce: vol. 2, para. 291), a symbol 'is a Representamen, whose Representative character consists precisely in its being a rule that will determine its interpretant' (Peirce: vol. 2, para. 292) so that for Peirce all words, sentences, books and other conventional signs are symbols. He brings together these three kinds of representamen in the following example: 'Take for instance, "it rains". Here the icon is the mental composite photograph of all the rainy days the thinker has experienced. The index is all whereby he distinguishes *that day*, as it is placed in his experience. The symbol is the mental act whereby (he) stamps that day as rainy' (Peirce vol. 2, para. 438).

A return can be made to our 'Digger story' to flush out these distinctions. But given the problematic status of Firstness, let's work back from Thirdness through Secondness to Firstness, from the symbol through its index to the icon. If the 'Digger story' is treated as a symbol of or for childhood, generally speaking, then we might interpret it in the way we regard Clive King's *Stig of the Dump*: as a narrative typifying a child's developmental stage in which children imagine they have a friend, perhaps even one for which a place has to be prepared or 'saved' at the meal table. Lou, Barney's sister, suggests just this possibility in that book: 'Stig's just a pretend-friend, isn't he Barney?'.[31] In relation to the indexical nature of the 'Digger story', we faced a difficulty: Who is or was Digger? In this kind of context Peirce remarks: 'If the Secondness is an existential relation, the index is genuine. If the Secondness is a reference, the index is degenerate' (Peirce: vol. 2, para. 283). In interpreting this quotation, Carl Hausman has pointed out that we need to distinguish, in the use of the term 'reference', a) 'reference as a relation' and b) 'reference as a relatum', 'a referent different from the relation of reference'.[32] For Peirce the index is degenerate because condition a), the first condition, is satisfied but the second condition, b), may not be because the referent may have no physical existence. So Peirce, in the next sentence, remarks: 'All degenerate seconds may be conveniently termed internal, in contrast to external seconds, which are constituted by external fact, and are true actions of one thing upon another' (Peirce: vol. 1, para. 365).

Earlier, we experienced difficulty in separating a sense of pure Firstness from what was called the Firstness of Thirdness. Similarly, we will have difficulty in separating the notion of icon from the iconic. The former Peirce defines logically: icons are simply 'non-relative characters' (Peirce: vol. 1, para. 564) whose 'significant virtue' is due simply to 'Quality' (Peirce: vol. 2, para. 92). On this account an icon 'is a representamen which fulfils the function of a representamen by virtue of a character which it possesses in itself, and would possess just the same though its object did not exist' (Peirce: vol. 5, para. 73). He cites the shape of a centaur which might be embodied in a statue as such 'whether there be a centaur or not' (Peirce: vol. 6, para. 73). On this account, a feeling 'red' 'is necessarily an Icon' (Peirce: vol. 2, para. 254) whereas a painting is not.

The treatment of the iconic seems to apply more to making sense of icons representing the Firstness of Thirdness where the idea of similarity or analogy comes to the fore: now an icon becomes a 'diagrammatic sign' which exhibits a similarity or analogy to the subject of discourse' (Peirce: vol. 1, para. 369): 'a sign may be iconic, that is, may represent its object mainly by its similarity, no matter what its mode of being... Any material image, as a painting, is largely conventional in its mode of representation' (Peirce: vol. 2, para. 276).

So, in making a comparison between contemplating a picture and hearing the 'Digger story', the latter can be regarded iconically or, if its felt quality is dwelt upon, experienced as an icon, just as a painting can, as a material image 'in itself, without legend or label' (Peirce: vol. 2, para. 276). Whilst the status of 'Digger' is regarded as a likeness to Stig in Clive King's story or some other fictional friend, we have an iconic representation, since Digger is real enough for the child, even if, as the subsequent questioning of him might show, Digger has no actual existence.[33] Once the status of Digger is ignored, however, and attention is paid instead to the nature of his activities 'the mind floats in an ideal world and does not ask or care whether it be real or not'.[34] Having focused on this theory of signs – Peirce's distinctions were picked up in chapter VII – we now need to examine another facet of his philosophy relating to his theory of categories, his semiotic theory of cognition, noting again how further theorising emerges from his semiotic. So the reference of the Christian Trinity is 'the *son of God*' but the place of the interpretant and the ground seems sometimes uncertain. In the 1866 *Lowell Lecture XI* 'Divine *Logos* or word' issuing from 'Paternity' is the former, whereas the '*Ground*, being that partaking of which is requisite to any communion with the Symbol' corresponds to 'the Holy Spirit'.[35] Yet it seems more obvious to identify God the Father with Firstness, the Son with Secondness and the Holy Ghost with Thirdness![36]

Our examination of the 'Digger story' has shown how far it can be regarded symbolically, indexically or iconically, how inferences can be drawn and how processes of reasoning can be used in our thinking about it. Indeed in his 1869 review *English Doctrine of Ideas*, Peirce credits the psychologist Wundt with the claim 'that every train of thought is essentially inferential in its character, and is, therefore, regulated by the principles of inference'.[37] As we have already appreciated, whereas his father was interested in mathematics, where necessary conclusions can be drawn deductively, the son Charles was more interested in the science by which conclusions can be drawn. In Peirce's own time, mathematical logic – deductive in character – was distinguished from the inductive, which characterised the logic of science. It was this latter kind of logic which became the focus of his attention where he distinguished induction proper from retroduction. These forms of inference along with deduction, explored in chapter V, can be briefly described.

Peirce defined retroduction, which he sometimes called hypothesis or abduction, as a form of inference 'where we find some very curious circumstance, which would be explained by the supposition that it was a case of a certain general rule' (Peirce: vol. 2, para. 624–625).

In this country's interior, fish-fossils are found today	Result
If sea, where fish once swan, washed over this country's interior, then we would find fish-fossils today	Rule
∴ Sea, where fish once swam, washed over this country's interior	Case

Such reasoning probably lay at the heart of Sherlock Holmes's method as Thomas Sebeok's *You Know My Method*[38] illustrated. Yet such inferences, evoking possibilities, may be false!

This youngster is speaking of a man in the forest called Digger	Result
If this youngster saw an actual man in the forest then he would speak of a man called Digger	Rule
∴ This youngster saw an actual man in the forest	Case

On the other hand, induction, meant to apply for all actual experiences, 'is the inference of the rule from the case and the result' (Peirce: vol. 2, para. 622).

This boy is a young child	Case
This boy is an inventor of fictional friends	Result
∴ All young children are inventors of fictional friends.	Rule

Finally, the most familiar form of inference, deduction, applies 'general rules to particular cases' (Peirce: vol. 2, para. 620), which may or may not be embodied in experience, to yield a result.

All young children are story–creators.	Rule
This boy is a young child	Case

∴ This boy is a story-creator. Result

One thing we can appreciate now, in examining these figures of inference, is that the result manifests something of a qualitative nature, that is to say, Firstness; the case, because it points to something actual, invokes Secondness; the rule exemplifies Thirdness.

In this somewhat extended account of the first stage of Peirce's intellectual development it has been shown how his interest in mathematics could have generated his doctrine of categories his theory of signs and his hunt for different kinds of inferences. Within his biographical development, things didn't quite happen in this linear fashion since throughout his whole life he worked to clarify these major achievements individually in his own enquiries. Yet, throughout that first stage, we saw how the exacting methods of mathematics and experimental science, forwarded by his father proved so crucial in his son's theorising, as did his critical and sceptical attack towards other philosophers and their philosophies. So, for example, his father was a Unitarian and regarded the views of the mystic Emmanuel Swedenborg seriously. The latter's ideas were introduced to both father and son by Henry James Sr., father of William and Henry James. But although Charles, in his 1870 review of *The Secret of Swedenborg*, indicated its 'spiritual nutriment', his acquired 'cool-headedness' judged that Henry James Sr. had 'no right to call his work philosophy' since Swedenborg's doctrines 'are incapable of being established by reason.'[39] Though his father's mystical spiritualism may have provided then the necessary dynamo for Charles's philosophical writings, it was never allowed to dominate their contents. His father, however, on the death of Charles's brother Benjamin Mills Peirce at the age of twenty-six – subsequent to his not only burning the candle at 'both ends, but at every other point on its surface'[40] – in that very year, warned Charles against his obsession with logic. The son responded by devoting himself to science, so as to pursue his logical interest,[41] as will be seen as the second stage of his intellectual development (1871–1884) unfolds.

Indeed, it was about 1870 when, with the Darwinian

positivistic thinker Chauncey Wright, William James and Nicholas St John Green, the Cambridge Metaphysical Club was formed. Because Green emphasised the significance of applying Bain's definition of belief 'as that upon which a man is prepared to act' (Peirce: vol. 5, para. 12), Peirce called him 'pragmatism's grandfather'. But in the way Peirce treats belief, in his first birth certificate *The Fixation of Belief*, his statement of Bain's theory is haunted by the categories. Belief involves 'the feeling of believing' which is 'a calm and satisfactory state', secondly 'it appeases the irritation of doubt' and thirdly it sustains 'in our nature some habit which will determine our actions' (Peirce; vol. 5, para. 371–372; vol. 5, para. 397). The categories became clear when these relations are established: 'Doubt is an uneasy and dissatisfied state (Firstness) from which we struggle to free ourselves (Secondness) and pass into the state of belief (Thirdness)'.[42] Unlike Bain's definition of belief Peirce's treatment stresses patterns of acting or habits – rather than actions or mere reactions – setting all this within what Peirce now terms 'inquiry' (Peirce: vol. 5, para. 372–375).

The significance of inquiry is developed in pragmatism's second birth certificate *How To Make Our Ideas Clear*. Now predictions about an object in the way it can be used practically are emphasised. Indeed, Peirce credits Kant with the insight concerning 'an inseparable connection between rational cognition and rational purpose' (Peirce: vol. 5, para. 412) in the use of his term 'pragmatisch'. This insight led Peirce to regard pragmatism, unlike James its later populariser, as a doctrine of meaning – Thirdness – not one in regard to action alone: 'Consider what effects, which might conceivably have practical bearings, we conceive the object of our conceptions to have. Then our conception of these effects is the whole of our conception of the object' (Peirce: vol. 5, para. 402). And this doctrine is placed in the context of the argument between Catholics and Protestants over transubstantiation; whether or not bread and wine become literally Christ's body and blood in the taking of the sacrament. Without facing that dispute – examined as has been done elsewhere[43] in terms of the adequacy of Peirce's pragmatic doctrine – we can return to our 'Digger story' to illustrate two of its directions.

Consider two objects in terms of a child's conceptions: Digger and Phil the butcher. The butcher owned a shop along the road where the boy lived. The pragmatic doctrine suggests two important ideas in relation to what was called a degenerate and a genuine index earlier. It defines an object's existence in dispositional, observational terms: 'If you go down the street you will see the butcher's shop and if it is not Sunday or before nine o'clock or after five in the evening, you are likely to see Phil, perhaps chopping off the head of a pig on the counter.' Similarly: 'When you next see Digger you can ask him something: can he get you one of those lions' heads?' This kind of operational approach proved enormously influential in legitimising theoretical ideas in science and in clarifying what counts as scientific proof. Secondly, the pragmatic maxim hints at what is included within Peirce's *logic of abduction* – namely the stages of inquiry, explored in chapter VI – for settling the truth of the matter. But there is a reservation on this claim, clarified in pragmatism's third birth certificate, Peirce's 1878 paper *The Doctrine of Chances*. This article emphasises the importance of a community of investigators in scientific inquiry – a claim examined in chapters II and IV of this book – but also a particular danger. This danger emerges with those interpreters who, like Deledalle, think that the truth and a sense of reality emerge as the result of a common enterprise sustained by 'the community of scientists all working on the same investigation'.[44]

Against Deledalle's claim Peirce specifically warns against the danger of what can be labelled 'Distorted Thirdness': our own perversity 'and that of others may indefinitely postpone the settlement of opinion'; '...it might even conceivably cause an arbitrary proposition to be universally accepted as long as the human race should last' (Peirce: vol. 5, para. 408). Consider the arbitrary proposition 'that man only acts selfishly' (Peirce: vol. 5, para. 382) or the exaggerated claims made for 'the beneficial effects of greed' (Peirce: vol. 6, para. 284). Only if our cognitive interest is unlimited, embracing the whole community to include '...all races of beings with whom we can come into immediate intellectual relation' can 'Distorted Thirdness' be avoided. This claim provides the rationale for his assertion that 'logic is rooted in the social principle' set out in the form of three regulatives: '...interest in an indefinite community, recognition of the possibility of that interest being made supreme, and hope in the unlimited continuance of intellectual activity.' These three 'indispensable requirements of logic' are identified with 'that famous trio of Charity, Faith and Hope', 'the finest and greatest of spiritual gifts' according to St Paul (Peirce: vol. 2, para. 654–655). In chapter II, these claims were related to his scientific research in his pendulum experiments. In addition, his photometric researches sought to discover a more exact determination of standard yards and metres using diffraction gratings, a technique known as 'the spectrum meter'. And it was for these international activities that he was elected to the National Academy of Sciences in 1877 which Peirce hoped, in correspondence with his father, was awarded as a result of his work on logic, viewed as a scientific contribution.[45] In the fall of 1879 he began lecturing in logic at Johns Hopkins University until his dismissal in 1884.

Earlier you may have been unhappy with the attempt to distinguish Phil the butcher from Digger simply in terms of statements or descriptions of what you, as an inquirer, might do in regard to their possible existence. Shouldn't a philosopher have something to say about objects or persons existing as such? Peirce expressed this difficulty precisely in 1885: 'The actual world cannot be distinguished from the world of imagination by any description. Hence the need of pronouns and indices, and the more complicated the subject the greater the need for them.' Some way then has to be found for referring to an object to indicate a 'thisness'. You know this to be a sign, a sign called an index. It was while Peirce was at Johns Hopkins University, working with his student O H Mitchell, that he learnt the true significance of this sign: 'The introduction of indices into the system of algebra or logic is the greatest merit of Mr Mitchell's system. He writes F_1 to mean that the proposition is true of every object in the universe, and F_u to mean that the same is true of some object. This distinction can only be made in some such way as this' (Peirce: vol. 3, para. 363).

Let's approach this from another angle. Your unhappiness may be caused because you want to say that things are real independent of how their reality is represented in language or otherwise expressed. This claim is two-pronged. Consider the generalisation: 'All butchers cut off the heads of their animal carcasses'. This may or may not be so, but Peirce makes its truth, in his Pragmatistic period, dependent upon the inquiries of investigators rather than by one person, and even if the nature of ultimate reality can't be settled – 'that which is thought in the final opinion' (Peirce: vol. 8, para. 12) 'which would finally result from investigation' (Peirce: vol. 5, para. 408) – the truth about it can prove acceptable to such a group of investigators at a particular time[46] since it is likely it will continue to be verified subsequently, if true; if not it will have been falsified. But surely the reality of universals is different from that of individuals. Thirdness's reality must be different from that of Secondness, mustn't it?

In 1885, Peirce responds that individuals must be real or exist in a way different from the way in which universals or generalisations are real. Reality is the knowable: existence is the experienceable. Individual objects are real because they exist as 'dynamical objects'; their existence has nothing to do with the way they are represented to consciousness, but dynamic objects are never fully determinate, never fully knowable since, in their case, there is not just one truth claim to test about them, as there is in the case of a universal claim, but, rather, many truth claims: 'Phil is grey-eyed, sentimental, kind, hardworking, fond of children etc. etc.' Moreover because an individual exists through time, s/he may manifest different actualities at different times: 'Phil is unkind, lethargic today, cross with the children he meets etc. etc.' So to determine absolutely, his existence would require an infinity of such descriptive statements, making different claims about him at diverse times. But such individuals can be represented by indexical expressions: 'the man over there'; by names in ordinary language, 'Phil', 'the butcher'; by variables of quantification in logic.[47] Hence Apel's conclusion: 'the individual can be identified in a situation as real, but cannot be designated through symbols (i.e. signs for general concepts), yet can be so designated through indices and can in this way be made the subject of a true or a false statement...'[48]

This conclusion has an important consequence. No longer need Peirce press his idealistic claim that reality is to be identified with the final outcome of investigation by inquirers. He can now become more realistic. The truth of generalisations, e.g. 'All butchers chop off the heads of animal carcasses' is now made to depend on individual existences which can either verify or falsify the reality of such generalisations. Of course it is true, as the seven-year-old son once said to his father, 'You can only recognise something if you know what you are looking at',[49] so that Secondness still seems entwined with Thirdness. But any conception of the object, an 'immediate object' in one's representations, can still be challenged, by the way the actual 'might react, or have reacted against my will' (Peirce: vol. 3, para. 613) as a 'dynamical object', generating thereby the possibility of a different representation, a different 'immediate object'. Such brute actualities are examples of Secondness. They challenge the demand for intelligibility, whereas in the case of individual objects and processes – as represented – Thirdness is as much necessary as Secondness so as to account for their ongoing reality made intelligible through further inquiry or evolving semiotic activity.

Though the beginning of this third stage of his intellectual development (1885–1902) is connected with his university dismissal, it may really have begun with the death of his father in 1880. If this stage is marked by Peirce's turn to realism, that event might explain his increasing interest in an evolutionary metaphysics. In that year he had attended his father's lecture on evolutionary cosmology published later in *Ideality in the Physical Science*. After his father's death his ideas may well have influenced his son's writings in his *Design and Chance* of 1884, where Charles speculated on the evolutionary origins of natural laws; his 1887 *A Guess at a Riddle*, where he suggested how his categories can be seen as active in the universe thereby viewed as underpinning the different branches of science and philosophy; his *Monist* papers (1891–1893), where he developed his metaphysical views more fully.

This time of his life seems haunted then by the non-presence of his father: on the one hand a lack of self-control, perhaps encouraged by his father's over-indulgence towards him as a boy, meant that he suffered from an excess of individualism. Notice within his own theorising that value claims only arise in relation to further intellectual activity at this time. Such considerations may not be irrelevant to his dismissal.[50] On the other hand, as you may have noted already in some of the examples which Peirce related to scientific inquiry, he sought to bring religion and science together in a greater epistemological whole. With the death of his father, his spiritual dynamo may have become more active. Previously, whereas his father forwarded a mysticism seeking to relate mind to nature in experience, his son sought such a relation through the processes of scientific inquiry by a community of experimenters.[51] But his realism was pushing him more in his father's direction. Meanwhile, his own personal prospects grew bleaker: he moved into premises he called Arisbe, Milford in Pennsylvania in 1857 where he stayed with his second wife until his death and his Coast Survey appointment ended in 1891 so that his only source of revenue came from his private practice as a chemical engineer or, possibly. from the publication of his papers.

We have already seen how Peirce tried to establish the existence of individual entities independently of the reality of generalisations. Now he turned to the reality of those generalisations themselves – Thirdness itself. Two changes can be recorded on this front. In an 1889 entry for the *Century Dictionary* he credited the idea of ideal–realism to his father[52] which became influential in his advocacy of Galileo's *Il Lume Naturale*: 'It is certain that the only hope of retroductive reasoning ever reaching the truth is that there may be some natural tendency toward an agreement between the ideas: which suggest themselves to the human mind and those which are concerned in the laws of nature' (Peirce: vol. 1, para. 81). With respect to his father's demand that such reasoning still had to be confirmed by empirical science, his son's logic of abduction – the stages of inquiry: retroduction, deduction and induction – provided the key.

The second development can be detected by such articles as his 1903 *The Reality of Thirdness* where Charles Peirce is concerned to show that 'whatever is truly general refers to an indefinite future', 'cannot be fully realised' since it is 'a potentiality' not an actuality (Peirce: vol. 2, para. 148) in experience. Real potentiality is more definite than logical possibility – if we put two seeds in different soils their 'future difference is then a *reality*, already' (Peirce: vol. 6, para. 349); it is not referring to 'utter nothingness' to talk about a potential being (Peirce: vol. 1, para. 218) even if it is less definite than an actual event taking place in the present. When an experimenter speaks of a phenomenon, he is referring to 'what. *surely will* happen… in the living future' (Peirce: vol. 5, para. 425). So 'to say that of the *real* objective facts some *general* character can be predicted, is to assert the reality of a general' (Peirce: vol. 6, para. 100).

Such shifts within his conception of philosophy generated wider cosmological speculations. These include his conjectures about tychism, anancasm and agapasm – touched on in chapter XI – made possible by the different way he came to regard his categories.

Peirce became more dependent upon William James during the fourth and final stage of his life (1902–1914). James provided financial assistance; publicised pragmatism as a doctrine originating with Peirce rather than with himself, in his Californian address of 1898 *Philosophical Conceptions and Practical Results*;[53] enabled Peirce to deliver lectures such as those at Harvard in the Spring of 1902 – *Pragmatism as a Principle of Right Thinking* – where the final lecture presented pragmatism as the logic of abduction with no connection made directly with metaphysics (Peirce: vol. 5, para. 180–212). None the less, Peirce wanted to distinguish his position more clearly from that of James. It wasn't just that James was interested in forwarding a philosophy – pragmatism – which might 'be acceptable to everyone' whilst, at least for his 1989 Harvard Lectures, Peirce was more interested in formal logic.[54] Rather, Peirce wanted to distinguish his own doctrine 'pragmaticism', from James's use of his earlier doctrine so as to be able 'to kiss his child' 'pragmatism' goodbye '…and relinquish it to its higher destiny' in the hands of William James himself and F C S Schiller (Peirce: vol. 5, para. 414).

It is during this period, however, where we can see the son

distancing himself more from his father's achievement. By 1896, Peirce viewed his father's definition of mathematics as 'the science which *draws* necessary conclusions' (Peirce: vol. 4, para. 239), though satisfactory as a necessary condition, as not sufficient since it omits the framing of hypotheses from which necessary conclusions can be drawn.[55] His new view of mathematics is triggered by his attempt to draw a natural classification of the sciences. So in 1903, mathematics is placed with phenomenology above the normative sciences aesthetics, ethics and logic and the latter is now defined as 'the theory of self-controlled, or deliberate, thought' which must appeal to ethics for its principles, just as ethics must appeal, in its turn, to aesthetics. Rather than logic being part of semiotics, because all 'thought is performed by means of signs, logic may be regarded as the science of the general laws of signs. It has three branches: 1. Speculative Grammar, or the general theory of the nature and meanings of signs, whether they be icons, indices, or symbols; 2. Critic, which classifies arguments and determines the validity and degree of force of each kind; 3. Methodeutic, which studies the methods that ought to be pursued in the investigation, in the exposition, and in the application of truth. Each division depends on that which precedes it' (Peirce: vol. 1, para. 191).

In his correspondence to Lady Welby in 1909, we can see another direction in which Peirce attempted to distance himself from his father's influence: 'If I had a son, I should instil into him this view of morality, and force him to see that there is but one thing that raises one individual animal above another, – self-mastery; and should teach him that the will is free only in the sense that, by employing the proper appliances, he can make himself behave in the way he really desires to behave.' This passage can be read as a criticism of the way his father, Benjamin Peirce, had reared his son. But in addition to that point there is another one. This is a much more significant one, emerging out of this critique. He repeats his often made claim that logic is 'the Ethics of the Intellect' in that it is the science of 'bringing Self-Control to bear to gain Satisfactions' – But what guides the ethics? The answer to that question was explored in chapters XI and XII of this book. The problematic explored there is rendered here too since Peirce suggests, in answer to that question, two possible answers.

They are provided by *hedonism – in referring to the satisfaction of personal desires – and/or *decisionism, what someone 'will desire if he sufficiently considers it'. In answer to the further question 'And what might that desire be?' we have again a concern with the individual's life: 'to make his life beautiful, admirable', 'the freedom to become Beautiful' since there is 'no freedom to be or to do anything else.'[56]

It might be claimed that from 1911 Peirce distanced himself too from his father's admiration for Swedenborg's philosophy, especially in relation to the problem of evil.[57] If true, such a distancing might have two important implications: firstly, for Peirce's conception of realism and, secondly, with respect to the very issue we have just been exploring. It might be argued, subsequent to all this, that Peirce did have reservations about his 'objective idealism', that 'matter is effete mind', 'inveterate habits becoming physical laws' (Peirce: vol. 6, para. 25). For Peirce this meant that there was a 'continuity between the characters of mind and matter, so that matter would be nothing but mind that had such indurated habits as to cause it to act with a peculiarly high degree of mechanical regularity, or routine' (Peirce; vol. 6, para. 277). I don't see, as Joseph Brent does,[58] that his emphasis upon Secondness in 1902, in any way undoes his 'Schelling fashioned idealism' of 1892 (Peirce: vol. 6, para. 102): 'Do I mean that the idea calls new matter into existence? Certainly not. That would be pure intellectualism, which denies that blind force is an element of experience distinct from rationality, or logical force' (Peirce: vol. 1, para. 220). Rather, what Peirce is doing in this passage is to clarify the status of Thirdness, that Thirdness requires Secondness in order to be manifested in the world. But such remarks might raise a question: 'I may be asked what I mean by the objects of (a) class *deriving their existence* from an idea?' His concern is whether or not any class has 'a defining character' and whether or not it 'is as "natural" or "real" as another' (Peirce: vol. 6, para. 220).

The second implication is in regard to Peirce's ethical fudge, indicated earlier. What governs ethics? It appears to be an ideal, an ideal which 'by modifying the rules of self-control modifies action, and so experience too – both the man's own and that of others, and this centrifugal movement that rebounds in a new centripetal movement and so on' (Peirce: vol. 5, para. 402 no.3/12). This tells us nothing. It is merely descriptive of a process unless Peirce means to identify such an ideal with whatever occurs, no matter how far in the long run, within experience. Such an implication of this 1906 remark is carried by what he says in his 'A Neglected Argument for the Reality of God' in 1908 where what is taken to exist is regarded as some kind of perfection, that is to say man should 'bless God for the law of growth with all the fighting it imposes upon him – Evil, i.e. what it is man's duty to fight, being one of the major perfections of the Universe'; evil is part of 'the secret design of God' (Peirce: vol. 6, para. 479).[59]

Such claims are of course due to the influence of Swedenborg, the thinker his father so much admired. In his son's 1893 paper *Evolutionary Love*, God is compared to evil, as light can be to darkness: 'a luminary can light up only that which otherwise would be dark'. God's 'Creative Love... must be reserved only for what intrinsically is most hostile and negative to itself' (Peirce: vol. 6, para. 287) If so, isn't the latter necessary for the former's existence? Again, in 1905, we are told that Henry James Sr.'s *Substance and Shadow* provides 'the problem of evil' with its 'most beautiful and satisfactory solution' (Peirce: vol. 8, para. 263). In 1906 we are once more informed that in James's book we can find there 'the obvious solution to the problem of evil' (Peirce: vol. 6, para. 507). But such a view of creation trivialises evil and thereby renders problematic any notion of the aesthetic ideal being identified with what is being worked out in evolutionary terms, evolutionary love in the name of agapism or not![60]

Except for their possible effect upon Peirce's own form of life, these metaphysical speculations in no way detract from the major achievements of his philosophy yielded to a student exploring his inquiries. The reason for that judgement is quite clear. Peirce's metaphysical speculations are of derivative interest in the broad sweep of his cognitive investigations. The latter begins with Peirce's fascination with mathematics, a fascination related to his concern not so much with 'how things actually are, but how they might be supposed to be' (Peirce: vol. 5, para. 40). And that fascination yields, by 1868, his logic of relations which convinced him that One, Two and Three are more than count words like 'eeny, meeny, mony, mi' but 'carry vast though vague ideas'. And why stop at the Monad, Dyad amid Triad? 'The reason is that while it is impossible to form a genuine three by any modification of the pair, without introducing something of a different nature from the unit and the pair, four, five, and every higher number can be formed by mere complication of threes' (Peirce: vol. 1, para. 362–363).

By this route Peirce provided 'a premetaphysical way of thinking'[61] about anything at all. His approach avoids either saying *what there is* or simply *saying* what there is before inquiry begins. Out of this approach, he gains a phenomenology establishing his categories 'quality', 'relation' and 'representation', in other words Firstness, Secondness and Thirdness. Three kinds of representation can then be distinguished through the use of these categories: 'likenesses or icons', 'indexes' and 'symbols'. But these kinds of representation constitute one of the trivium of conceivable sciences namely formal grammar. But there are also two others: logic and formal rhetoric. Common to all these three conceivable sciences is a general demarcation between different kinds of symbols, namely terms, propositions and arguments, whilst in relation to logic we have correspondingly three kinds of inference: hypothesis or retroduction, induction and deduction. Whereas for the earlier Peirce, following his father, the logic of mathematics was deductive and thereby demarcated from the logic of science – regarded as retroductive and inductive in character – after his father's death Peirce became more interested in the logic of scientific inquiry, the logic of abduction – marked by the stages retroduction, deduction and induction – explored in chapter VI of this book. In this way the logic of mathematics disappears into the logic of science itself which, in its turn becomes identified with semiotics, the sign-theory of cognition.[62] Now we can see why Peirce claimed 'metaphysics may be deduced from cognition simply'.[63]

As we have already seen, his theory of categories also enabled him to distinguish three normative sciences – aesthetics, ethics and logic – which themselves lie between the two philosophical sciences phenomenology and metaphysics so that philosophy itself becomes identified with phenomenology, the normative sciences, along with metaphysics, in that logical order. Indeed, even if Peirce only became interested in aesthetics from about 1893[64] it was in this fourth and final stage of his life when his ideas in this area became more significant.[65] At the same time he was still working on science and mathematics dealing with such topics as 'secundal arithmetic, three-valued logic, map projections and the earth's ellipticity'. A remark in his correspondence to Lady Welby in 1908 might indicate his rationale for all this: 'it has never been in my power to study anything – mathematics, ethics, metaphysics, gravitation, thermodynamics, optics, chemistry, comparative anatomy, astronomy, psychology, phonetics, economics, the history of science, whist, men and women, wine, metrology, except as a study of semeiotic'[66] Now, as we have seen, his earlier position – in which logic was but a branch of semiotics – had changed for, as he put it in 1901, thought in itself is 'essentially of the nature of a sign' (Peirce: vol. 5, para. 594; vol. 5, para. 259–263), so that logic was seen as semiotics in the way the sciences had become classified. Indeed at the time of his death, through cancer in 1914, he was still working on his *System of Logic, considered as Semeiotic* which he regarded in 1910 as 'a *theory of inquiry*, intended to 'show the real nature of any inquiry's validity, and the degree thereof, and to consider how to build up a solid structure of science'.[67]

NOTES

[1]G W F Hegel's Introduction to his *Phenomenology of Spirit*, p.46, sect. 73.

[2]Paul Weiss, 'The Essence of Peirce's System' *The Journal of Philos.*, vol. 37, no.10, pp.253–264; J Feibleman, *An Introduction to the Philosophy of Charles S Peirce* (1970).

[3]Murray C Murphey, *The Development of Peirce's Philosophy.*

[4]Karl-Otto Apel, *Charles S Peirce: From Pragmatism to Pragmaticism.*

[5]Douglas R Anderson, *Strands of the System: The Philosophy of Charles Peirce.*

[6]R Tursman, *Peirce's Theory of Scientific Discovery* (1987), Max Fisch Peirce, *Semeiotic and Pragmatism.*

[7] 1.3 stands for vol. 1, para. 3, not the page no. 3 of the *Collected Papers of Charles Sanders Peirce*, ed. by C Hartshorne, P Weiss and A W Burks (1921–58), vols. 1–8, Harvard UP.

[8]N E Boulting, 'Charles Sanders Peirce (1839–1914), Peirce's Ideal of Ultimate Reality and Meaning Related to Humanity's Ultimate Feature as Seen through Scientific Inquiry' in *American Philosophers' Ideas of Ultimate Reality and Meaning*, pp.28–45.

[9]Bert P Helm, 'Axiologic Rather Than Logic: Priority Of Aesthetic And Ethical Over Semiotic Logic of Scientific Discovery: A Comment On Boulting's Presentation of Peirce's Idea of Ultimate Reality and Meaning', in *American Philosophers' Ideas of Ultimate Reality and Meaning* pp.45–47.

[10]N E Boulting, 'Grounding The Notion of Ecological Responsibility: Peircian Perspectives' in *Religious Experience & Ecological Responsibility*, pp.119–142.

[11]For a detailed biography see Joseph Brent's splendid *Charles Sanders Peirce: A Life*.

[12]J L Eposito, 'Peirce's Early Speculations on the Categories' *Proc. of the C S Peirce Bicent, Int. Congress Grad. Studies*, pp.343–346.

[13]Max Fisch, (1986), *Peirce, Semeiotic and Pragmatism*, op. cit., p.382.

[14]Peirce quoted by Max H Fisch in his Introduction to the *Writings of Charles S Peirce*, vol. 1, (1857–1866), p.xxiv.

[15]'Peirce always regarded himself primarily as an epistemologist – or, as he would put it, a logician (but he regarded logic epistemologically rather than in a purely formal or mathematical way).' Joseph M Ransdell, 'The Epistemic Function of Iconicity in Perception', *Peirce Studies in Peirce's Semiotic: A symposium by Members of the Institute for Studies in Pragmaticism*, p.51.

[16]Max Fisch (1986), *Peirce, Semeiotic and Pragmatism*, op. cit., p.331.

[17]Max Fisch (1986), *Peirce, Semeiotic and Pragmatism*, op. cit., p.331.

[18]This example is based on a suggestion in Joseph Ransdell's 'Firstness, Secondness, Thirdness in Language' in *The Signifying Animal: The Grammar of Language & Experience*, p.171.

[19]'Ronald Jebson, 61, appeared before Brent magistrates court, north-west London, accused of murdering Susan Blatchford, 11, and Gary Hanlon, 13, who disappeared from their homes in Enfield, north London, in March 1970. 'After 11 weeks of intense searching police found their bodies in a copse in Epping Forest, just 30 minutes walk from their homes.' Maurice McLeod 'Man charged with "babes in the wood" murders', the *Guardian*, Wednesday, 29 March, 2000, p.5.

[20]Max H Fisch, Introduction to the *Writings of Charles S Peirce*, op. cit. p.xxvi.

[21]A N Whitehead, *Process & Reality*, p.174, 175.

[22]Joseph Ransdell, 'Semiotic Objectivity' in *Frontiers in Semiotics*, sect. II.

[23]Joseph Brent, *Charles Sanders Peirce*, op. cit., p.13.

[24]A N Whitehead, *Science and the Modern World*, p.83.

[25]'The one intelligible theory of the universe is that of objective idealism, that matter is effete mind, inveterate habits becoming physical laws' (Peirce; vol. 6, para. 25). 'In fact, habits, from the mode of their formation necessarily consist in the permanence of some relation, and therefore on this theory, each law of nature would consist in some permanence…' (Peirce vol.1, para. 415).

[26]N E Boulting, 'Charles Sanders Peirce (1839–1914), op. cit.

[27]Charles Hartshorne, 'The Relativity of Non–relativity: Some Reflections on Firstness' in *Studies in the Philosophy of C S Peirce*, pp.215–224, p.216.

[28]Charles Hartshorne, 'Charles Peirce's 'One Contribution to Philosophy' and his most Serious Mistake' in *Studies in the Philosophy of Charles Sanders Peirce – Second Series*, pp.455–474 p.463.

[29]Charles Hartshorne, 'The Relativity of Non–relativity, op. cit., p.223.

[30]Klaus Oehler, 'A Response to Habermas' in *Peirce and Contemporary Thought: Philosophical Inquiries*, pp.267–271, 268.

[31]Clive King, *Stig of the Dump*, p.16.

[32]Carl R Hausman, 'Metaphors, Referents and Individuality' in *Journal of Aesthetics & Art Criticism*, 42, pp.181–195, p.186, *Metaphor and Art*, pp.91–92.

[33]In response to his sister's remarks, Barney replies 'No, he's really true!' Barney describes Stig thus: 'He's a sort of boy. He just wears rabbit skins and lives in a cave. He gets his water through a vacuum cleaner and puts chalk in his bath. He's my friend.' (Clive King, *Stig of the Dump*, pp.16–17). For Barney, then, Stig is totally real; he exists.

[34]C S Peirce, 'Trichotomic' (1888) in *The Essential Peirce: Selected Philosophical Writings*, vol. 1 (1867–93) chapt. 20, p.282.

[35]C S Peirce, 'Lowell Lecture XI' in C S Peirce, 'Trichotomic', op cit., p.503.

[36]Joseph Brent, *Charles Sanders Peirce: A Life*, op. cit., p.332. Max Fisch quotes Peirce who outlined a draft in 1907 for his account of pragmatism: 'A Sign mediates between the *Object* and its *Meaning* – Object the father, sign the mother of meaning. Fisch comments: 'That is, he might have added, of their son, the Interpretant'. See the Introduction to *Writings of Charles S Peirce*, vol. 1,Max H Fisch, op. cit., above p.xxxii. Peirce could have said 'Sense' rather than 'Meaning', the former the mother of meaning.

[37]C S Peirce, 'The English Doctrine of Ideas' (1869) in *Writing of Charles S Peirce: A Chronological Edition*, vol. 2, (1867–1871), chapt. 30, p.307.

[38]T A Sebeok and J Sebeok-Umiker, *You Know My Method: A Juxtaposition of Charles S Peirce & Sherlock Holmes* (1979), Semiotica 26.

[39]C S Peirce, 'The English Doctrine of Ideas', op. cit., pp.436, 438.

[40]Joseph Brent, *Charles Sanders Peirce*, op. cit., p.77.

[41]Max Fisch (1986), *Peirce, Semiotic and Pragmatism*, op. cit., p.315.

[42]The interjection of the categories is Fisch's suggestion. Max Fisch (1986), *Peirce, Semiotic and Pragmatism*, op. cit., p.96.

[43]N E Boulting, 'Charles Sanders Peirce (1839– 1914), op. cit., pp.33–34.

[44]G Deledalle, *Charles S Peirce: An Intellectual Biography* (1990), John Benjamin Publishing Company, p.22.

[45]Max Fisch (1986), *Peirce, Semiotic and Pragmatism*, op. cit., pp.316–317, 379.

[46]'There is nothing then, to prevent our knowing outward things as they really are, and it is most likely that we do thus know them in numberless cases, although we can never be absolutely certain of doing so in any special case.' (Peirce: vol. 5, para. 311) This is the basis of Peirce's fallibilism.

[47]Manley Thompson, 'Peirce's Verificationist Realism' *The Review of Metaphysics*, September 1978, vol. 32, no.1, issue no.125, pp.76–98, pp.78, 80.
[48]Karl-Otto Apel, *Charles S Peirce: From Pragmatism to Pragmaticism*, op. cit., p.136.
[49]'Just as a designation can denote nothing unless the interpreting mind is already acquainted with the thing it denotes, so a reagent can indicate nothing unless the mind is already acquainted with its connection with the phenomenon it indicates' (Peirce: vol. 8, para. 368 no. 23).
[50]Max Fisch 'Peirce at the Johns Hopkins University', Max Fisch (1986), *Peirce, Semiotic and Pragmatism*, op. cit., pp.35–78.
[51]Joseph Brent, *Charles Sanders Peirce: A Life*, op. cit., p.18.
[52]Joseph Brent, *Charles Sanders Peirce: A Life*, op. cit., pp.33, 205, 344.
[53]Max Fisch (1986), *Peirce, Semiotic and Pragmatism*, op. cit., pp.283–285.
[54]R J Roth, 'Is Peirce's Pragmatism Anti–Jamesian?' *Int. Philos. Quart.* (1965) vol. V, no.9, November, pp.541–563.
[55]Max Fisch (1986), *Peirce, Semiotic and Pragmatism*, op. cit., p.333.
[56]C S Peirce, *Semiotic and Significs: The Correspondence between Charles S Peirce and Victoria Lady Welby*, p.112; see also Joseph Brent (1993), *Charles Sanders Peirce: A Life*, op. cit., pp.49.
[57]Beverley Kent, 'Peirce's Esthetics: A New Look', Trans. of the *C S Peirce Soc.* (1976), vol. 12, no.3, pp.263–283, pp.276–277.
[58]Joseph Brent, *Charles Sanders Peirce: A Life*, op. cit., p.344.
[59]C S Peirce, 'he should see the innocents who are dearest to his heart exposed to torments, frenzy and despair, destined to be smirched with filth, and stunted in their intelligence, still he may hope that it be best for them, and will tell himself that in any case the secret design of God will be perfected through their agency;' (Peirce: vol. 6, para.479).
[60]For a contrary view, see Carl R Hausman, 'Eros and Agape in Creative Evolution: A Peircian Insight' *Process Studies*, vol. 4, no.1, Spring 1974, pp.11–25.
[61]Joseph Ransdell (1980), 'Firstness, Secondness, Thirdness in Language', op. cit., p.150.
[62]Max Fisch, (1986), *Peirce, Semeiotic and Pragmatism*, op. cit., pp.324–325, 392. See also Sandra B Rosenthal, 'Pragmatic Experimentalism and the Derivation of the Categories' in *The Rule of Reason: The Philosophy of Charles Peirce*, pp.120–138, p.123.
[63]C S Peirce, 'Notes on Logic Book' (1873) in *Writings of Charles S Peirce A Chronological Edition*, vol. 3 (1872–1878), chap, 39, p.108.
[64]M O Hocutt, 'The Logical Foundation of Peirce's Aesthetics', *Aesthetics and Art Criticism*, 1962, vol. 21, pp.157–166.
[65]Douglas R Anderson, *Creativity and the Philosophy of C S Peirce* (1987) Amsterdam, N Nijhoff.
[66]C S Peirce, *Semiotic and Significs*, op. cit., pp.85–86.
[67]Quoted by Max Fisch, (1986), *Peirce, Semeiotic and Pragmatism*, op. cit., pp.313, 392.

BIBLIOGRAPHY

Adorno, Theodor W. *Aesthetic Theory*, (1970) trans. by C Lenhardt, New York, Routledge and K Paul, 1986; trans by R. Hullot-Kentor, London: The Athlone Press 1999.

——, Theodor, *Negative Dialectics*, New York, Continuum Publishing, 1979

——, Theodor, *Prisms*, Cambridge MA, MIT Press, 1982

——, Theodor, *Minima Moralia*, London, Verso, 1988

Anderson, Douglas R, *Creativity and the Philosophy of C S Peirce* (1987) Amsterdam, M Nijhoff

——, Douglas R, *Strands of the System: The Philosophy of Charles Peirce*, (1995) Indiana, Purdue, UP

Apel, Karl-Otto, 'The Common Presuppositions of Hermeneutics and Ethics: Types of Rationality Beyond Science and Technology', in *Research in Phenomenology*, vol. IX, ed. by J Sallis, Humanities Press, 1979, pp.35–53

——, Karl-Otto, 'Types of Rationality Today: the Continuum of Reason between Science and Ethics', in *Rationality Today*, ed. by T F Geraets, Ottawa UP, 1979, pp.307–350

——, Karl-Otto, *Towards a Transformation of Philosophy*, trans. by G Adey and D Frisby, London, Routledge and K Paul, 1980

——, Karl-Otto, *Charles S Peirce, From Pragmatism to Pragmaticism*, trans. J M Krois, Amherst, Massachusetts, UP, 1981

Appleton, Jay, *The Experience of Landscape*, (1975) revised edition, New York, John Wiley, 1996

Appleyard, Bryan, 'A Flaw in the British Nature', The *Sunday Times*, 12 January 1997

Applied Environment Research Centre Ltd., *North Kent Marshes Study* Consultant's Final Report presented and published by KCC, May 1992, pp.4, 38 and sects. 4.24 and 27(i) and 6.9–11

Aristophanes, *The Birds* in *The Birds and Other Plays*, trans. by D Barrett, New York, Penguin, 1978.

Aristotle, *Nicomachean Ethics*, trans. by W D Ross, Oxford,

Clarendon Press 1921
——, *The Politics*, ed. S Everson (1988), Cambridge UP, 1990, sects. 1325, b9–10 and 1252 b4, 1254a, 36–38
Baker, Alan R H, 'On Ideology and Landscape', Introduction to *Ideology and Landscape in Historical Perspective* ed. by A R H Baker and C Biger, Cambridge UP, 1992
Bartley III, W W, *The Retreat to Commitment*, A A Knopf, 1962
Batten, L A, (et al.) (eds.) *Red Data Birds in Britain*, a joint publication by the Nature Conservancy Council and RSPB, T and A D Poyser, London 1990
Bender, B, 'Stonehenge: Contested Landscapes' in B Bender's (ed.) *Landscape: Politics and Perspectives*, Oxford, Berg, 1995, p.261
Benjamin, Walter, 'The Work of Art in the Age of Mechanical Reproduction' in *Illuminations*, New York, Schocken Books, 1978, pp.217–251
——. Walter, 'Central Park' in *New German Critique*, Winter 1985, vol. 34, pp.32–58
——, Walter, 'On the Mimetic Faculty' and 'A Small History of Photography' in *One-Way Street and Other Writings*, trans. by E Jephcott and K Shorter, London, New Left Books, 1979, pp.160–388, 240–257
——, Walter, 'Some Motifs in Baudelaire' in *Charles Baudelaire: A Lyric Poet in the Era of High Capitalism* (1939) trans. by Harry Zohn, New York, Verso, 1989, p.148
Berlin, Isaiah, 'On Man of Ideas and Children's Puzzles' in BBC 2's series *Men of Ideas: Creators of Modern Philosophy*, *The Listener* 26 January 1978 p.110
Bernstein, Richard J, 'Towards a More Rational Community' in *Proceedings of the C S Peirce Bicentennial International Congress*, Graduate Studies of Texas Tech. Univ., ed. by K Ketner et al., Texan Tech. Press, 1981, pp.115–120
Boersch, A, 'Landscapes: Exemplar of Beauty', *Brit. Journal of Aesthetics*, vol. 11, no.1, Winter 1971, pp.81–95
Bolen, E G, and W L Robinson, *Wildlife Policy and Management*, Prentice Hall (3rd ed.), 1995
Botwinick, A, (1983), *Hobbes and Modernity*, America UP
Boulting, N E, (1988), 'Conceptions of Human Action & the

Justification of Value Claims' in *Inquiries into Values*, ed. by S H Lee, Lewiston, Mellen Press, pp.173–193

Boulting, N E, (1989), 'Hobbes and the Rationality Problem in *Hobbes: War among Nations*, ed. by T Airaksinen and Martin Bertman, Aldershot: Gower Press pp.179–189

Boulting, N E, (1990), 'Edward Bullough's Aesthetics & Aestheticism Features of Reality To Be Experienced' *URAM*, vol. 13, no. 3, September, pp.55–85

Boulting, N E, (1992), 'Senses of Liberal Education in Relation to the Problem of Rationality', *Philos. of Educ. Soc. of Great Britain Conference: Papers of the 26th Annual Conference Institute of Education,* University of London

Boulting, N E, (1994A), 'Peirce's Idea of Ultimate Reality and Meaning Related to Humanity's Future as Seen through Scientific Inquiry' in *American Philosophers' Ideas of Ultimate Reality & Meaning*, ed. by A J Reck (et al.), Toronto UP, pp.28–45

Boulting, N E, (1994B), 'Review of Thomas E Hosinski "Stubborn Fact & Creative Advance: An Introduction to the Metaphysics of A N Whitehead"', Trans. of *C S Peirce Soc.* Fall, vol. XXX, no. 4, pp.1081–1091

Boulting, N E, (1995A), 'Does Any Form of Nationalism Offer a Perspective on Ultimate Reality and Meaning?' *URAM*, vol. 18, no. 3, September, pp.192–211

Boulting, N E, (1995B), 'Between Anthropocentrism and Ecocentrism' in *Philosophy in the Contemporary World*, vol. 2, no. 4, Winter, pp.1–8

Boulting, N E, (1996A), 'Grounding The Notion of Ecological Responsibility: Peircian Perspectives' in *Religious Experience & Ecological Responsibility* ed. by D A Crosby and C D Hardwick, New York, Peter Lang, pp.119–142

Boulting, N E, (1996B), 'The Emergence of the Land Ethic: Aldo Leopold's Idea of Ultimate Reality Meaning' *URAM*, vol. 19, no. 3, September, pp.168–188

Boulting, N E, (1996C) 'On Endymion's Fate: Responses to the Fear of Death' in *Rending & Renewing the Social Order*, ed. by Y

Hudson, (*Studies in Social & Political Theory* vol. 17) Lewiston, Mellen Press pp.367–385

Boulting, N E, (1998), 'Sartre's Existential Consciousness: Implications for Subjectivity' *Philos. in the Contemp. World*, vol. 5, no.4, Winter pp.11–23

Boulting, N E, (1999), 'Necessity, Transparency, and Fragility in Simone Weil's Conception of Ultimate Reality and Meaning', *URAM*, vol. 22, no. 3, September, pp.223–246

Boulting, N E, (1999), 'The Aesthetics of Nature' in *Philos. in the Contemp. World*, vol. 6, nos. 3–4, Fall–Winter, pp.21–34

Bourassa, S C, *The Aesthetics of Landscape*, New York, Belhaven Press, 1991

Brent, Joseph, *C S Peirce: A Life*, especially chapt. 6, 'The Wasp in the Bottle', Indiana UP, Bloomington, 1993

Buck-Morss, Susan, Morss's, *The Origin of Negative Dialectics*, Hassocks, Sussex, Harvester Press, 1979

Buck-Morss, Susan, *The Dialectics of Seeing* (1989), Cambridge, Mass., MIT Press 1991

Budiansky, S, *Nature's Keepers: The New Science of Nature Management*, London, Weidenfeld and Nicolson, 1995

Bullough, E, *Aesthetics: Lectures and Essays*, edited and Introduction by E M Wilkinson, London: Bowes and Bowes 1957

Calabrese, Omar, 'Some Reflections on Peirce's Aesthetics from a Structuralist Point of View' in *Peirce and Value Theory*, ed. by Herman Perret, J Benjamins Publishing Company, Amsterdam, Philadelphia, 1994 pp.143–151

Callicott, J B, 'The Wilderness Idea Revisited: The Sustainable Development Alternative' in *Reflections on Nature: Readings in Environmental Philosophy*, ed. by L Gruen et al., Oxford UP, 1994, pp.252–265

——, J B and K Mumford (1997), 'Ecological Sustainability as Conservation Concept', *Cons. Biol.* vol. 11, pp.32–40

Carroll, Noel, 'On Being Moved by Nature: Between Religion and Natural History' in *Landscape, Natural Beauty and the Arts*, ed. by S Kemal and I Gaskell, Cambridge UP, 1993, pp.244–266

Carson, Allen, 'The Aesthetics of Art and Nature' in *Landscape, Natural Beauty and the Arts*, edited by Salem Kemal and Ivan Gaskell, Cambridge UP, 1993 pp.228–243

Clarke, K, *Civilization*, BBC/John Murray, London 1971

Cobham Resource Consultants, Avalon House, Marcham Rd., Abingdon, Oxon OX14 lUG, *The Kent Thames Gateway Landscape: A Landscape Assessment and Indicative Landscape Strategy* prepared for Kent County Council and the Countryside Commission, September 1994 pp.45, 57, 58, p.28, R4; p.28, R6, p.45, R1a

Cosgrove, Denis, *Social Formation and Symbolic Landscape* (1984), London, Croom Helm

——, Denis, and Daniels, S, 'Iconography and Landscape', Introduction to *The Iconography of Landscape: Essays on the Symbolic Representation, Design and Use of Past Environments*, ed. by D Cosgrove and S Daniels, Cambridge UP, 1988, p.1

Countryside Commission, *Landscape Assessment: A Countryside Commission Approach* (CCD18), Cheltenham, UK, Countryside Commission, 1987

Countryside Commission, *The Cotswold Landscape* (CCP294), Cheltenham, UK, Countryside Commission, 1990

Countryside Commission, *Assessment and Conservation of Landscape Character: The Warwickshire Landscapes Project Approach* (CCP332), Cheltenham, UK, Countryside Commission, 1991

Countryside Commission, *Environmental Assessment: The Treatment of Landscape and Countryside Recreation Issues* (CCP326), Cheltenham, UK, Countryside Commission, 1991

Countryside Commission, *Landscape Assessment Guidance* (CCP423), Cheltenham, UK, Countryside Commission, 1993

Countryside Commission, *Conservation Issues in Local Plans* (CCP485), London, English Heritage, 1996

Countryside Commission and English Nature, *The Character of England: Landscape, Wildlife and Natural Features* London, December 1996

Craig, John L, 'Managing Bird Populations: For Whom And At What Cost?' in *Pacific Conservation Biology* (1998) vol. 3, pp.172–82

Darwin, Charles, *The Voyage of the Beagle*, London, Everyman's Library, Dent

——, Charles, *The Origin of Species*, (1859), New York, Penguin, 1974

Deledalle, G, *Charles S Peirce: An Intellectual Biography* (1990), Elmsford, NY, John Benjamin Publishing Company

Department of Environment and Welsh Office, *The Countryside and the Rural Economy* (PPG7), London, HMSO, January 1992

Department of Environment and Welsh Office, *The Countryside – Environmental Quality and Economic and Social Development* (PPG7 revised) London, HMSO, February 1997

Deuseur, Hermann, 'Charles S Peirce's contribution to Cosmology and Religion', *Religious Experience and Ecological Responsibility*, ed. A Crosby and C D Hardwick, New York, Peter Lang, 1996

Dewey, John (1934), *Art as Experience*, New York, Capricorn Books, 1958.

Duncan, J S (et al.), 'Ideology and Bliss: Roland Barthes & the Secret Histories of Landscape' in *Writing Worlds*, ed. by T S Barnes (et al.), Routledge, 1992.

DuPuis, E M and P Vandergeest, (eds.) Introduction to *Creating the Countryside*, Philadelphia, Temple UP, 1996.

Elton, Charles, *Animal Ecology and Evolution*, Oxford, Clarendon Press, 1930

Eposito, J L, 'Peirce's Early Speculations on the Categories' *Proc. of the C S Peirce Bicent, Int. Congress Grad. Studies*, ed. by K L Ketner et al., no. 23, September, 1981 pp.343–346

Feibleman, J, *An Introduction to the Philosophy of Charles S Peirce* (1970), Cambridge MA, MIT Press

Fisch, Max H, in his Introduction, *Writings of Charles S Peirce,* vol. 1, (1857–1866), ed. by Max H Fisch (et al.), 1982, Bloomington, Indiana UP

——, Max, *Peirce, Semeiotic and Pragmatism*, ed. by K L Ketner and C J W Kloesel (1986), Bloomington, Indiana UP

Gibbs, G, (1988) *Learning by Doing: A Guide to Teaching and Learning Strategies*, London, Further Educ. Unit;

Goudie, Andrew, *The Human Impact on the Natural Environment*, 4th Edition, Oxford, Blackwell, 1993, pp.71–76

Gould, S J, *Life's Grandeur*, J, Cape, London, 1996

Green, Nicholas 'Looking at the Landscape: Class Formation and the Visual' in *The Anthropology of Landscape: Perspectives on Place and Space* ed. by E Hirsch and H O'Hanlon, Clarendon Press 1995, pp.31–42

Gregory, Derek, 'The Crisis of Modernity' in *New Models in Geography*, vol. 2 ed. by R Peet and N Thrift, London: Unwin Hyman 1989, pp.376–377

Grossman, Morris, 'Interpreting Peirce' trans. of *C S Peirce, Society*, Winter 1985, vol. XXI, no.1, pp.109, 114

Guha, R, 'Radical American Environmentalism and Wilderness Preservation: A Third World Critique' in *Reflections on Nature: Readings in Environmental Philosophy*, ed. by L Gruen et al., Oxford UP, 1994, pp.241–252

Habermas, Jurgen, 'A Reply to My Critics' in *Habermas – Critical Debates*, ed. by J B Thompson and D Held, London, Macmillan 1982

——, Jurgen, *Communication and the Evolution of Society*, trans. by T McCarthy, Boston, Beacon Press, 1979

——, Jurgen, *Knowledge and Human Interests*, (1968), London, Heinemann, 1978

Hargrove, Eugene C, 'Weak Anthropocentric Intrinsic Value' in *The Monist* vol. 75, no. 2, April 1992, pp.183–207

Hartshorne, Charles, 'The Relativity of Non–relativity: Some Reflections on Firstness' in *Studies in the Philosophy of C S Peirce*, ed. by P P Wiener and F H Young, Cambridge, Harvard UP, 1952, pp.215–224

——, Charles, 'Charles Peirce's One Contribution to Philosophy and his most Serious Mistake' in *Studies in the Philosophy of Charles Sanders Peirce – Second Series*, ed. by E C Moore and R S Robin, Mass. UP, 1964, pp.455–474

——, Charles, 'The Aesthetic Matrix of Value', chapt. XVI of *Creative Synthesis and Philosophic Method*, La Salle, Illinois, Open Court, 1970

——, Charles, *Creative Synthesis and Philosophic Method*, La Salle, Illinois, Open Court, 1970

——, Charles, 'Why Study Birds?', *Virginia Quarterly Review*, vol. 46, no.1, Winter 1970, pp.133–140

——, Charles, *Born to Sing*, Indiana UP, Bloomington, 1973, chapt. 2

——, Charles, 'Beyond Enlightened Self–Interest: The Illusions of Egoism' in *The Zero Fallacy and Other Essays in Neoclassical Philosophy*, ed. by M Valady, Chicago, Open Court, 1997

Hausman, Carl R, 'Eros and Agape in Creative Evolution: A Peircian Insight', *Process Studies*, vol. 4 no. 1, Spring 1974, pp. 11–25

——, Carl, 'Metaphors, Referents and Individuality' in *Journal of Aesthetics & Art Criticism*, 42, Winter 1983, pp.181–195; *Metaphor and Art*, Cambridge UP, 1989

—— Carl, 'Metaphorical Reference and Peirce's Dynamical Object', the Appendix to *Metaphor and Art*, Cambridge and New York, Cambridge UP, 1989, pp.209–23

——, Carl, *Charles S Peirce's Evolutionary Philosophy*, Cambridge University Press, 1993

—— Carl, 'Charles Peirce's Evolutionary Realism' in *Classical American Philosophy*, ed. by Sandra B Rosenthal, Carl R Hausman, Douglas R Anderson, Urban Champaign, Illinois UP, 1999

——, Carl, *Charles Peirce's Process Realism*, [a manuscript to be published]

Hayman, Peter, *The Mitchell Beazley Birdwatcher's Pocket Guide*, Peter Hayman, London, Mitchell Beazley, 1982.

Heffernan, James D, 'The Land Ethic: A Critical Appraisal' *Environ. Ethics*, vol. 4, no. 3, Fall 1982 pp.245–7, 246

Hegel, G W F, *Philosophy of Mind*, trans. by A V Miller, New York, Oxford, Oxford UP, 1972

——, G W F, Introduction, *Phenomenology of Spirit*, trans. by A V Miller, New York, Oxford UP, 1977

Heidegger, Martin, 'Building Dwelling Thinking' (1951) in *Poetry, Language, Thought*, New York, Harper, Colophon Books, 1971 chapt. IV

——, Martin, *Being and Time*, trans. by J Macquarrie and E Robinson, Oxford, Blackwell, 1978, pp.100, 416, pp.468–469

Hobbes, Thomas, *De Cive: English version* – a critical ed. By H Warrender, Oxford, Clarendon Press 1983, chapt. 1, sect. 7,

'Philosophical Rudiments concerning Government and Society'

Hobbes, Thomas, *Leviathan* in *The English Works of Thomas Hobbes*, ed. by Sir William Molesworth (London 1839), London, John Bohn, 1966

Hirsch, Eric, 'Landscape: Between Place and Space', Introduction to *The Anthropology of Landscape* ed. by E Hirsch and M O'Hanlon, Oxford, Clarendon Press 1995

Hocks, R A, *Henry James and Pragmatistic Thought*, North Carolina UP, 1974

Hocutt, M O, 'The Logical Foundation of Peirce's Aesthetics', *Aesthetics and Art Criticism*, 1962, vol. 21, pp.157–166

Horkheimer, Max and Theodor W Adorno, *Dialectic of Enlightenment*, trans. by J Cumming, New York, Continuum 1982

Hungerland, I and G R Vick, 'Hobbes's Theory of Signification', *Journal of the History of Philosophy*, October 1973, vol. XI, no.4, pp.459–482

Hunt, J D, *The Figure in the Landscape: Poetry, Painting and Gardening in the Eighteenth Century*, Baltimore, John Hopkins UP 1989

Huxley, J, *Animal Language*, New York, Grosset and Dunlap, 1964

Ihde, Don, *Technics and Praxis*, Boston, D Reidel Pallas, pbk, 1979

Jahanbegloo, R, *Conversations with Isaiah Berlin*, Paris, Peter Halban, 1991

James, Henry, *The Spoils of Poynton*, New York, Penguin 1986

Johnson, L, *A Morally Deep World*, Cambridge UP, 1991

Jonsen, A R (with S Toulmin), *The Abuse of Casuistry*, Berkeley, California UP, 1988 p.251, pp.61–71

Kant, Immanuel, *The Critique of Judgment*, trans. by J C Meredith, Oxford, 1982

Kent, Beverley, 'Peirce's Esthetics: A New Look', trans. of the *C S Peirce Soc.* (1976), vol. 12, no. 3, pp.263–283, pp.276–277

Kent Wildlife Trust, 'Landfill – Success, Threats, Opportunities and Respectability' in *Wildlife News*, Autumn/Winter 1997, p.2

King, Clive, *Stig of the Dump*, London, Puffin Books, 1963

Kolb, David A, 'On Management and the Learning Process' in *Organisational Psychology: A Book of Readings*, second edition, ed.

by D A Kolb, I M Rubin, J M McIntyre, Englewood Cliffs, New Jersey, Prentice – Hall Inc. 1974 pp.27–42

Lakatos, Imre, 'Falsification and the Methodology of Scientific Research Programmes' in *Criticism and the Growth of Knowledge*, Imre Lakatos and Alan Musgrave, Cambridge UP, 1970, sect. 2

Lee, Desmond, *Introduction to Plato: The Republic*, New York, Penguin, 1974

Leopold, Aldo, *A Sand County Almanac and Sketches Here and There*, 1949, Special Commem. Ed., Oxford UP, 1987, pp.47, 225

——, Aldo, *The River of the Mother of God & Other Essays*, ed. by S L Flader and J B Callicott, Madison, Wisconsin UP, 1991

LPAC, *Draft Supplementary Planning Advice on Waste*, London Planning Advisory Committee (Policy No. W10), July 1997

Lukacs, George, *The Theory of the Novel*, London, Merlin Press, 1978

Lyotard, Jean-François, 'The Postmodern Condition', (1984) quoted in *The Postmodern History Reader* ed. by Keith Jenkins, New York and London, Routledge 1997, sect. 2, p.36

Marx, Karl, 'The Rise and Downfall of Capitalism' in 'Grundrisse' included in *Karl Marx: Selected Writings* ed. by D McLellan (1977), New York, Oxford UP, pp.362–365

——,Karl, 'The Communist Manifesto' in 'Grundrisse included in *Karl Marx: Selected Writings*, ed. by D McLellan (1977), New York, Oxford UP, p.221–247

——,Karl, 'Commodities' in 'Capital' included in *Karl Marx: Selected Writings*, ed. by D McLellan (1977), New York, Oxford UP, p.421

——, Karl, *Economic and Philosophic Manuscripts of 1844* (1959), London, Lawrence and Wishart, 1977

McCarthy, T, *The Critical Theory of Jurgen Habermas*, Cambridge, Polity Press, 1984

McDaniel, Jay, 'Physical Matter as Creative and Sentient' in *Environmental Ethics*, vol. 5, Winter 1983 pp.291–317

Meine, Curt, *Aldo Leopold: His Life and Work*, Madison, Wisconsin UP, 1988

Merleau-Ponty, Maurice, *Sense and Nonsense* trans. by H L Dreyfus, Evanston,Northwestern UP 1964

Misak, Cheryl, 'A Peircian Account of Moral Judgments' in *Peirce and Value Theory*, ed. by H Parrett, John Benjamin Publishing Company, Amsterdam, Philadelphia, 1994 pp.39–48

Murphey, Murray C, *The Development of Peirce's Philosophy*, Cambridge, Mass., Harvard UP, 1961

Nannery, L, 'The Problem of Two Lives in Aristotle's Ethics' in *International Philosophy, Quarterly*, September 1981, vol. 21, no. 3, pp.277–293

Nature Conservation Topic Group, 'Nature Conservation Topic Paper' prepared for the *North Kent Marshes Initiative*, Medway Estuary and Swale Management Plan; Topic Papers, Part 1, presented to assist the KCC's Draft Estuary Management Plan for Summer 1997, p.4

Newby, P T, 'Towards an Understanding of Landscape Quality', *Brit. Journal of Aesthetics,* vol. 18, no.4, Autumn 1978, pp.345–355

Oehler, Klaus, 'A Response to Habermas' in *Peirce and Contemporary Thought: Philosophical Inquiries*, ed. by K L Ketner, New York, Fordham UP, 1995, 267–271

Olwig, K R, 'Sexual Cosmology', chapt. 10 of *Landscape: Politics and Perspectives*, ed. by B Bender, Oxford, Berg, 1995, pp.35, 310, 328, 307–343, 331, 328

Orians, Gordon H, 'An Ecological and Evolutionary Approach to Aesthetics' in *Landscape, Meanings and Values*, ed. by E C Penning-Rowsell and D Lowenthal, London, Allen and Unwin 1986, chapt. 2, p.9

Paley, A (Estuary Project Officer), 'Coastal Processes and Flood Defence Topic Paper' prepared for the *North Kent Marshes Initiative* Medway Estuary and Swale Management Plan; Topic Papers, Part 1, presented to assist the KCC's Draft Estuary Management Plan for Summer 1997, p.4

Parrett, H, 'Peircean Fragments on the Aesthetic Experience', in *Peirce and Value Theory*, ed. by H Parrett, Amsterdam, Philadelphia, John Benjamin Pub. Co., 1994 pp.179–190

Passmore, J A, 'The Dreariness of Aesthetics', (1951) in *Contemporary Studies in Aesthetics*, ed. by F J Coleman New York, McGraw-Hill, 1968 pp.427–443

——, J A, *Man's Responsibility for Nature*, (1974), London, Duckworth, 2nd edition, 1980

Pearce, Mike, 'Dumping on Democracy' in *Sheppey Gazette*, Wednesday, 14 October 1998

Peirce, Charles S, *The Collected Papers of Charles S Peirce*, (vols. I–VI ed. by C Hartshorne and P Weiss, vols VII–VIII, ed. by A Burks), Harvard UP, 1931–1958

——, Charles S, 'The Law of Mind' in *Philosophical Writings of Peirce* selected, edited and introduced by Justus Buchler, New York, Dover 1955, chapt. 25,

——, Charles, S, *Semiotic and Significs*, ed. by C S Hardwick, Bloomington, Indiana UP, 1977, pp.85–86, p.112

—— Charles S, 'The Place of our Age in the History of Civilization', 21 November 1863, *Writings of Charles S Peirce – A Chronological Edition*, vol. 1, 1857–1866, Max H Fisch et al. eds. Indiana University Press, Bloomington, 1982

——, Charles, S, 'Trichotomic' (1888) in *The Essential Peirce: Selected Philosophical Writings*, vol. 1 (1867–1893) ed. by N Houser and C Kloesel, Bloomington, Indiana UP, 1982, chapt. 20

——, Charles, S, *Writings of Charles Peirce*, vols 1–6 (1857–90), Max H. Fisch and the Peirce Editorial Projects (eds.) Bloomington, Indiana UP, 1982–2000

——, Charles, S, 'The English Doctrine of Ideas' (1869) in *Writing of Charles S Peirce: A Chronological Edition*, vol. 2, (1867–1871), Bloomington, Indiana UP, 1984 chapt. 30

——, Charles, S, 'Notes on Logic Book' (1873) in *Writings of Charles S Peirce: A Chronological Edition*, vol. 3 (1872–1878) Bloomington, Indiana UP, 1986 chapt, 39

——, Charles S 'Lowell Lecture XI' in *The Essential Peirce: Selected Philosophical Writings*, vol. 1 (1867–1893) ed. by N Houser and C Kloesel, Bloomington, Indiana UP, 1992

——, Charles S, 'Elementary Account of the Logic of Relations' 1986 in *Writings of Charles S Peirce: A Chronological Edition (1854–1856)*, vol. 5 ed. by Max H Fisch et al., Bloomington, Indiana Univ. Press 1993

Peters, R S, *Hobbes*, (1956), Penguin, 1967.

——, R S, 'Worthwhile Activities' in *Ethics and Education*, London, Allen and Unwin, 1966, chapt. V

——, R S, (ed.) 'The Justification of Education' in *The Philosophy of Education*, New York, Oxford UP 1975

Piaget, Jean, *The Moral Judgement of the Child*, trans. by M Gabain, London, 1932

Plato, *The Republic of Plato* trans. with an introduction by F M Cornford, Oxford, Clarendon Press 1961

Popper, Karl R, *Open Society and its Enemies*, (1943) vol. 2, Princeton University Press, p.230

——, Karl R, *Conjectures and Refutations*, New York, Harper Torchbooks, 1963

——, Karl R, *Objective Knowledge*, Oxford, Clarendon Press, 1972

Porteous, J D, *Environmental Aesthetics*, New York, Routledge, 1996

Price, K, 'The Truth about Psychical Distance, *Journal of Aesth. & Art Criticism*, vol. 35 (Summer 1977), pp.411–423

Rader, Melvin, *Wordsworth: A Philosophical Approach* Clarendon Press 1967

Ransdell, Joseph M, 'The Epistemic Function of Iconicity in Perception', *Peirce Studies, No. 1: Studies in Peirce's Semiotic: A Symposium by Members of the Institute for Studies in Pragmaticism*, ed. by J E Brock (et al.), Institute for Studies in Pragmaticism, 1979, pp.51–66

——, Joseph, 'Firstness, Secondness, Thirdness in Language' in *The Signifying Animal: The Grammar of Language & Experience*, ed. by I Rauch and G F Carr (1980) Bloomington, Indiana UP, pp.135–185

——, Joseph, 'Semiotic Objectivity' in *Frontiers in Semiotics* ed. by John Deely (et al.), (1986) Bloomington, Indiana UP, pp.236–254

Read, E S, 'Knowers Talking About The Known', *Synthese*, vol. 12, 1992

Rolston, Holmes, 'The Wilderness Idea Reaffirmed' in *Reflections on Nature: Readings in Environmental Philosophy*, ed. by L Gruen et al., Oxford UP, 1994, pp.265–278

——, Holmes, 'Does Aesthetic Appreciation of Landscapes Need to be Science-Based?' *Brit. Journal of Aesthetics*, vol. 35, no.4, October 1995, pp.374–386

Rorty, R, 'Philosophy as Science as Metaphor and as Politics' in *The Institution of Philosophy: A Discipline in Crisis*? ed. by A Cohen and M Dascal, Open Court, 1989

Rosen, Stanley (1968), *Plato's Symposium* Yale UP, 1987

Rosenthal, Sandra B, 'Pragmatic Eprimentalism and the Derivation of the Categories' in *The Rule of Reason: The Philosophy of Charles Peirce*, ed. by J Brunning (et al.), Toronto UP, 1997 esp. pp.120–138

Roth, R J, 'Is Peirce's Pragmatism Anti-Jamesian?' *Int. Philos. Quart.* (1965) vol. V, no.9, November, pp.541–563

Rousseau, Jean-Jacques, *The Confessions*, trans. and with an introduction by J M Cohen, New York, Penguin Books, 1963

——, Jean-Jacques, *Emile or On Education*, trans. with an introduction by Allan Bloom, New York, Penguin Books, 1979

Sagoff, Mark, 'Ethics and Economics in Environmental Law' in *Earthbound: New Introductory Essays in Environmental Ethics*, ed. by T Regan, Random House, 1984 pp.147–77

Santayana, George, 'The Sense of Beauty: Being the outlines of Aesthetic Theory' (1896), in *The Works of George Santayana*, vol. II co-ed. by W G Holzberger and H J Saatkamp Jr., Intro. by A C Danto: Critical Ed, MIT, 1988

Sauer, James, 'Discourse, Consensus and Value: Conversations About The Intelligible Relation Between The Intelligible Relation Between The Private and Public Spheres' in *Political Dialogue: Theories and Practices*, ed. by S L Esquith, Amsterdam: Atlanta GA, 1996 pp.143–66

Schelling, F W J, *The Philosophy of Art*, trans. by D W Stott, Minnesota UP, Minneapolis

Schiller, F, *On Aesthetic Education of Man in a Series of Letters*, ed. by E W Wilkinson and L A Willoughby, New York, Oxford UP, 1986

Sebeok T A, and J Sebeok-Umiker, *You Know My Method: A Juxtaposition of Charles S Peirce & Sherlock Holmes* (1979), Semiotica 26

SERPLAN, *Waste: It's Reduction, Re-Use and Disposal* Waste Planning Guidance (RPC 2266), The London and South East Regional Planning Conference, 1992

SERPLAN, *Developing the Waste Planning Guidelines: Advice on Planning for Waste Reduction, Treatment and Disposal in the South-East 1994–2005*, (RPC 2700) The London and South East Regional Planning Conference, December 1994

Shakespeare, William, *Measure for Measure*, Act II, Scene 2, lines 117–120

Spencer, H, *On Education* (1854–1859), Cambridge UP

Spinoza, B, *A Political Treatise* introduction and trans. by R H M Elwes, New York, Dover, 1951

Stearns, I, 'Firstness, Secondness and Thirdness' in *Studies in the Philosophy of Charles Sanders Peirce*, ed. by P P Weiner and P H Young, Cambridge, Mass, Harvard UP, 1952

Stock, Brian, 'Reading, Community and a Sense of Place', in *Place: Culture: Representation*, ed. by J Duncan and D Ley, London and New York, Routledge 1993, chapt.16 .

Straus, Leo, 'The City and the Man', Chicago UP, 1964, chapt. 2

Svetlana Alpers, *The Art of Describing: Dutch Art in the Seventeenth Century* (1983), New York, Penguin, 1989

Swanwick, Carys, of Land Use Consultants, *Landscape Assessment: Principles and Practice,* a report to The Countryside Commission For Scotland, Kirkcaldy, Inglis Allen 1991

Taylor, D W, (et al.) (eds.) *The Birds of Kent: A Review of Their Status and Distribution*, Meresborough Books on behalf of the Kent Ornithological Society, 1984

Thomas, J, 'The Politics of Vision' and 'The Archaeologies of Landscape' in *Landscape: Politics and Perspective* ed. by B Bender, Oxford, Berg 1995, pp.19–48

Thompson, Manley, 'Peirce's Verificationist Realism' *The Review of Metaphysics*, September 1978, vol. 32, no.1, issue no.125, pp. 76–98

Tilley, Christopher, *A Phenomenology of Landscape* (1996) Oxford, Berg

Tomin, Julius, 'Pursuit of Philosophy' *Hist. of Pol. Thought*, vol. V, no.3, Winter 1984, pp.527–542

——, Julius, 'Plato's First Dialogue' in *Ancient Philosophy* (17), 1997, pp.31–45

Tucker, Emma, 'Where Rubbish is a Way of Life', *Financial Times*, weekend 1 /2 July, 2000, p.xxii

Tursman, R, *Peirce's Theory of Scientific Discovery*, (1987), Bloomington, Indiana UP

Ulrich, R S, 'Views Through a Window May Influence Recovery from Surgery', *Science* vol. 224 (1984), pp.420–421

——, R S, 'Stress Recovery During Exposure to Natural and Urban Environments' *Journal of Env. Psych.* vol. 11 (1991) pp.201–30

——, R S, 'Biophelia, Biophobia and Natural Landscapes' in *The Biophelia Hypothesis*, ed. by S R Kellert and E O Wilson, Washington, Island Press, 1993

Warnock, Geoffrey J, *English Philosophy Since 1900*, New York, Oxford UP, 1958

Watkins, J W N, 'Decision and Belief' in *Decision Making*, R Hughes (ed.) London, BBC, 1967

Weil, Simone, *The Need for Roots* (1949) London, Routledge and Kegan Paul, 1978, p.164

——, Simone, *Gateway to God*, UK: Collins, Fontana Books, 1982

——, Simone 'Philosophy', *Formative Writings*, (1929–1941), ed. and trans. by D T McFarland, and W Van Ness, Amherst, Massachusetts UP, 1987 pp.283–288

——, Simone, *Gateway to God*, (1952), Glasgow: Fontana Bks.,, 1982

——, Simone, *Gravity and Grace*, London, Routledge, Ark paperbacks, 1988

——, Weil, S.W and McGill, I (1989), *Making Sense of Experiential Learning*, Open UP

Weiss, Paul, 'The Essence of Peirce's System', *The Journal of Philos.*, vol. 37, no. 10, 9 May 1940, pp.253–264;

Weitemeier, Hannah, *Yves Klein 1928–1962, International Klein Blue*, Cologne, Germany, B Taschen 1995

Whitehead, Alfred North, 'The Romantic Reaction' in *Science and the Modern World*, New York, Macmillan Press, 1926, chapt. V

——, Alfred North, *Science and the Modern World*, New York, Macmillan Press, 1926

——, Alfred North, *Symbolism: Its Meaning and Effect*, 1827 New York, Capricorn Books, 1959

——, Alfred North, *Process and Reality*, ed. by D R Griffin and D W Sherburne, New York, The Free Press 1979

——, Alfred North, *Adventures of Ideas*, New York, Macmillan, 1933

——, Alfred North, *Modes of Thought*, G P Putnam, Capricorn Books, 1958

——, Alfred North, 'Remarks' (1936) in *The Interpretation of Science: Selected Essays*, edited with an introduction by A H Johnson, New York, Bobbs Merrill, 1961, p.211

Williams, Raymond, *The Country and the City*, London, Chatto and Windus, 1973

Wilson, E O, *Sociobiology: The New Synthesis*, Harvard UP, 1975

Wilson, Jeffrey, *The Old Telegraph*, Chichester, Fillimore and Co, 1976

Winch, Peter, *The Idea of a Social Science*, Routledge and Kegan Paul, 1959

Wittgenstein, Ludwig, *Philosophical Investigations*, Oxford, Blackwell, 1963

Wood, H W Jnr, 'The United Nations World Charter for Nature' in *Ecology Law Quarterly*, 1985, pp.977–96

Woodcock, M, and R Perry, *Birds* London, Collins Gem Guides, 1980.

Zeman, J J, 'The Esthetic Sign in Peirce's Semiotic', *Semiotica* 19, 1977, pp.241–258

SUBJECT INDEX

A

B

C

D

E

G

H

K

T

U

W

NAME INDEX

D

E

F

G

H

I

J

K

L

M

N

O

P

R

S

T

U

V

W

Z

TERMS INDEX

A

B

C

D

E

F

S

T

U

V

Printed in the United Kingdom
by Lightning Source UK Ltd.
9665200001B/1-10